The Lost World of Fossil Lake

Snapshots from Deep Time

THE LOST WORLD
of FOSSIL LAKE

LANCE GRANDE

With photography by
Lance Grande *and* John Weinstein

The University of Chicago Press | Chicago and London

LANCE GRANDE has been doing paleontological fieldwork in the Fossil Butte Member of southwestern Wyoming for more than thirty years and is one of the world's foremost authorities on this amazing local- ity. He is also a curator at the Field Museum of Natural History, Chi- cago, where he conducts research on fishes, paleontology, geology, and evolutionary biology. He is the award-winning author of more than one hundred books and scientific articles, including *Gems and Gem- stones: Timeless Natural Beauty of the Mineral World*, also published by the University of Chicago Press.

The University of Chicago Press, Chicago 60637
The University of Chicago Press, Ltd., London
© 2013 by Lance Grande
All rights reserved. Published 2018.
Printed in China

22   3 4 5

ISBN-13: 978-0-226-92296-6 (cloth)
ISBN-13: 978-0-226-92298-0 (e-book)

Library of Congress Cataloging-in-Publication Data

Grande, Lance, author, photographer.
  The lost world of Fossil Lake: snapshots from deep time / Lance Grande; with photography by Lance Grande and John Weinstein.
    pages cm
  Includes bibliographical references and index.
    ISBN 978-0-226-92296-6 (cloth: alkaline paper) — ISBN 978-0-226-92298-0 (e-book) 1. Fossils—Green River Formation. 2. Paleontology—Green River Formation. 3. Paleobiology—Green River Formation. 4. Green River Formation. I. Weinstein, John (John S.), 1955– photographer. II. Title.
    QE747.G74G73 2013
    560'.45630978782—dc23   2012029631

♾ This paper meets the requirements of ANSI/NISO Z39.48-1992 (Permanence of Paper).

Frontispiece: The dark-colored fossils from the Fossil Butte Member are preserved on flat light-colored limestone slabs. These slabs often resemble portraits or "snapshots" of the creatures that lived in Fossil Lake 52 million years ago. This slab includes fishes (†*Mioplosus labracoides*), a snail (*Viviparus* sp.), and a small bird skeleton with feathers preserved (†*Primobucco mcgrewi*), illustrating the rich fossilized diversity locked in the limestone. Total length of upper fish is 135 millimeters and the slab is specimen number FMNH PA783 from near the K-spar tuff bed above the 18-inch layer at FBM Locality A.

To

Professor Robert E. Sloan

The whole art of teaching is only the art of awakening the natural curiosity of young minds for the purpose of satisfying it afterwards.

ANATOLE FRANCE, *The Crime of Sylvestre Bonnard* (1881)

The direction which education starts a man, will determine his future life.

PLATO, *The Republic* (380 BC)

Contents

Preface

Back to the beginning of what brought me here. As a paleontologist and evolutionary biologist, I have followed many different paths and worked on a variety of scientific issues. But a common thread through much of my scientific career has been my interest in the fossil assemblage of the Fossil Butte Member (FBM) of the Green River Formation. The FBM contains some of the world's most beautifully preserved fossils. It is revered as a paleontological mother lode by professional scientists and is considered to be an aesthetic treasure by amateur fossil collectors, natural-art dealers, and even interior decorators. The U.S. National Park Service hails it as a monumental source of national patrimony. The FBM quarries are some of the most productive freshwater fossil localities in the world. Literally millions of complete animal and plant fossils have been excavated from these deposits over the last 150 years. The well-preserved, lifelike poses of these fossils on flat limestone slabs make the fossils seem like snapshots of extinct life taken by some long-ago time traveler. The FBM contains an entire 52-million-year-old ecological assemblage, from microscopic pollen and algae to 12-foot-long palm fronds and 13-foot-long crocodiles. Hundreds of extinct species of plants and animals abound. Fishes and other aquatic animals are particularly abundant, because the FBM deposits were produced within a lake. But non-aquatic animals and plants are also present, since many of them fell into the lake from the air or washed into it from the shore. It is, for example, one of the best sources of well-preserved fossil birds and bats in the world. With such a well-preserved, broadly representative fossil record, it is easy to visualize these fossils as a vibrant community of living organisms.

Over the history of Fossil Lake, sediment built up on the lake bottom to depths of hundreds of meters in thickness, resulting in pressures necessary to turn sediment into rock. A combination of regional mountain uplift and erosion over the last 40 or 50 million years eventually brought the fossiliferous rock layers back to the earth's surface. The FBM limestone now sits on the tops of buttes in the high mountain desert of southwestern Wyoming, where it has been mined for over 140 years by professional paleontologists and commercial fossil quarriers alike. This fossiliferous rock is used by biologists, paleontologists, geologists, amateur collectors, interior decorators, and even specialty tile manufacturers. The FBM provides the most comprehensive picture of Eocene life that we know of, surpassing even the beautifully preserved assemblages of the middle Eocene Messel Formation in Germany and the late Eocene Florissant Formation in Colorado (Gruber and Micklich 2007; Meyer 2003).

It was the aesthetics of these very fossils that first drew me to seriously consider science as a profession. I was an economics student in my junior year at the University of Minnesota in the mid-1970s when I saw my first fossil fish from the FBM (a †*Knightia eocaena*). The fossil was given to me by my friend Hans Radke, who had purchased it from a rock shop in Wyoming. It was a beautiful piece that required little effort to imagine as a living creature, even given its unimaginably great age. The fossil reawakened an attraction to fishes, anatomy, and deep history that I had developed as a child. I took it to Professor Robert E. Sloan in the Department of Geology of the University of Minnesota for identification. He studied it carefully for a moment, and then his eyes lit up. Looking at me, he said, "I won't tell you what this is, but if you take my vertebrate paleontology course, you will be able to identify it yourself." I took the bait and enrolled in his course. In the class we looked at many fossils, were taught the importance of drawing them on paper to fully "see" them, and learned to put extinct fossil species into a greater evolutionary context. I was so captivated by the course that I eventually switched my educational focus from business to paleontology and never looked back. I have never regretted that decision, and over the years I have felt very fortunate to be able to make a living doing something I enjoy and still find fascinating. Science can be difficult and extremely time-consuming, and if one did not enjoy doing it, it would probably be cruel and unusual punishment. But it can also be intensely satisfying and absorbing once you begin to make new discoveries of lasting value and achieve successful results. It then becomes an enjoyable exploratory journey rather than mere work. In the words of the famous Chinese social philosopher Confucius, "Choose a job you love, and you will never have to work a day in your life." I also learned a valuable lesson from my geology professor Dr. Sloan about the power of a good teacher to inspire his or her students. And it is today's students who will help determine the future of our social culture, our intellectual progress, and to some degree our success as a species.

Since my time as a graduate student, I have had numerous research projects and publications focusing on a variety of biological, geological, and even philosophical topics; but many of these endeavors connected to some aspect of the fossils from the Fossil Butte Member, where it all began for me. The exceptional preservation and diversity of organisms fossilized there meant that my most major scientific projects would usually include at least some species from the FBM. And I begin on that note of great personal attachment, and with the desire to once again synthesize what I have learned about the organisms preserved in the FBM. I hope to show the scientific importance and rich beauty of this unique fossil assemblage, and maybe in the process even attract a few new students to the field of paleontology.

Lance Grande
The Field Museum
Chicago, January 2013

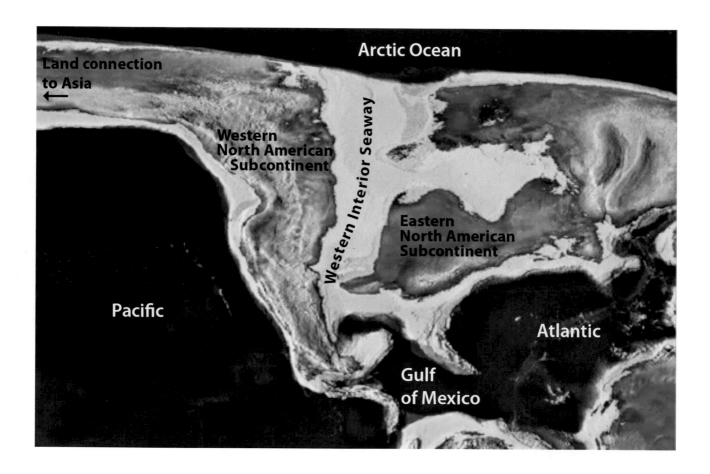

Arctic Ocean

Land connection
to Asia
←

Western
North American
Subcontinent

Western Interior Seaway

Eastern
North American
Subcontinent

Pacific

Atlantic

Gulf
of Mexico

FIGURE 1

Map of North America during the Late Cretaceous
(approximately 90 million years before present)
showing the maximum extent of the Western Interior
Seaway. This seaway completely divided the North
American continent into eastern and western subcontinents for more than 25 million years, from the
end of the Cenomanian to the end of the Cretaceous.
Map modified after Scotese (in prep.).

In the Beginning

Every story must have a beginning, and this one starts in the middle of the Cretaceous period, about 90 million years ago. By that time, rising sea levels had created the great Western Interior Seaway, dividing the North American continent into distinct eastern and western subcontinents that would remain separated for the next 25 million years. This immense marine waterway was 762 meters (2,500 feet) deep and hundreds of miles wide in places. It connected the Arctic Ocean in the north to the Gulf of Mexico in the south, and it served as an impenetrable barrier to the dispersal of most freshwater and terrestrial plants and animals between the eastern and western North American subcontinents (fig. 1). The effects of this long continental division—together with a solid land connection between the western subcontinent and Asia—greatly influenced the early development of the North American fauna and flora. The BIO-GEOGRAPHIC influence from Asia was still very strong in the Eocene Green River Formation of western North America (MacGinitie 1969; Grande 1994b).

Time Changes Everything

From the Late Cretaceous into the Early Cenozoic, Earth and its inhabitants went through a severe transformation. A huge impact from the Chicxulub asteroid off the coast of Mexico's Yucatán peninsula caused widespread, massive extinction at the end of the Cretaceous, wiping out an estimated 75 percent of all living species on the planet. The great seaway disappeared due to plummeting sea levels and mountain uplift; and along with it, all pterosaurs, marine reptiles, ammonites, and dinosaurs became extinct, except for a single surviving dinosaurian lineage: birds. The post-Cretaceous world had gone through tremendous change in its physical topography and suffered a profound loss of biodiversity; but now in the Early CENOZOIC, Earth's BIOTA was in the process of renewing itself in a different form. The age of mammals arrived on land with a vengeance and overtook the long rule of reptiles. Flowering plants and broad-leafed trees diversified, as did the animals that pollinated and fed on them. Birds and bats were also becoming more diverse and now controlled a sky devoid of pterosaurs. Teleost fishes (the

FIGURE 2
Geologic time scale for 100 million to 30 million years before present, showing the approximate ages and durations of the Green River lakes and various events discussed in this book. Ages of Green River lakes revised since Grande (1980, 1984), based on radiometric dates from Smith, Carroll, and Singer (2008), Smith et al. (2010), and Bowen, Daniels, and Bowen (2008). History of Western Interior Seaway after Scotese (2004). MYBP = millions of years before present. * = point in time at which the middle unit of the FBM was deposited by Fossil Lake.

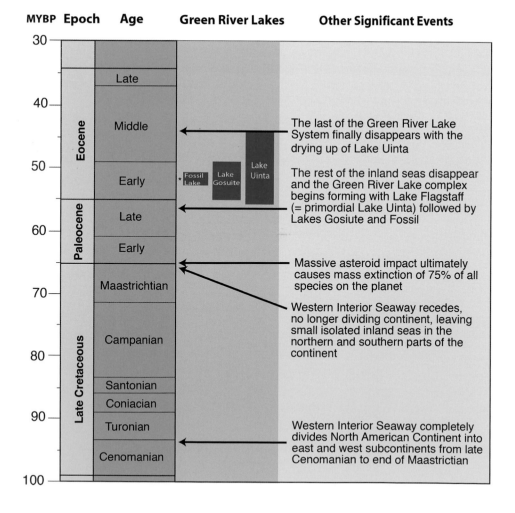

group containing 96 percent of all living fishes today) expanded their dominance in both marine and freshwater environments, and numerous freshwater lake systems formed in the western part of a North American continent no longer divided by a seaway.

One of the most remarkable western lake systems to appear was the Early Cenozoic Green River Lake System (figs. 2 and 3). This great lake complex was comprised of three lakes at its peak; Lake Uinta, Lake Gosiute, and the subject of this book: Fossil Lake. The earliest appearance of this lake system began in the late Paleocene as a result of regional uplift. As the mountains rose over time, the areas between them warped downward into basins that collected runoff from the surrounding highlands. The lake system eventually formed from thousands of years of drainage and runoff and persisted for a very long time (figs. 2 and 3). This was a true great lake system with one of the longest durations of any known lake system on Earth. One of its lakes, Lake Uinta, lasted well over 10 million years. There are no North American lake systems today that have anything close to the maturity reached by the Green River Lake complex. The North American Great Lake System that exists today is only about 10,000 years old.

FIGURE 3
Approximate areas of three Green River lakes during four different intervals of geologic history, modified after Grande (1984, 1994b), with sedimentological data from Buchheim, Cushman, and Biaggi (2012) and Eugster and Hardie (1975), and with age data from Smith et al. (2010). MYA = million years ago.

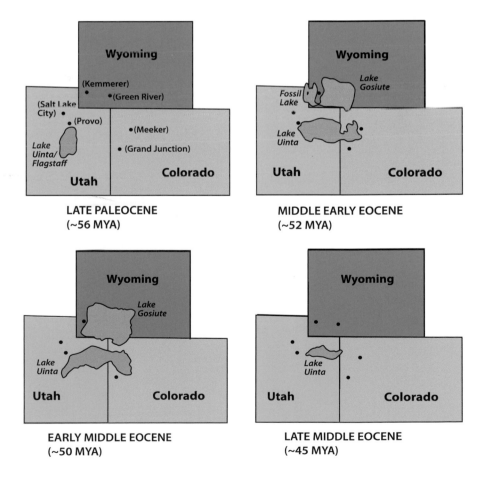

Dating techniques for the Green River Formation sediments continue to be refined, and time estimates used here represent revisions of previous work (e.g., Grande 1984, 1994b). Most of the dates for the beginning and end of each Green River lake are based on correlations of INDEX FOSSILS and STRATIGRAPHY rather than ABSOLUTE DATES. Within the Green River Formation, only a few absolute dates (based on radiometric data) have been determined, but one that is well

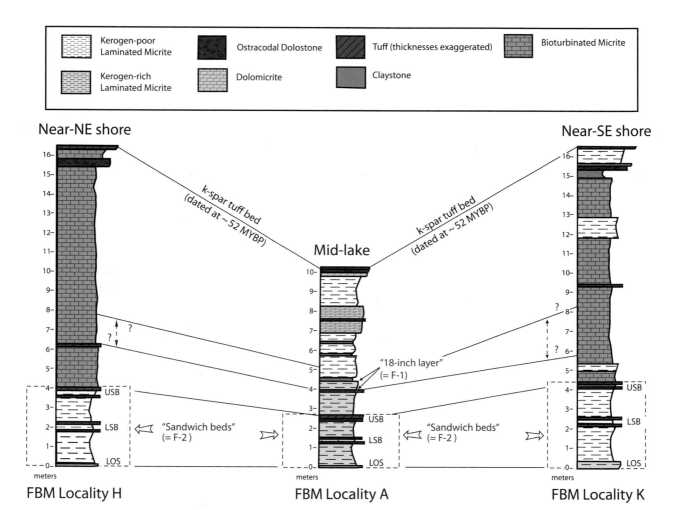

FIGURE 4

Comparative stratigraphic sections of the FBM (*sensu* Buchheim 1994a) at FBM Localities H, A, and K (roughly plotted on fig. 7). Note the relative positions of the 18-inch layer and the sandwich beds. FBM Localities A–N are discussed and plotted on maps in appendix A. The K-spar tuff layer is a volcanic ash bed that has been dated at approximately 52 million years before present (Smith, Carroll, and Singer 2008; Smith et al. 2010). The base of the FBM starts just below the lower oil shale (LOS), and the top of the FBM is between 3 and 5 meters (10 and 16 feet) above the K-spar tuff bed. Note the shorter stratigraphic column in the mid-lake locality relative to the two nearshore localities, indicating that there was a lower sedimentation rate in the mid-lake region of Fossil Lake. Modified after Grande and Buchheim (1994), based mostly on fieldwork by Buchheim. USB = upper sandwich beds; LSB = lower sandwich beds; LOS = lower oil shale; MYBP = million years before present. All measurements taken with the LOS unit of Buchheim (1994), as the base.

established is for the K-spar tuff bed near the top of the FBM (fig. 4) dated at about 52 million years before present (data repository of Smith et al. 2010). There is still much that needs to be done to definitively date the origin and extinction dates of the three lakes.

Although the Green River Lake System persisted over a period of millions of years, it was not a continuously stable freshwater environment for all of that time. Geological evidence indicates that all three of its lakes went through drying phases in which shorelines fluctuated wildly and salinity concentrations rose to levels lethal to most freshwater fish species. The few species of fishes and other freshwater aquatic organisms that survived such periods must have retreated up the tributaries that fed the lakes in order to survive. Some probably just became extinct, because the aquatic diversity of freshwater fishes achieved within the FBM is never seen again in the Green River Lake System.

The long-buried SEDIMENTARY ROCK left from the Green River Lake System, particularly the Fossil Butte Member of southwestern Wyoming deposited by the middle phase of Fossil Lake, provide a window into the deep past—an unprecedented comprehensive picture of the North American biota transitioning between the Mid-Cretaceous and the present day. This book will focus only on the paleontology of the Fossil Butte Member of the Green River Formation (here abbreviated as the FBM). A few older post-Cretaceous formations (e.g., the Tongue River Formation) as well as older members of the Green River Formation (e.g., the Flagstaff Member) show sporadic occurrences of fossils or an abundance of specific types of fossils (Estes 1976; La Roque 1960), but the FBM provides the earliest comprehensive look at a post-dinosaur community in graphic, visually revealing detail. It provides natural portraits of an entire community, long since extinct. A few younger members within the Green River Formation show great diversity of insect and plant fossils, but they lack the overall biotic diversity known from the FBM.

The Larger Stratigraphic Context: The Relationship of the FBM to the Other 13 Members of the Green River Formation

The source of the FBM, Fossil Lake, was part of a unique PALEOGENE lake system of three lakes that produced the Green River Formation. The great longevity of the Green River Lake System (particularly Lake Uinta) made the Green River Formation one of the largest documented accumulations of lake sediments in the world. This enormous FORMATION extends over an area of more than 65,000 square kilometers (25,000 square miles) and over much of its geographic expanse it is 600 meters (2,000 feet) or more in thickness.

Today the Green River Formation is divided into 14 different subdivisions called MEMBERS (fig. 5). Each of the three lakes that created the formation had a different duration (figs. 2 and 3). Lake Uinta—whose lithified sediments are located in the Uinta Basin of Utah and the Piceance Basin of Colorado—was the largest of the three Green River lakes. It was the first to form and the last to disappear, lasting at least 12 million years, and is today represented by six members. The earliest stage of its history was originally referred to as a separate lake called Lake Flagstaff. Later geologists found that the Lake Flagstaff history was continuously connected with that of Lake Uinta, and thus considered it to be synonymous with Lake Uinta. Fouch (1976) subsequently downgraded the Flagstaff "Formation" to "Member" status within the Green River Formation (e.g., now the Flagstaff Member of the Green River Formation).

Typically, Lake Uinta was LAGOONAL to very shallow LACUSTRINE, represented today by mudstones, sandstones, siltstones, and shales (Baer 1969). The Flagstaff Member of Lake Uinta formed in what is now central Utah during the late Paleocene. Eventually, in the latest part of the Paleocene, Lake Uinta dried up in its southern end while expanding eastward into what is now Uinta Basin in Utah and Piceance Basin in Colorado. At the same time, Lake Gosiute and Fossil Lake were forming in Wyoming. The sedimentary rock left by Lake Uinta represents one of the thickest documented accumulations of lake sediments in the world, with thicknesses greater than 2,100 meters (7,000 feet) in places (Cashion 1967). The shales left by Lake Uinta are rich in fossil fuels, containing an estimated 1 trillion barrels of oil (Pitman, Pierce, and Grundy 1989). Lake Uinta deposits (from youngest to oldest) consist of the Parachute Creek Member, the Garden Gulch Member, the Douglas Creek Member, the Anvil Point Member, the Cow Ridge Member, and the Flagstaff Member (fig. 5). The sedimentary rocks

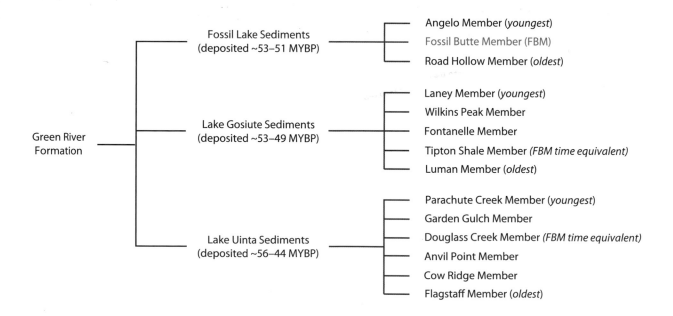

Lake and Member divisions of the Green River Formation

from Lake Uinta do not contain a very diverse vertebrate fauna, and fishes are far less common than in the FBM of Fossil Lake or the Laney Member of Lake Gosiute, but the Lake Uinta deposits contain some of the most productive plant and insect fossil localities within its Parachute Creek Member, near Douglas Pass, Colorado, and Bonanza, Utah (e.g., MacGinitie 1969).

Lake Gosiute, whose LITHIFIED sediments are found in Green River Basin directly east of Fossil Basin, was the second largest of the Green River lakes. Its deposits formed five different members in the Green River Formation (from youngest to oldest): the Laney Member, the Wilkins Peak Member, the Fonta-nelle Member, the Tipton Shale Member, and the Luman Member (fig. 5). Longer lived than Fossil Lake, but shorter lived than Lake Uinta, Lake Gosiute is thought to have been shallow with a fluctuating shoreline throughout its history. During periods of contraction, the lake often became saline, and the fishes retreated to freshwater tributaries, marshes, and ponds that fringed the lake (Buchheim and Surdam 1981). There were also periods during deposition of the Laney Member Lake Gosiute deposits when the lake was freshwater, indicated by abundant ic-talurid catfishes (Ictaluridae) and suckers (Catostomidae), two PRIMARY FRESH-WATER FAMILIES of fishes. Ictalurid catfishes and suckers, which are bottom dwellers, are curiously absent from the FBM deposits, probably due either to ecological conditions (Fossil Lake possibly being less hospitable to bottom dwell-ers than Lake Gosiute) or differing age (the FBM deposits are about 3 million years older than the Laney Member deposits, perhaps indicating that ictalurids and suckers did not penetrate the Green River Lake System until after Fossil Lake had become saline or dried up). The fish assemblage of the 49-million-year-old Lake Gosiute deposits (the Laney Member) was very different from that of the 52-million-year-old Fossil Lake deposits (the FBM) (Grande 1994b, table 2).

The third lake represented by the Green River Formation was Fossil Lake, part of which is the subject of this book. At its peak, Fossil Lake also extended into the northeastern corner of what is now Utah and the southeastern corner of what is now Idaho (fig. 7). This lake has also been referred to as "Unnamed Green River Lake West of Gosiute Lake" by Bradley (1948) and "Fossil Syncline Lake" by McGrew (1975). Fossil Lake is represented in the Green River Forma-tion by three members (from youngest to oldest): the Angelo Member, the Fos-sil Butte Member, and the Road Hollow Member (Buchheim and Eugster 1998; Buchheim, Cushman, and Biaggi 2012). The Road Hollow Member represents the earliest formation of Fossil Lake and is the thickest of the three members that contain deposits from Fossil Lake. It is the time equivalent of part of the Luman Member of Lake Gosiute (Buchheim, Cushman, and Biaggi 2012). It originated in the southern part of Fossil Basin as a floodplain that gradually migrated north-ward and deepened as the lake evolved. The Angelo Member is the youngest member and records the latest phase and eventual demise of Fossil Lake. It is the time equivalent of the Wilkins Peak Member of Lake Gosiute (Buchheim, Cush-man, and Biaggi 2012). In the Angelo phase, the lake became saline, and there

←⋯ FIGURE 5
Branching diagram showing the context of the Fossil Butte Member (in blue) within the entire Green River Formation. The FBM is only 1 of 14 MEMBERS, but it has the most diverse fossil assemblage of any of them. The approximate FBM time-equivalent members in Lake Gosiute and Lake Uinta are also indicat-ed. There is great variation in thickness, geographic area, age, and duration of deposition among the members of the Green River Formation.

was a general disappearance of fishes and other aquatic organisms. The rest of this book and all the fossils illustrated in it will focus on the thinnest but most fossiliferous of the three members, the Fossil Butte Member (FBM), which was sandwiched between the Road Hollow and Angelo Members. The FBM ranges from only about 22 meters in thickness in the nearshore deposits to about 12 meters in thickness in the mid-lake quarries, and it represents only about 9 percent of the stratigraphic thickness of the Fossil Lake deposits. For more information on the stratigraphy of the Fossil Lake members, see Buchheim, Cushman, and Biaggi (2012).

Sweet Spot in the Eocene: The Relatively Short but Productive Duration of the FBM

The FBM is only a tiny slice of the Green River Formation. To put it in perspective, the Green River Lake System lasted at least 12 million years, and Fossil Lake appears to have lasted for less than 2 million of those years. And of that 1 to 2 million years, the FBM represents less than 9 percent of Fossil Lake's stratigraphic record (Buchheim, Cushman, and Biaggi 2012, fig. 2). Thus, the FBM probably represents only a few tens of thousands of years at most, which in the context of GEOLOGIC TIME is a blink of an eye.

The FBM represents Fossil Lake in its largest, deepest, and freshest stage. The southeastern end of the FBM connects to the Wilkins Peak Member of Lake Gosiute in places, indicating a brief, narrow connection between Fossil Lake and Lake Gosiute in the early Eocene (Buchheim, Cushman, and Biaggi 2012). If fishes from Fossil Lake were able to enter Lake Gosiute via this connection, they were later wiped out in Gosiute during deposition of the Wilkins Peak Member (fig. 5) when Lake Gosiute dried and became hyper-saline and toxic to fishes. The later Gosiute fish assemblage of the Laney Member is a very different one than that of the FBM assemblage from Fossil Lake, with only a few overlapping genera (†*Atractosteus*, †*Amia*, †*Knightia*, †*Diplomystus*, †*Asineops*, and †*Phareodus*). (Dagger symbols indicate extinct groups or taxa.) Although Grande (1994b, table 2) listed the percoids †*Priscacara* and †*Mioplosus* as rare occurrences in Laney Member deposits, rechecking the original reports of these genera from the Laney indicate that these reports are in error. It appears that percoids are absent from most, if not all, of the Laney Member localities.

The known aquatic biodiversity of the FBM exceeds that of the other 13 members of the Green River Formation combined. In part this was probably due to the variety of well-preserved habitats in these deposits. It is also due to the intense level of fossil excavation from the Fossil Butte Member over the years. Grande and Buchheim (1994, 34) estimated that over half a million fossil fishes were excavated by commercial quarries from 1970 through 1994—mostly complete to nearly complete skeletons. Another million incomplete or damaged fishes were excavated and discarded during the same period. I estimate that now over

200,000 specimens are excavated annually. After more than a century of collection, millions of fossils have been recovered from the FBM, so even proportionately rare elements of the overall fauna and flora (e.g., birds, bats, etc.) are now known by numerous specimens. And the diversity continues to climb. It wasn't until 1998 that a member of the pike and pickerel family (Esocidae) was discovered in the FBM. Out of a million-plus fishes that have been excavated over the last century from the FBM, this unique specimen is the only one of its kind (fig. 78). Large sample size is a definite advantage for deciphering past biodiversity based on fossil localities. But regardless of sample size, the biodiversity of aquatic vertebrates in the FBM is unmatched in any other part of the Green River Formation, because even the diversity of common fish species is high.

Within the FBM, some of the most productive fossil localities are located within a relatively narrow zone of CALCIMICRITE limestone including the "18-inch layer" and the underlying "sandwich beds" (described on pages 10–15), together representing what is probably only a few thousand years of the lake's history at most. Other highly productive zones are the so-called "mini-fish beds" near the K-spar tuff bed (figs. 4, 74, and 83, *bottom*). It is the K-spar tuff that has been RADIOMETRICALLY DATED to be about 51.97 years old (Smith et al. 2010, data repository). As explained on page 8, all of the fossils illustrated in this volume are from the FBM, and nearly all are from either the 18-inch layer or the sandwich beds of this member. Green River Formation fossils from members outside of the FBM (some of which are millions of years younger or older than the FBM) were excluded from this book to provide a more contemporaneous community picture. It is the snapshot-like aspect of a relatively contemporaneous extinct biota that continues to attract my interest to the FBM. In fact, the FBM contains such a diversity of fossil organisms that it is hard to do it complete justice in a single volume, so there are references cited throughout this text for further reading.

The importance of the FBM as a national heritage was formally recognized in 1972 with the establishment of 8,200-acre Fossil Butte National Monument, 10 miles west of Kemmerer, Wyoming. This monument protects significant sections of the FBM on federal lands as a national heritage. There are many important FBM fossils deposited in museums around the world, but the largest of these collections is the one at the Field Museum in Chicago. The visitor center at Fossil Butte National Monument also has a small museum with one of the finest displays of any national park and includes many important FBM specimens (fig. 8B, *bottom*, on p. 21).

From Lake Muck to Limestone: The FBM Matrix Forms

Because the mountains surrounding Fossil Lake contained significant amounts of limestone ($CaCO_3$), the streams flowing down their slopes became super rich in dissolved calcium carbonate before reaching the lake. As these streams filled the lake basin, Fossil Lake became SUPERSATURATED with calcium carbonate (the principal component of limestone). Like air supersaturated with moisture resulting in rain, the water column supersaturated in calcium carbonate would periodically "rain" calcium carbonate, which accumulated on the lake bottom as a lime ooze. For such a thing to have occurred, the lake must have been somewhat alkaline, with a pH of at least 8 to 8.5. As lake-bottom sediments accumulated to hundreds of feet thick during the early part of the Eocene, the enormous weight of the upper sediments compressed and DEWATERED the lower sediments, eventually turning them into a limestone. The extreme compression that converted the ooze into limestone also helped flatten the fossilizing organisms along the BEDDING PLANE (figs. 5, 217). Starting in the middle Eocene, the lake dried up and the region continued to rise in elevation, reversing the process of deposition to one of erosion. Over the next 30 to 40 million years, the region was eroded to a series of buttes, exposing portions of the Fossil Lake limestone (i.e., the FBM of the Green River Formation). Erosion continued around the buttes, creating the huge valleys between them. The maps in figure 3 show the hypothesized borders of all of the Green River Formation's PALEOGENE lakes at four points in time, but most of the lake deposits were eroded away long ago. Today all that remains of Fossil Lake lies near the tops of the isolated buttes that were spared from erosion (figs. 11, 213, *bottom*).

Three Mother Lodes of the FBM: The 18-Inch Layer, the Sandwich Beds, and the Mini–Fish Beds

The two layers within the FBM that have produced the most celebrated "mother lodes" of fossils are the 18-inch layer (also referred to as the "F-1" or "black-fish" layer) and the underlying "sandwich bed" layer (also referred to as the "F-2," "split-fish," or "orange-fish" layer where it outcrops near the shoreline in the region of Thompson Ranch). More recently a third highly fossiliferous layer near the K-spar tuff beds has been mined for fossils, the so-called "mini-fish beds" (e.g., fig. 74). Each of these layers has provided a rich source of fossils for a number of independent fossil quarrying operations (see FBM Localities A–M in appendix A). The 18-inch layer quarries around Fossil Butte were mined for nearly a century before quarrying operations in the nearshore sandwich beds or the mini–fish beds began in earnest. The rock exposures of the FBM were prominent features of the topography surrounding the town of Fossil and its railroad station (fig. 8A, *bottom left*). The mid-lake 18-inch layer averages about a half meter in thickness (thus the common name "18-inch layer"). In contrast, the nearshore sandwich beds are several meters in thickness (fig. 4). The sandwich beds also occur in the mid-lake quarries below the 18-inch layer, but the fishes there are darker in color with less of an orange tint than in the nearshore localities. The sandwich beds in the mid-lake quarries are also thinner than in the nearshore quarries due to higher sedimentation rates near shore (Buchheim 1994b).

The LAMINATED limestones of these highly fossiliferous layers split cleanly along bedding planes, producing slabs of rock with very flat surfaces. The brown to orange-brown fossils embedded in the lighter-hued limestone slabs are so beautifully preserved that they sometimes resemble wildlife portraits. Their beauty is also appreciated in the art world, and these fossils are considered "natural art" by many art galleries, interior decorators, architects, and collectors. The 18-inch and sandwich layers are not only packed full of well-preserved fossils; they are also among the easiest rock layers within the FBM from which to remove large quantities of easy-to-prepare fossils safely and efficiently. Today about 40 square kilometers of 18-inch and sandwich bed layers remain on a mixture of private, state, and federal lands.

The reason that the 18-inch layer around Fossil Butte is sometimes called the "black-fish layer" is because they are often preserved with black bones and dark brown scales, and are darker in color than those from the nearshore sandwich beds of the Thompson Ranch region. The 18-inch layer deposits are also strongly demarcated in cross section with alternating light and dark bands, or laminations (fig. 6A, *left*). The light bands are pure carbonate rock, and the dark bands consist of KEROGEN, an organic-rich material from decomposed microorganisms. The couplets of light and dark laminations were previously thought to represent annual cycles of Fossil Lake called VARVES by some geologists, resulting from

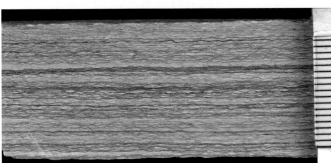

annual (i.e., seasonal) water turnover in the lake (e.g., Grande 1984, 183, 184). If the couplets had been annual markers, they would indicate that the 18-inch layer was deposited over a period of several thousand years. But more recently, work by Buchheim (1994b) has shown that the frequency of lamination couplets within equivalent stratigraphic levels varies significantly between the mid-lake and the nearshore deposits of the FBM; so at least some of them cannot possibly be varves. In addition, the remarkable preservation of large skeletons with all bones in normal position suggests a much more rapid burial process. The couplets may actually represent more frequent events such as periodic storms, major variations in lake inflow, periodic die-offs of microorganisms, or other common environmental events that happened more than once per year. We will probably never know for sure what the exact duration was for the deposition of all the 18-inch layer sediments, but the 18-inch and underlying sandwich layers (fig. 4) together likely represent a period lasting as long as several thousand years or as short as a few hundred. We can tell that this period was one of relative ecological stability in Fossil Lake, because so many of the lake's aquatic species are present from the lowest part of the sandwich beds to the uppermost part of the 18-inch layer. The sandwich bed period of deposition also represents the maximum-area extent of Fossil Lake.

The excellent preservation of the lamination couplets in the mid-lake 18-inch layer and the rarity of bottom-dwelling species there (e.g., stingrays, DECAPODS, and mollusks) suggest that Fossil Lake may have had a stratified water column at its center, as described on page 358, with well-oxygenated surface water (epilimnion) and stagnant, lethal bottom waters (hypolimnion). Bottom-dwelling fishes other than †Notogoneus could not tolerate the stagnant bottom waters in the mid-lake regions and thus were rare or absent there. Organic material and dead animals that sank from the upper regions would have been relatively protected from scavenging long enough to become buried and well preserved as fossils. The 18-inch layer is conformably "capped" on its upper and lower surface by tough, water-resistant oily shale and oily limestone layers 10 to 15 centimeters (4 to 6 inches) thick. These layers probably helped preserve the fossiliferous rock sandwiched between them from severe groundwater weathering.

The typical nearshore sandwich layer is several meters in thickness. Bottom-dwelling organisms (e.g., stingrays, crayfishes, shrimps, and gastropods) are more common in this layer than in the 18-inch layer. The nearshore sandwich beds of the Thompson Ranch region (FBM Localities H, I, J, and M) are also sometimes referred to as the "split-fish beds" because of the manner in which many of the layer's fishes have been exposed, particularly in the early years of quarry operations. Smaller fishes, in particular, often "split out" in these beds as a "part and counterpart" (fig. 6B) as well as other fossils (figs. 30, 115, *top*; and 138). Splitting out is usually not a good thing for the fossils, because it usually damages the fossil, pulling part of it on to each of the two split surfaces. Early in the history of commercial collecting, split fishes were thought to be a boon for

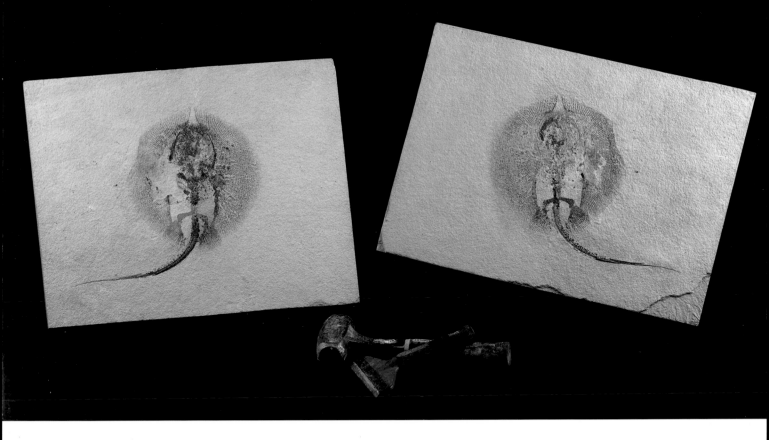

business because it was a "two-for-one," and for rarer pieces, each side could
have its missing pieces painted in. In recent years, commercial preparation has
become much more sophisticated and detailed, and for anything but the most
common species, splits are not desirable nor are the heavily restored pieces. In
fact, if a rare specimen like a bird, mammal, or turtle is found as a split, the part
and counterpart are sometimes glued back together and the piece is prepared
from one or both of the unexposed sides (e.g., fig. 116).

The sandwich bed fossils of the nearshore Thompson Ranch region are usually lighter in color (as is the rock), have an orange to orange-brown hue (fig. 6A,
upper right), and do not have the prominent dark bands or laminations that the
18-inch layer has (fig. 6A, *lower right*). Along with the much lighter-colored rock,
the plant and insect fossils from the Thompson Ranch sandwich beds are often
colorless impressions, as are most of the bird feathers. Some of the commercial
quarriers have recently started mining the sandwich beds in the mid-lake regions
where the fossils are of a darker color (e.g., at FBM Locality A).

The different FBM quarry sites provide comparative paleontological samples
contrasting the nearshore versus farther offshore aquatic communities of Eocene
Fossil Lake. The known species of the mid-lake 18-inch layer are largely the same

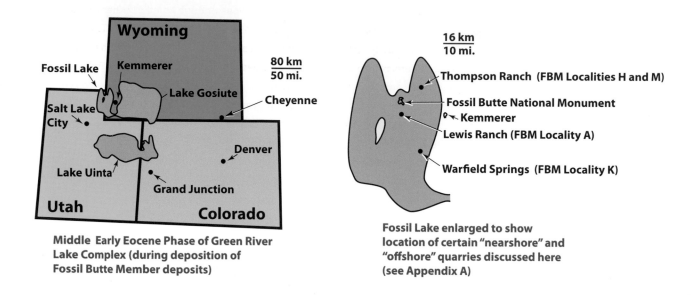

Middle Early Eocene Phase of Green River Lake Complex (during deposition of Fossil Butte Member deposits)

Fossil Lake enlarged to show location of certain "nearshore" and "offshore" quarries discussed here (see Appendix A)

FIGURE 7

Map showing rough location of the three major FBM localities discussed here (Localities A, H and M, and K). See appendix A for a more complete listing of recent FBM quarry locations. Locality A is one of the many 18-inch layer quarries, and Locality H is one of the many sandwich bed quarries. The Warfield Springs locality appears to be a different lithological expression of the sandwich beds, with an identical time horizon, but with different colored rock and fossils due to different ecological conditions in that region of the lake.

as those from nearshore sandwich beds. The main diversity and paleoecological differences between the nearshore and mid-lake fossils are in relative proportion of species and size classes (table 1, pages 176–77). In the mid-lake 18-inch deposits, stingrays, crayfishes, shrimps, *Goniobasis* snails, baby fishes for many FBM species, mooneyes (*Hiodon*), trout-perches (†*Amphiplaga*), and the mollusk-eating gar †*Masillosteus* are much rarer than in the sandwich beds. The nearshore sandwich beds produce more mammals, birds, schools of juvenile fishes, and young turtles, and have far fewer †*Priscacara* and a complete absence of †*Notogoneus*. We can see the effects of the different ecological habitats within Fossil Lake on the distribution of its inhabitants. For example, nearshore habitats were generally better suited as nurseries for young aquatic organisms as they are today. Well-oxygenated bottom waters near shore were better suited for stingrays and other bottom-dwelling organisms. And, of course, flightless land animals fell into the lake at its edge rather than its center.

The Warfield Springs locality on the southern end of Fossil Lake (fig. 7, FBM Locality K), which has been mined for decades, contains vertebrate fossils with black-colored bone in a soft, chalky-white limestone. This locality is the southern extension of the sandwich beds, but the bone color more closely resembles that of the mid-lake 18-inch layer deposits (e.g., fig. 111). This unit was thought to have been adjacent to a deltaic environment. More recently, the nearshore deposits from the midwestern side of Fossil Lake have also been mined (FBM Locality L). The sandwich bed fossils from this locality also more closely resemble the color of the mid-lake 18-inch layer fossils than the Thompson Ranch fossils (e.g., fig. 143B).

The 18-inch and sandwich bed layers are close enough to each other in geologic time that they represent a relatively contemporaneous sample. There is also another extremely fossiliferous layer located near the K-spar tuff bed (fig. 4), sometimes referred to as the "mini-fish beds," that produces extremely large quantities of common fish species (†*Knightia*, small †*Diplomystus*, and small †*Mioplosus*). This layer also produces the greatest abundance of the paddlefish †*Crossopholis*, although even here †*Crossopholis* is a proportionately rare element of the fish fauna. This layer is mined by some of the commercial quarries to supply vast quantities of small, inexpensive †*Knightia* and small †*Diplomystus* to rock and fossil shops around the world. The †*Knightia* in these beds are usually quite small, hence the name "mini-fish beds" sometimes used by the local quarriers for this layer (not to be confused with the Lake Gosiute Laney Member beds to the east, also called mini-fish beds, which consist primarily of the small clupeid †*Gosiutichthyes*; see Grande 1982b).

The World of Fossil Lake as Illustrated in the FBM: A Transitional North American Biota

Early Eocene southwestern Wyoming was as different from today's world as it was from its Cretaceous predecessor. Fifty-two million years ago, southwestern Wyoming contained lake-basin lowlands with a warm, wet, subtropical environment surrounded by highlands and mountains containing pine and other more temperate highland flora. There were active volcanoes to the north of Fossil Lake, as evidenced by ash beds (K-spar tuff) that are thickest in the north part of the lake, thinning to the south. Fossil Lake was home to a large freshwater lake system and a diverse community, including many already ancient families and orders on their way to extinction, as well as early representatives of what would diversify over time into major families and orders of animals and plants living today. There were palm trees, balloon vines, crocodiles, alligators, monitor lizards, boa constrictors, parrots, and a whole community of tropical and subtropical and warm-temperate organisms. Sea level was still somewhat higher than it is today, and the warmer, more humid climate was not only regional but global. Today, in contrast, the area is a high mountain desert region with little standing water, where the snowy season begins in September and the winter temperatures plunge to well below 0 degrees F. The fossils of the FBM provide a firm testament to nature's dictum: *time changes everything*. One broader aspect of this story is that major changes in climate result in major extinctions, and over geologic time there is no stopping it. The warmer, wetter environmental ecosystems of Wyoming and the thousands of interconnected species that they supported are now entirely extinct. Who knows what the long-range result will be from today's rapidly changing global climate?

Looking at the FBM as an Extinct Community

The diversity and abundance of fossilized contemporaneous species throughout the 18-inch and underlying sandwich layers allow us to study not only the species biodiversity, but also interactions that took place among the extinct species within the community. It is the best picture of life in the early Eocene that has ever been discovered. We can see species that were eaten by other species and preserved as stomach contents or fossilized in the moment they were being swallowed by a predator. We can see leaves with chew marks and the insects that might have been eating them. We can see which fish species were schooling, and we can compute ratios of predator versus prey species. We can study size-classes of fishes in mass mortalities, or see which species and size-classes were typically nearshore dwellers and which were typically mid-lake dwellers. For many fish species, we can find complete growth series ranging from hatchlings to large adults. Even the process of reproduction is revealed by fossil stingrays: some preserved in mating position, some with embryos still inside, and others next to newborn offspring. We also find fossilized fecal matter (coprolites) with bone and other materials indicating food items.

In the figure captions, I will provide specimen catalog numbers for fossils illustrated in this book. This will not only provide readers with a reference number for each piece, but it will also identify the institution that has the specimen. There is a key to the institutional abbreviations on page 393. In some rare cases for extremely important specimens, pieces from private collections are illustrated, but for most of these there are casts deposited at the Field Museum (FMNH) or another public repository. Before describing and illustrating the many types of fossils in the FBM, I will briefly discuss the history of mining and excavation in these deposits.

Fossils from the FBM

*History,
Controversy,
and
Quarry Life*

The early FBM quarriers of the 1870s through the 1950s focused primarily on the 18-inch layer. It was not until the 1960s that the nearshore sandwich layers and the FBM mini–fish beds began to be quarried significantly. Although there are other zones within the FBM that have been mined for fossils, the 18-inch layer and the sandwich beds have been responsible for producing the vast majority of all specimens collected from the FBM during the last 150 years. They have been the most desirable zones to work because of the diversity of fossils there, and because they have been the easiest layers from which to extract, transport, and prepare fossils.

Many different people have worked in the FBM over the last century and a half. Some used the FBM for scientific research while others simply mined the fossils for commercial reasons. Some of these people were charismatic with intriguing names like "Pap" Wheeler, "Stovepipe" Smith, and "Peg-leg" Craig. Some left a good record of their history, while others left little or no personal records. Some were early scientists, while others were just people trying to make a living. Most were pioneers in one way or another.

A Rich Cast of Characters: Early Scientific Work

Much of the earliest scientific collecting in the FBM was done by U.S. Geological and Geographical surveyors of the West in the decades following the Civil War. The material was sent back east to early paleontologists to be described and named in scientific publications. The very first descriptions of FBM fossils were by Fielding Bradford Meek (1817–1876) in 1870. These were mollusk specimens from "Fossil Hill" (today known as Fossil Butte). Meek was a merchant whose leisure hours were devoted to collecting rocks and fossils. When his businesses ultimately failed, he went to work for various state and federal geological and geographical surveys and started publishing on the invertebrate fossils from those surveys. During his career, he published over 100 papers. Although Meek's 1870 paper contained the first published descriptions of FBM specimens, it was not the first to contain a description of a species that occurs in the FBM. The first species that can be found in the FBM had already been named by Joseph Leidy 14 years earlier from a different locality to the east.

In 1856 Joseph Leidy (1823–1891) described the fossil herring †*Clupea humilis*, from the Laney Member of the Green River Formation, based on a specimen given to him by a Dr. John Evans. The name was eventually changed to †*Knightia eocaena,* in part because the name †*Clupea humilis* was found to have already been used for a different fish in Europe. Although the specimens that Leidy studied were not from the FBM, †*Knightia eocaena* was later also found to be represented in the FBM and in fact is the most common fish species there. Thus, Leidy indirectly made one of the earliest contributions to understanding the FBM fossil fish assemblage.

Called by some as "the last man who knew everything" (Warren 1998), Leidy was a professor of anatomy at the University of Pennsylvania and is known as the "Father of American Vertebrate Paleontology," as well as the "Father of American Parisitology." He was one of the earliest supporters of Charles DARWIN and lobbied successfully to have him elected to the Academy of Natural Sciences in Philadelphia. Leidy was known as a gentleman who had broad talent and expertise in many areas, including research, teaching, and administration. He published over 220 scientific papers on paleontology.

One of Leidy's students, Edward Drinker Cope (1840–1897), published the first scientific DESCRIPTIONS of fossil vertebrates collected from the FBM. Cope was a paleontology prodigy with little formal college training, but he was one of the most productive vertebrate paleontologists and zoologists that ever lived. In less than 40 years, he produced over 1,400 publications. He financed much of his fieldwork with his own money, and in 1895 he sold his fossil collection of about 13,000 specimens (including about 150 FBM specimens) to the American Museum of Natural History in New York. Cope's famous 1885 monograph, sometimes referred to by vertebrate paleontologists as "Cope's bible" (Cope 1885b), as well as earlier publications (e.g., Cope 1877) describe many fishes from the 18-inch layer of the FBM, which Cope refers to as the "Twin Creek" beds (now known as the Fossil Butte sites). Cope's work in the FBM focused on the fishes. Unlike many of the prolific paleontologists of his day, Cope did much of his own fieldwork; although it is not clear if he collected the FBM specimens himself or if they were instead collected by fossil collectors that he commonly hired for some localities. He described and named about 30 species of fishes from the FBM. Only 14 of those species are still recognized today as valid species because many of them were found to be JUNIOR SYNONYMS. Cope's archrival Othniel Charles Marsh (1832–1899) beat him to the punch on one of these species: a stingray (see page 32).

After Cope, there were a few other early scientists who included FBM specimens in their studies (e.g., Alpheus S. Packard [1880], on the decapods; Samuel H. Scudder [1890], on insects; Charles R. Eastman [1900], on a bird and

gars; Léo Lesquereux [1883] and Roland W. Brown [1937], on plants; Charles W. Gilmore [1938], on a snake), but no comprehensive study has focused on the FBM in particular. More recently, one of my colleagues, Paul Buchheim, has focused on the STRATIGRAPHY, SEDIMENTOLOGY, and geology-based PALEOECOLOGY of the FBM. He has spent many years in the region reconstructing Fossil Lake's boundaries through time with careful stratigraphic fieldwork.

Most early paleontologists who studied the FBM fossils did not collect the fossils themselves. They usually received them either from survey geologists or hired contract collectors, or they purchased them from commercial fossil quarriers. Although much has been written about Leidy, Cope, Marsh, and other professional paleontologists of the late nineteenth and early twentieth centuries, little has been said about the commercial and contract collectors who were often the ones who physically excavated the fossils studied by the professional scientists.

Commercial Fossil Excavations of the FBM through History

The birth and demise of the town of Fossil, Wyoming, is integral to the history of the early fossil quarriers in the FBM. Many of the earliest full-time fossil quarriers, like Robert Lee "Peg-leg" Craig and "Pap" Wheeler, lived there. The town's very name refers to the FBM fossils that are found so abundantly preserved in the surrounding rock exposures. The town of Fossil began in 1881 as a major railroad station for servicing steam-driven trains of the Oregon Short Line Railway. Settlers began arriving in 1884, attracted by the train stop, the presence of water, ranching, oil speculation, and fossil mining. The first settlers to set up a full-time residence in the town of Fossil were Richard H. Lewis and his wife, Susanna. Their descendants are still prominent figures in the area today, and they own an impressive amount of fossil-rich land that they lease to a number of commercial fossil operations. Although at the start of the twentieth century it looked as though Fossil might be a major Wyoming city one day, the town's population peaked early in the twentieth century at 151 total residents. Shortly after that, with the advent of the diesel engine in the 1930s and the appearance of more convenient servicing stations elsewhere, the train station at Fossil became much less important. This was the beginning of the end for the town, and like the thousands of fossilized species that once inhabited Eocene Fossil Lake, the town of Fossil eventually became extinct and ceded its status as a regional population center to the town of Kemmerer, about 10 miles to the east. Today there are only two inhabited buildings remaining in the town of Fossil: one a residence for some of the remaining Lewis family members (who came to own what is left of the town of Fossil), and the other a residence for Carl and Shirley Ulrich. There is a very telling sign at the border of Fossil today that says:

FOSSIL
City Limits — Pop. 2 ± 4

FIGURE 8B

Today the town of fossil has only five residents: Carl and Shirley Ulrich and Richard, Roland, and Betty Lewis. *Top left:* This humorous sign at the border attests to the town's now-defunct status. *Top right:* Photograph of the Lewis family taken in 2009; *from left to right:* Roland, Betty, and Richard L. Lewis. The west end of Fossil Butte shows in the background. *Bottom:* The 4,600-square-foot Fossil Butte Monument Visitor Center, beautifully nestled among the natural vegetation with the west end of Fossil Butte in the background. Many fine FBM specimens are on exhibit in this facility.

One of the earliest commercial collectors was a Mr. A. Shoomaker, a conductor for the Union Pacific Railroad with a sideline of collecting fossils. He collected a number of FBM fishes and other pieces in the 1870s and then sold the collection to O. C. Marsh at Yale University in 1877. In his letter to Marsh of October 9, 1877, Shoomaker offered the collection to him for "[cost plus] 25% from the proceeds for my trouble" (letter in the archives of the Yale Peabody Museum). This is about all that is known about Mr. Shoomaker. I could not even find his full first name.

"Pap" Wheeler was a legendary pioneer of the region during the late nineteenth century, who also had an early commercial operation in the FBM. An article in the *Kemmerer Gazette* (February 1, 1928) described him as "a famous fossil digger, over 70 years of age, [who] lived in a cabin on Fossil Hill. In 1889 in company with a man named Cutch he started walking to La Barge to spend the winter and was frozen to death. His body was never found." It is uncertain how the reporter knew Wheeler froze to death if his "body was never found," but clearly the life of an early fossil "digger" in the FBM was a hard one.

Chas "Stovepipe" Smith is said to have been a commercial quarrier during the early and mid-1880s on what is now Fossil Butte in the National Monument (then called Fossil Hill), but I could find no details of his exploits. Ike Smith quarried the FBM for the first few years of the twentieth century, and the Smith Hollow quarry where he worked (FBM Locality B) was named after him. Irvan Porter and his brother Dan dug fossils in the FBM from 1912 through 1917 on Fossil Butte and FBM Locality B and sold fossils to various museums and collectors, but no detailed information about this family is known.

Since the late nineteenth century, there have been a number of family "dynasties" that have had multiple generations involved in the FBM fossil quarrying business. One of the first was the Haddenham family, with a long although sporadic presence working the FBM from the 1880s through the 1960s. There are many fine Haddenham specimens in major museums around the world (e.g., fig. 50, *top*, and fig. 118, *right*). William James Haddenham's family immigrated to Wyoming from London, England, in 1882. In 1888 they moved to the town of Fossil and "engaged in digging fossils and were very successful" (*Kemmerer Gazette*, 1928). That year Wyoming was still two years away from becoming the 44th state of the Union and was still part of the great Western Territories. The region around Fossil was "very wild" then, and according to George Haddenham (son of William J.) in the above-cited *Gazette* article, "There were many Indians and wild game. The Indians would build a town of tents in one night and when they would get enough 'fire water' they would get very disagreeable. They would trade buck skins and beads for food, especially bread, as meat was plentiful." A few years later the Haddenhams relocated to the town of Green River, but three following generations of Haddenhams (David C., another son of William J.; David F., son of David C.; and Robert, son of David F.) continued over the next six decades to come back and quarry the FBM for fossils. David C. Haddenham (fig. 9, *top left*) spent a total of 50 years quarrying fossils on Fossil Butte, beginning in

FIGURE 9

Early commercial quarriers who worked in the FBM. All photos courtesy of Arvid Aase and the Fossil Butte National Monument except where indicated as otherwise. *Top left:* David C. Haddenham (1881–1962) and son David F. Haddenham (1905–1985) in their FBM quarry on Fossil Butte. Photo taken somewhere around 1930. *Top right:* Robert Lee Craig (1866–1938), commonly known as "Lee Craig," preparing a †*Notogoneus* that he collected from the 18-inch layer of his FBM quarry in Smith Hollow (FBM Locality B). Photo circa 1925, courtesy of the Utah State University. *Bottom left:* Lee Craig shown in the tented wagon where he lived in the valley below his FBM quarry during many of his digging seasons. Photo circa 1927. *Bottom right:* Lee Craig in his quarry in 1927. Craig is the man with the hat.

the early 1900s, when he occasionally worked with another early worker of the FBM, Robert Craig.

Robert Lee Craig (1866–1938) was the first full-time commercial fossil quarrier for which there are many significant records (fig. 9, *top right* and *bottom*). Lee Craig, as his friends called him, built a reputation as "the leading hard-rock fisherman of the country" (*Kemmerer Gazette*, 1926) and was one of southwestern Wyoming's most picturesque pioneers. Even *National Geographic* magazine carried an article about him and his quarrying activities (Powell 1934). From 1897 to 1937, Craig worked various 18-inch layer localities on Fossil Ridge, including Smith Hollow (FBM Locality B), and on Fossil Butte. Craig lost his right leg in a Wyoming coal-mining accident several years before he started his fossil quarrying career. Some writers have since given him the moniker "Peg-leg" Craig for the wooden leg he had to wear after the accident, but that nickname was applied to him mainly by these later writers rather than his contemporaries. He was actually known by local residents of his day as "Judge Craig," because he was a justice of the peace and an election judge for the nearby town of Kemmerer. Besides digging fossils, Craig was said to have enjoyed playing poker, shooting pool, and chewing tobacco.

Life was not easy for Craig. He carved his own wooden leg pegs out of barrel staves (e.g., fig. 9, *top right*). He lived the summer each year in a tent that he pitched about a quarter mile downhill from his quarry or in a tent-covered wagon. Each morning at 7:00 a.m. he would push a wheelbarrow about 300 feet up a steep dirt path to the quarry to work the 18-inch layer near the town of Fossil. Using hammers, hatchets, hacksaw blades, and chisels, he split the rock looking for fossils and carried his finds down the hill in his wheelbarrow. There were no tractors or bulldozers in the early days, so dynamite or black powder was used to remove the 20- to 40-foot-thick layers of rock overlying the 18-inch layer. Back at camp, most of his evenings and rainy days were spent cutting out the fossils and preparing them with a pocket knife. Craig's primary business was mail-order, and he sold pieces to the Smithsonian, Yale, Harvard, Princeton, and the University of Michigan, just to name a few of his institutional clients. He also sold to foreign museums, wealthy collectors, and passing travelers. Craig would even sell a few pieces locally at the train stop in the town of Fossil and in the town's only grocery store. During the summers, his life in the camp tent was quite spartan, and as Lyle Bradley put it (1987), "Craig stacked up poorly on personal habits." Near the end of the digging season, he would send most of his pieces out east and get his main compensation so he could rent a house to wait out the cold winter weather. When spring came, he would move back into his tent by the quarry. But he was said to always keep a clean suit to wear for when he filled the role of election judge or justice of the peace.

Craig had the strange idea that the fossils in his quarry were 750,000 years old when he started his quarrying activity. There is no known explanation why he thought that, but he marked his age estimates on the specimen labels that he provided to museums and other buyers. He once related to a visitor, "I prepared a fish for the Smithsonian Institution in Washington and labeled it 750,032 years old. They wrote me and asked why the 32 years? I told them I had been digging for 32 years." Some of the famous pieces that Craig excavated include a large gar and a snake (illustrated in figs. 53, *top*, and 107, *bottom*). Most of his business, of course, was with fossil fishes.

Craig often worked alone in his quarry, although he occasionally hired part-time help. It was the seasonal help that ultimately did him in. During his last year of full-time quarrying, two unscrupulous men Craig had hired for the summer of 1935 changed his life for the worse. They came to his camp one night, tied him up, beat him, and set his tent on fire in an attempt to rob him of cash they thought he was hiding. Craig lived a marginal existence, though, and the only money to be had at the time was the few dollars he had in his pockets. Craig managed to crawl out of the burning tent and was taken to the Kemmerer hospital, where he was treated. Although he partially recovered and tried to continue his work in the fossil fish beds, he never fully recovered and was in ailing health from that point on. His assailants were never captured. Craig died in 1938 in the Kemmerer hospital at the age of 72 and was buried in the paupers' end of the Kemmerer cemetery next to a cattle rustler. His headstone was a simple concrete block with the name "Lee" cast into it. Many decades later, Craig's headstone was replaced with a custom-made marker in granite with a carving of an FBM fish (†*Priscacara*) and the words "Father of Fossils, Robert Lee Craig" carved into it. The new headstone was financed by the president of the American Association of Paleontological Suppliers (Bill Mason), ceremoniously presented to the mayor of Kemmerer, and installed in 2001. The original headstone is now in the collection of Fossil Butte National Monument.

The commercial presence in the FBM continued well beyond Craig's passing and to the present day. The later generations of the Haddenham family continued their quarrying activity off and on until the late 1960s. Various others for whom there are no records and for whom even the names have been lost worked the FBM during the early half of the twentieth century. But it is not until the 1940s, 1950s, 1960s, and early 1970s that we see the early presence of three new family quarrier dynasties in the FBM: the Ulrich, Tynsky, and Hebdon clans.

The Ulrich family is probably the best known of the present-day commercial quarriers (present day as of 2010, that is). Carl and Shirley Ulrich began working the FBM in 1947 around the vicinity of Fossil Butte and eventually moved to land leased from the state of Wyoming, where their work crews still quarry today (FBM Locality E). My own first experience digging in the FBM began with a week on the Ulrich state commercial quarry many years before I had finished graduate school and become a professional paleontologist. Although I had to buy each fish that I found, I learned a few early tricks of the quarrying trade from Carl and his work crews. Their years of experience working the FBM rock had taught them techniques that they freely relayed to me, many of which I still use today with my own quarry crews from the Field Museum.

Carl Ulrich, born in 1925, started working in the 1940s as an underground coal miner south of Diamondville and as a part-time card dealer in the casinos of Kemmerer and Jackson, Wyoming. As a child he had dug for fossils on Fossil Butte with David Haddenham. Shirley Ulrich was the daughter of parents who worked for the coal mine in the days when the mine held its workers in virtual servitude, and she had no desire to follow in their footsteps. Early in their relationship when things started to get serious, Shirley told Carl in no uncertain terms that he had to give up card dealing if he wanted to marry her. Carl chose Shirley over the casino, and a lifelong partnership began, in more than one way.

In the late 1940s, the Ulriches started looking for FBM fossils as a hobby but managed to sell a few pieces, which eventually convinced them to make a go at the commercial fossil business. The prospect of digging fossils for a living ultimately seemed much more attractive than coal mining or card dealing. Things started slowly in the beginning but eventually blossomed into an extremely successful enterprise that still thrives today. The Ulriches were awarded some of the very first of the commercial fossil quarrying permits from the state of Wyoming in the late 1960s. They have been working the same FBM quarry for decades (FBM Locality E). Through their permit, they are allowed to keep and sell most of the fossils that they find, but as with all the state commercial quarries, they are required to turn some of the rarer species of vertebrate fossils over to the Wyoming Geological Survey (e.g., birds, bats, turtles, amphibians, crocodilians, and mammals).

Carl and Shirley's son Wallace also engaged in the FBM fossil business and had his own quarry on private land for a brief time (FBM Locality B). Currently, Wally works stone from his parents' quarry and fashions much of it into tabletops and other interior decorative stone pieces in which the fossil-bearing stone is hardened and polished. Carl and Shirley's daughter Gail and their grandson Paul also occasionally pitch in to help. Michael Snively, hired in 1990, is currently managing much of the Ulrich fossil enterprise and may carry the core business into the future.

The Ulriches were among the first of the twentieth-century commercial quarriers to take special care and use modern tools in the preparation of many of their fossils, particularly the rarer pieces (e.g., the †*Cyclurus* in fig. 59). They were strong advocates of protecting federal lands from commercial exploitation and helped campaign for the creation of a national park to preserve significant portions of the FBM. In 1972 the park became a reality with the creation of Fossil Butte National Monument (fig. 8B, *bottom*). For the Ulriches and the other legal commercial quarriers, this preserve served a dual purpose. First, it showed altruistic foresight in preserving a significant part of this national heritage for future generations. Second, it eliminated a lot of potential competition and

FIGURE 10

Representatives of three late twentieth-century family dynasty operations in the FBM. *Top:* Carl and Shirley Ulrich standing in front of their house in the town of Fossil. Photo taken in July 2010. You can also see their quarry (FBM Locality E) far in the distance in the upper left of the photo. *Bottom left:* James E. Tynsky working in the small †*Knightia* mass-mortality beds near the K-spar tuff layer in his quarry at FBM Locality A in July 2010. *Bottom right:* Rick Hebdon in his sandwich bed quarry at FBM Locality H in July 2009.

uncontrolled outside exploitation. Everyone benefited from this initiative in the end.

The Ulriches' business profile and political recognition grew through the late twentieth century. In 1960 President Eisenhower purchased a large †*Diplomystus dentatus* from Carl Ulrich to present as a gift to Japan's Emperor Hirohito from the American people. During the George H. W. Bush presidency, Secretary of State Jim Baker would occasionally buy FBM fossils from the Ulriches to present to foreign officials. Carl and Shirley were also friends with Casper, Wyoming, resident Dick Cheney before he became vice president of the United States, and they grew up with early Kemmerer resident Edgar Herschler, who would later become governor of Wyoming from 1975 to 1987. Wally Ulrich is a past state chairman of the Wyoming Republican Party, and in 2010 he was appointed for a time as the acting state geologist for the state of Wyoming by Governor Dave Freudenthal. The breadth and depth of the Ulrich family's involvement in state and federal politics over time is remarkable and has earned them the respect of many public officials with regard to their fossil enterprise and cultural networking.

Carl and Shirley built a very successful business out of their FBM fossil operations, and their pieces are in museums and private collections all over the world. In 1979 they moved from the town of Kemmerer to the defunct town of Fossil, where they built a large house at the base of Fossil Butte just outside the park on the road to the Fossil Butte National Monument visitor center. From this beautiful estate and prime location (fig. 10, *top*), huge windows look out on the main butte of the monument, and the famous Twin Creek runs just outside their back door. (The famous paleontologist E. D. Cope used "Twin Creek" as his name for the Fossil Butte localities.) The Ulriches also have a fossil-preparation facility and sales floor located in the building, which is visited by many tourists each day. With the emigration of the Ulriches to Fossil, the population was once again on the rise, boosted to a grand total of five residents. (Kemmerer is not in any danger of losing its regional prominence to the town of Fossil any time soon.) The Ulrich family has clearly made its mark in the history and commercialization of FBM fossils.

Another "dynasty operation" in the FBM is the Tynsky family—four generations of fossil quarriers who have leased both state and private lands. In the 1960s, Sylvester "Switz" Tynsky and his two sons Robert D. (Bob) and James A. (Jim A.) opened the very first of the nearshore sandwich bed quarries. It was located on a 40,000-acre sheep ranch belonging to a friend of Bob's, Ronald Thompson. At first other quarriers were skeptical about whether the Tynskys would find many fossils at this locality because it was so far north of where the historic FBM quarries were. Development of the quarrying operations started slowly, but find fossils they did! Today their original site (FBM Locality H) is the most productive of all of the FBM quarries, and the Thompson Ranch is home to the most productive nearshore sandwich bed quarries. In the late 1980s, the lease for FBM Locality H was acquired by the Hebdon family (discussed below).

In the mid-1970s, Bob and Jim A. split the Tynsky operation between them, part for Jim A. and his son Duane, and the other part for Bob and his son James (Jim E.) and Jim's wife, Karen. Jim and Karen were eventually joined by their children Robbie and Stacey, adding a fourth generation of fossil diggers to the family. Jim and Karen's complementary skills and interests combined for a modest but successful trade. Tragically, Karen died of cancer in 1998 at the age of 49, and for many years to follow, Jim ran the business by himself, with his children pitching in from time to time. In 2009 he was joined by a new partner, Vicky White, whom he married in 2011. It is Jim E. Tynsky who has consistently been the most generous and helpful of all the quarriers to my own research career and to the Field Museum's teaching and field programs in the FBM.

In 1984 Jim E. Tynsky moved from FBM Locality H to a mid-lake 18-inch layer site on the Lewis family ranch, FBM Locality A, where he leased several acres. I have brought Field Museum crews to Jim's Lewis Ranch quarry each year for the last 26 years, building the world's greatest collection of FBM fossils at the Field Museum. Over the last eight years, I have also brought many classes of students for my University of Chicago Graham School course called "Stones and Bones" (a summer course in field paleontology for junior- and senior-level high school students from around the world). This experience has helped spark a number of students to pursue paleontology as a career. Jim E.'s overall contributions to the scientific and educational development of the FBM have been huge, but those contributions are of the sort that history often does not acknowledge and even some scientists do not fully understand. Perhaps these words will help his historical role be better remembered.

Another of the quarry dynasties still working in the FBM today is that of the Hebdon family. Virl and Shirley Hebdon, residents of Thayne, Wyoming, owned a large sheep ranch that included a creek and spring by the name of Warfield Springs. In 1970 their son Rick discovered fossil fishes eroding out of the creek bank, sparking the beginning of quarrying in the Warfield Springs area (FBM Locality K). Then Rick and his father, Virl, began mining fossils part-time, selling them to local rock shops. By 1983 Rick decided that the fossil business was much more lucrative and a whole lot more interesting than sheep ranching, and so he went into it full-time. He maintained ownership of the Warfield Springs quarry and added FBM quarries from both the mid-lake 18-inch layer and nearshore sandwich bed sites to his overall fossil operations (FBM Localities H and G). Hebdon currently works all three of these quarries with his wife, Tanya Hester. He employs 6 to 16 people per year, maintains a staff of fine preparators, and today has the largest and most broadly diversified of the commercial operations within the FBM.

One of the most recently formed dynasty operations quarrying in the FBM is the Lindgren family. Tom Lindgren (fig. 102) began working the sandwich beds of Thompson Ranch in 1978. He worked in the FBM sporadically in various sandwich bed and 18-inch layer quarries until 1985, when he began leasing FBM

Locality B from the Lewis family (discussed below). For the next 10 years, Tom continued leasing that site, and in 1998 he partnered with Scott Wolter of Minnesota to purchase this quarry from the Lewis family. For the next two years, Tom and Scott worked the quarry together, and in 2000 they sold the quarry to Green River Properties and the Green River Stone Company, which still owns the property today. Tom's three sons also eventually entered the business over the years, with David starting in 1997, Anthony (Tony) in 1998, and Adam in 2004. They worked in several different quarries, sometimes digging for "shares," sometimes subleasing small plots from other quarriers, and sometimes leasing quarries themselves. Adam found one of only three snakes known from the FBM (fig. 107, *top*) and currently works with his brother Tony on a site leased from the Lewis family. According to his father, David was the most prolific of his boys in finding fossil birds, turtles, and bats. I met David in 2010, and he was unusually enthusiastic about his work and a good fossil preparator. Although only in his early twenties, his body was heavily tattooed with images of his favorite finds, including a stingray, a †*Mioplosus* aspiration, an †*Amphiplaga*, a †*Diplomystus*, a †*Priscacara*, a dragonfly, a bat, and a turtle—all fossils he had excavated and prepared from the FBM. Tragically, he was killed in an automobile accident in 2011 at the age of 29. Tom, Tony, and Adam continue working in the FBM quarries of the Lewis Ranch and the Thompson Ranch.

The Lewis family (fig. 8B, *top right*), although never fossil quarriers themselves, were key to the development of today's FBM quarrying operations on private land. Richard H. Lewis and his wife, Susanna, moved to the town of Fossil in 1884 and were the first settlers to establish a home there. But as the town dwindled in the 1930s and 1940s, and eventually died in the 1950s, the Lewis family acquired various pieces of the town through purchase or filing for quiet title. Eventually, the Lewis family came to own the entire town of Fossil, although there was not much "town" to it once the railroad station moved to Kemmerer. They later sold one plot of land in Fossil to Carl and Shirley Ulrich, who built the beautiful residence referred to in figure 10 (*top*). The Lewis family also owns several thousand acres around Fossil Ridge. They were originally horse and cattle ranchers who intended to buy only the relatively lush valley property suitable for ranching; but due to an error by the surveyor, the Lewis family also ended up accidentally owning the tops of two large buttes. Much later, FBM fossils were discovered there, and the value of that once-worthless butte-top land would far exceed the value of the valley property. It was on the Lewis Ranch that the first privately owned 18-inch layer quarries were developed in the 1970s and 1980s. These quarries are still mined today by four or five different quarrying operations. The lease fees and share of fossil sales from the fossil quarrying operations have been much more lucrative than ranching for the Lewis family.

Fossils from the FBM can today be found all over the world, in rock shops, souvenir shops, and even art galleries. They consist mostly of the common varieties of fishes (†*Knightia*, †*Diplomystus*, †*Mioplosus*, †*Priscacara*, †*Cockerellites*, and †*Phareodus*). One of the FBM species, †*Knightia eocaena*, has the distinction of being the world's most common vertebrate fossil preserved as a complete skeleton, and I have personally seen it for sale in souvenir shops, flea markets, and street fairs of more than a dozen countries I have visited over the years. Art galleries prefer to display the larger and more unusual specimens, such as the showy palm fronds or well-prepared large plates with multiple fossils on it. The quality of preparation of commercial specimens ranges from superb to awful, so buyers should bring a magnifier. Rare pieces with high price tags should be carefully examined to make sure that they are not composites or heavily painted (see appendix D). At the Field Museum, like most scientific institutions, we prefer to get specimens that are unprepared, because we can usually spend much longer than commercial fossil dealers in preparing the pieces for display or scientific study.

There are probably few, if any, fossil deposits in North America that have been worked as extensively as the FBM quarries. The extreme relative rarity of many species in the FBM means that there would have been no other practical way to get a sufficient sample size of this fossil biota other than to have had the mass participation of many people. It is more or less a type of "citizen science." Permits for either scientific or commercial collecting in the FBM on state land can be obtained from the state of Wyoming, but most commercial quarrying currently takes place on private land. Cooperation and interactions between the commercial quarriers of the FBM and the professional scientists working with the FBM fossils have been exemplary for the most part. Many quarriers have donated extremely important material to scientific institutions (e.g., Jim E. Tynsky, Rick Jackson, Tom Maloney, Bob Kronner, Tom Lindgren). My own scientific career and science in general have benefited by the activities of these hardworking people.

Bone Wars and Other Conflicts Touching the FBM

It is amazing to me just how much passionate conflict has been associated to one degree or another with the FBM fossils. Perhaps this is inevitable for anything that involves a mix of ego, competition, notoriety, and money.

The first scientific DESCRIPTIONS of vertebrate fossils from the FBM came from Edward Drinker Cope and Othniel Charles Marsh, two nineteenth-century paleontologists famous for their fierce competitive "bone wars" over fossils of the American West. Both were pioneering vertebrate paleontologists during the Gilded Age of American history who used their wealth and influence to fuel their own expeditions and fossil acquisitions from amateur and commercial collectors. They were both remarkably productive, published an enormous number of papers, and named a large number of species. But each of the two iconic figures reportedly used less than ethical means to outcompete the other, including bribery, theft, slander, and even destruction of bones in the field to prevent the other from acquiring them. Each attacked the other in their scientific publications in an effort to ruin the other's social standing, scientific reputation, and ability to raise funding. The enmity and ruthless competition between them was said to have been largely responsible for Cope's mentor, the mild-mannered Joseph Leidy (page 18), eventually quitting the field of vertebrate paleontology, which he had originally helped found. In the end, both Cope and Marsh exhausted their own funds and ethical credibility in their attempts to gain paleontological supremacy. Today, thanks to the forgiving nature of historical memory, both men are recognized primarily for their vast scientific accomplishments.

A stingray from the FBM was involved in the Cope/Marsh feud. Both Marsh and Cope described the same species under different names, but Marsh's name, †*Heliobatis radians*, was described in 1877, two years prior to Cope's name, †*Xiphotrygon acutidens*, and therefore is the legal name by the nomenclatural rule of **PRIORITY**. Cope was aware of Marsh's previous description but either mistakenly thought his ray to be different than Marsh's species or dismissed Marsh's name out of spite. In a controversial tone that was typical between the two scientists, Cope described his ray without examination of the Marsh specimen, and in his published description he referred to Marsh's previous description as "a meager" work.

Another more recent conflict that has occasionally involved the FBM is between commercial/private collector interests and professional scientific interests. There are extremists on both sides of the issue of commercial fossil collecting with very strong opinions, including some commercial collectors who believe that they should be able to dig fossils anywhere they please, and a few professional paleontologists who believe that there should be no commercial development of fossils whatsoever. A good balance of resource use currently exists, with some of the FBM exposures protected through establishment of Fossil Butte National Monument and restrictions on state lands, while other exposures on private lands are available for responsible commercial development. Nevertheless, there have been a few instances of abuse by commercial collectors on one side and instances of overzealous protectionism on the other side.

The worst of the commercial and amateur collecting abuses is outright theft, with several documented cases of illegal removal of FBM fossils from Bureau of Land Management property and even from the national monument itself. In November 1993, a part-time commercial fossil digger and his friend broke into the Fossil Butte National Monument visitor center in an attempt to steal some of the fossils exhibited there (fig. 8B, *bottom*). They were not particularly skillful thieves, even signing the visitor book a few days prior to the break-in while they were casing the place for their heist. They caused some significant damage during the break-in and two years later were convicted in federal court for the crime.

Looters on state and private land often use clumsy tools to quickly remove pieces, destroying valuable material in the process. In my early days of working in the FBM, my field crew and I used to camp off-site by the Hams Fork River near

Kemmerer. One day after we had spent the morning exposing a particularly nice plate of large fishes in the quarry floor, we decided to wait until the next day to saw it out for removal. We threw a tarp over it and headed down the butte into town. When we returned the next day, all that remained in the quarry floor was a large hole with chisel marks around the edges. Someone had come into our quarry while we were gone and stolen the slab. That day we relocated our camp to the top of the quarry butte, and in the 25 years since then we have always camped on the quarry butte to ward off potential fossil poachers.

But some abuses have also been perpetrated by individuals trying to "protect" FBM deposits. In 1997 in the National Geographic documentary *Thieves of Time*, a melodramatic exposé of supposed illegal fossil dealing, a FBM stingray took center stage (as reported in the December 11, 1997, *Denver Westward News* by Eric Dexheimer). A Lincoln County sheriff who founded an anti-fossil-poaching campaign called "Operation Rock Fish" was receiving much attention in the press. He set up a sting operation for the National Geographic television documentary where he busted a fossil dealer in a Wyoming parking lot on camera and confiscated an "illegal" FBM stingray from him. Based on faulty information, the sheriff claimed that the fossil was illegally poached from public land. As the NGS show closes, the sheriff says to the camera, "This is a very good feeling that we're able to get this back. . . . It's like finding a lost child and returning it. It's home where it belongs. So we call it a win." Shortly afterward the "win" turned into a big loss when it was discovered that the stingray came from the private land quarry of Rick Hebdon (FBM Locality H). The sheriff's department then had to return the specimen to its rightful owner, an Italian fossil dealer. The piece was eventually auctioned off at the big annual fossil show in Tucson, Arizona. The proceeds from its sale were divided in half, with 50 percent going to the owner and 50 percent placed in a scholarship fund for student research in paleontology set up by the American Association of Paleontological Suppliers (organizational name changed to Association of Applied Paleontological Sciences in 2005), according to the president of that organization, Bill Mason.

The controversy between commercial development of fossil land and the professional scientists' desire to keep important specimens in the public domain is a complicated and often passionate one, and some of its key issues differ from locality to locality. The FBM deposits illustrate a case where a mutually beneficial equilibrium has been reached by most involved parties. Responsible commercial interests make a living excavating large numbers of these fossils from legal sites, while scientists such as me have access to a sample size far beyond any we could have possibly gotten by ourselves due to limitations of time and resources.

Even the metaphysical controversy between religion and science touches the FBM. Some believers of SPECIAL CREATIONISM consider the FBM (as well as the rest of the Green River Formation) as relevant to the "debate on the age of the earth" (e.g., Garner 1997) or as demonstrating that no evolutionary change has occurred (Yahya 2006). The Young Earth Creationists (today a Christian faith-based group) contend that Earth is only between 6,000 and 10,000 years old, following a literal interpretation of Genesis, rather than 4.5 billion years old, as indicated by science. They consider the FBM, with its millions of fishes located in a high mountain desert, to be evidence of the "great flood." Muslim writers, on the other hand (e.g., Yahya 2006), accept the geologic time scale and greater age of Earth, but claim that the FBM fossils are structurally identical to living species and that they therefore help prove that evolution has not occurred. (Comparative anatomy is a skill obviously lacking in such interpretations.) Other religious faiths have still different metaphysical interpretations of the extraordinary fossil assemblage as well. As a scientist, it is difficult to even know where to begin to respond to such faith-based views, other than to point out that creationism is not science. SPECIAL CREATIONISM is a specific form of fundamentalist religion of the type that once dictated that the sun and planets all revolved around Earth. The so-called "science vs. creationism" argument is one that has no end because science and religion are not really comparable entities. Religion is a *belief* whereas science is a *method*. Religious fundamentalism involves unquestioning acceptance of proposed "truths" as interpreted and communicated by spiritual leaders and communities, whereas science is a method that is forever challenging itself in the study of nature. Science is constantly creating and changing hypotheses, accepting those hypotheses supported by the most observable data, and falsifying hypotheses contradicted by EMPIRICAL data. This is not a specific criticism of either religion or science; it is merely a caution that they are not really comparable concepts.

The cultural history of the FBM fossil activity—whether by early pioneers, scientists, commercial collectors, or other parts of society—is rich and fascinating in its own right, but for now I will return to the more pragmatic aspects of the FBM and how we work there.

A Typical Day in the Field

Most of the FBM fossil quarries lie about 10 miles west of the coal-mining town of Kemmerer (est. 1897, pop. 2,651). Kemmerer and its neighboring town Diamondville (pop. 716) comprise the self-proclaimed "Fossil Fish Capital of the World," as touted on local road signs. The town of Fossil is now gone, and Kemmerer and Diamondville are home to many of the FBM quarriers. Kemmerer's fame also extends to housing the very first JCPenney store ("the Mother Store"), still there today. And as a Chicagoan, I am compelled to mention that Kemmerer is also known as "Little Chicago" for being a major supplier of moonshine liquor to Chicago and much of the midwestern region of the United States during Prohibition.

There are numerous full-time commercial FBM quarries, mostly within the 18-inch layer and sandwich bed horizons, and many are mapped in appendix A (FBM Localities A–M). The FBM fossils occur on private land, railroad land, state

FIGURE 11

The Lewis Ranch FBM Locality A, owned by the Lewis family of Fossil, Wyoming, and leased by commercial quarrier James E. Tynsky of Kemmerer. *Top:* The road up to the dig site. *Bottom left:* The 18-inch layer exposure along the side of the butte at FBM Locality A, as first discovered in 1983. Oily limestone capping layers above and below the layer make it resistant to weathering, which is why it was jutting out of the side of the butte here. *Bottom right:* A large bulldozer is used to remove about 6 meters (20 feet) of unproductive overlying rock ("overburden") while lighter tractors are used closer to the fossil layer so as not to damage it with the weight of the tractor.

land, Bureau of Land Management land, Forest Service land, and National Park Service land. Without special permits, it is not legal to dig FBM fossils anywhere except on private land with permission of the landowner.

Each summer my field crew—consisting of Field Museum preparators, volunteers, students, and myself—packs up the museum trucks and heads west to southwestern Wyoming for a couple of weeks of quarrying in the FBM. It is something I still look forward to after all these years, because every year is different and filled with the anticipation of exploration and discovery. We never know what will be under the next slab of limestone we pry up. All we know is that whatever it is, it hasn't seen the light of day for 52 million years.

We work in a five-acre section of the Lewis Ranch quarry leased to commercial quarrier Jim E. Tynsky (fig. 10, *bottom left*), where we have enjoyed digging privileges for over two and a half decades. During that time, my numerous team members and I have mined thousands of fossil fishes and a host of other fossil animals and plants out of the Wyoming butte, establishing the Field Museum as the world's largest, most comprehensive collection of FBM fossils.

The quarry lies in the high mountain desert about 2,165 meters (7,100 feet) above sea level. It takes some of our midwestern crew a day or two to adjust to labor at high altitude and low oxygen levels, but once our bodies have produced a few extra red blood cells, the excitement of the dig takes over. To hit the dry season and avoid the rains, we make the trip in late June and early July. The quarry cannot be efficiently worked in the rain, and lightning storms can be hazardous because our tents and vehicles are the highest points on the butte. Summer temperatures can swing widely, typically ranging from around 40 degrees Fahrenheit at night to 90 degrees in the heat of the day. There are no trees on the butte, so the only shade is from the trucks and rock exposures. Sunscreen, hats, and sunglasses are a must. Water must be drunk regularly to prevent dehydration in the dry mountain-desert air. Still, it is a sharply clean air we breathe and stunningly beautiful landscape that surrounds us. Antelope and deer roam the valley below us, golden and bald eagles soar overhead, and mountain bluebirds fly in and out of their nests in the quarry walls. In the evening, the stars are intense and sharply defined, and the moonlight is so bright you can read by it. The outdoor physical labor is spiritually cleansing for those of us who spend so much of our time at computer screens, microscopes, and office meetings for the rest of the year.

Life in the quarry begins at sunup. We are all camped at the top of the butte just above our quarry site. Although the cool mountain night air provides ideal sleeping temperatures in a warm sleeping bag, as soon as the sun hits the tent wall in the morning, the inside temperature rises from comfortable to intolerable. Sleeping late in the tent is not an option on sunny midsummer days. After a quick breakfast and coffee around the remains of the previous night's campfire, we head down to the quarry.

Fossil excavation is a mix of force and finesse. Bulldozers remove about 9 meters (30 feet) of OVERBURDEN to get near the top of the 18-inch layer (fig. 12,

FIGURE 12

Lewis Ranch FBM Locality A, after the upper layers of overburden have been removed. *Top:* Smaller tractors are used to dig to within a few inches of the tough upper capping layer. Then the last 2 to 2.5 centimeters (5 to 6 inches) of capping above the 18-inch layer are removed by hand in order to not damage the fossiliferous slabs with the weight of the tractor. Photo taken in 1986. *Bottom:* The 18-inch layer is lifted one thin layer at a time. Then we clean the surface of each slab with brooms and gasoline-powered leaf blowers, while looking for bumps on the limestone surface that indicate fossils beneath. Photo taken in 2009.

FIGURE 13

Lewis Ranch FBM Locality A. *Top left:* Once a fossil is discovered, it is outlined in pencil using custom-size templates to create uniform-size slabs for efficient packing. *Top right:* The limestone is cut along the pencil lines with a gas-powered saw fitted with a diamond masonry blade. *Bottom:* The layer around the cut slab is removed so the fossil-containing piece can be safely freed from the rock surface below.

FIGURE 14

Lewis Ranch FBM Locality A. *Top:* A cut slab removed from the rock surface below shows a nicely centered †*Diplomystus dentatus* still covered with a thin layer of rock. The fish skeleton under the surface is indicated by raised bumps running along the center of the slab. *Bottom:* Once cut, the uniformly sized slabs can be efficiently and safely packed for transport back to the Field Museum in Chicago.

top). Then hand-wielded sledgehammers, pry bars, and sturdy shovels are used to remove the next 13 or 15 centimeters (5 or 6 inches) of hard, oily, dark gray "capping rock." This is hard work that takes the first day or two of the trip and is dreaded by field veterans and is eye-opening for beginners. The tough capping rock smells like oil when you smash it, and it has helped protect the lighter-colored fossiliferous limestone layers from weathering over the millennia. Once the upper capping layer is removed, the finely laminated, lighter-colored layers of the 18-inch layer are exposed. We sweep the surface clean and check for signs of fossils in the angled sunlight that falls across the slab's surface and remove the fossils that we find. We work through the entire 18-inch layer one layer at a time, hammering sharpened flat shovels and strips of thin spring steel into the layer's edges to take advantage of naturally occurring separations along the bedding planes (fig. 12, *bottom*). If the rock is dry and splitting properly, it lifts up in large sheets between 2 and 5 centimeters (0.8 and 2 inches) in thickness. Once a slab is lifted, we check for fossils on its underside as well as on the newly exposed surface below. A very thin layer of rock (part of the MATRIX) usually covers the fossils, which can be both a blessing and a curse. The matrix will protect the fossils as they are sent back to the museum, but it can also hide small fossils if the sun is directly overhead. Angled sunlight is necessary to cast defining shadows from the underlying fossils on the light-colored rock (fig. 14, *top*). For this reason, work in the quarry is suspended between 11:30 a.m. and about 3:30 p.m., when the sun is near directly overhead.

Finding fossils is a different sort of operation in the nearshore sandwich bed quarries of Thompson Ranch (fig. 15). The fossiliferous layer is much thicker than the mid-lake 18-inch layer. The slabs also come up in thicker pieces that split more easily, and prospecting involves resplitting slabs with a hammer and chisel rather than taking up broad sheets of rock off the quarry floor. Large fishes with a thin covering of matrix can be found in the sandwich beds, just like the 18-inch layer, and once prepared these fishes are some of the most beautiful (e.g., fig. 6A, *top right*, and fig. 54).

When fossils are found in either the 18-inch layer or the sandwich beds, they are marked and cut out with a gasoline-powered, diamond-blade rock saw. The cut slabs are sized to fit carefully in custom-made crates for shipment back to the museum in Chicago (fig. 14, *bottom*). The locality is extremely productive. In two weeks' time, we can easily collect a couple of tons of cut slabs containing fossils. In 30 years' time, we have built a massive collection of this material in Chicago for use not only by me, but by scientists from all over the world.

FIGURE 15

Thompson Ranch FBM Locality H. This is a typical nearshore sandwich layer quarry (also called F-2). Unlike the mid-lake 18-inch layer quarries, the nearer-shore sandwich bed quarries are mined more commonly by putting the large blocks on edge and splitting them. Hence the name "split fish" sometimes given to fishes from this locality. *Top:* A broad exposure of the sandwich beds at Thompson Ranch. *Bottom left:* Searching for fossils in sandwich beds involves splitting a lot of rock. *Bottom right:* Splitting the slabs often exposes fossils, although this often results in damage to the specimen. Fortunately, this †*Diplomystus dentatus* split out cleanly.

Exposing the Record of
Past Life

*Fossil
Preparation*

For scientific study, it is not enough to have well-preserved fossils. The fossils must also be well prepared. Nature provides the opportunity and potential, but good fossil preparation enables scientific discovery. In the early days of Cope and Marsh, and Craig and Haddenham, it was enough to use a pocket knife and perhaps a bit of sandpaper to prepare fossils from the FBM. Today fossil preparation is taken to a much higher standard, resulting in a fossil with enough detail to easily include it in comparative studies of living species. The sheer beauty of a well-prepared FBM fossil is also something wonderful to behold.

So how do you prepare fossils from the FBM? It depends on which locality the specimen comes from to some degree, but there are standard tools for most jobs, including a binocular microscope, pin vises fitted with rods of tungsten or carbide steel sharpened to a needle point, fine air-powered engraving tools, and an abrasive machine (a micro-powder blaster that shoots a narrow stream of powder under adjustable pressure).

The ideal starting point for unprepared vertebrate fossils from the FBM is a slab with the fossil just under the surface; close enough to see the outline of all bones, but deep enough so none of the fish bones have split off onto the counterpart slab (fig. 16, *left*). Then the process involves the careful removal of rock from the surface of the bones and scales either by gently scratching it to dust with the pin vise, or by using the abrasive machine under a low pressure (fig. 16, *right*). If a needle-sharp hand tool is used, there should be a steady stream of gently forced air pointed at the fossil to keep the surface clean as the overlying matrix is scratched into dust. Otherwise, the dust will obscure the point at which you need to stop scratching to prevent damaging exposed bone. Research museums will spend several months on a single specimen if it is a critical research specimen. Modern commercial fossil preparation of rare pieces is also often of extremely high caliber (e.g., figs. 46, 54, *top*). But commercial preparation of more common

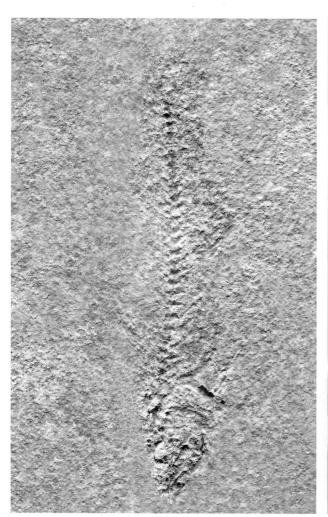

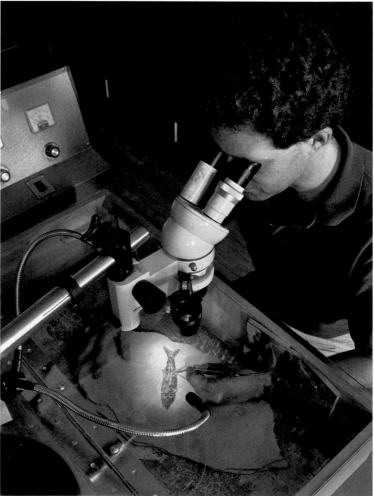

material is sometimes inadequate or even misleading for scientific purposes (see appendix D). Many commercially prepared specimens of the more common species are heavily damaged during preparation and subsequently colored and heavily restored (easy to detect with careful examination under a microscope) (see pages 385–86).

An abrasive machine (fig. 16, *right*) can also be useful for preparation of FBM vertebrate fossils if it is used carefully and on specimens with only a thin layer of rock covering the bones. An abrasive machine is a type of pinpoint powder blaster that shoots a very fine stream. One of the most widely used powders is powdered dolomite. The powder is ideally softer than the bone but harder than the rock encasing the fossil. This machine should be used on FBM fossils only under a binocular microscope and using low air pressure (e.g., 10 psi).

Leaves and insects from the FBM have no appreciable relief to them and are only discovered if they partially split out in the field. A pin vise must be used to finish preparing such tissue-thin pieces, not an abrasive machine. Once exposed, they should be protected from the bleaching effect of direct sunlight and dust, both of which will make fossil leaves and insects from the FBM fade.

Excavation and preparation are the first steps in deciphering Earth history from FBM specimens. Once fossils have been quarried and prepared, they must be identified and classified. Thus we move from the process of collection to that of research.

←⋯ FIGURE 16
Fossil preparation for scientific research. The ideal pieces for preparation are those with the skeleton still completely covered with rock, but close enough to the surface to see ridges indicating the bones beneath. *Left:* A good example is this unprepared †*Mioplosus* with the head facing down, which measures 140 millimeters (5.5 inches) long. *Right:* Preparation was done under magnification using needle-sharp tools, needle-fine engravers, and air-abrasive machines. A perfectly prepared †*Mioplosus* of this size is illustrated in fig. 83 (*top*).

EXPOSING THE RECORD OF PAST LIFE

Classification of Fossils and Their Place in the Web of Life

There is only one history of life on this planet, past and present. Whether you metaphorically call it an evolutionary tree or a web of life, there is an interconnection of evolutionary relationship that unites all of the organisms that ever lived here. So logically, all fossil species should be classified and organized within the same system that we classify living species. The FBM species discussed in this book are all systematically arranged within the LINNAEAN SYSTEM, an organizational system of NOMENCLATURE used for hundreds of years that is today so ingrained in biology that we give it little thought. Here I provide a short explanation of this system and its twofold significance: first as an organizational tool for formally naming and classifying natural groups of organisms (TAXA), and second as a rough expression of evolutionary relationships among taxa (PHYLOGENY).

There are many millions of species living today, plus many millions of extinct species represented by fossils. How do we keep the names and classifications of all fossil and living species organized so we can effectively study this unimaginable biodiversity? We use a system developed by Swedish botanist, physician, and zoologist Carolus Linnaeus (1707–1778). Just as a major library uses a Dewey decimal system to classify and locate a specific book among hundreds of thousands of volumes, biologists use the Linnaean system to efficiently classify and retrieve information about individual taxa among the millions of fossil and living taxa that are known.

The Linnaean system is arranged as a hierarchy (i.e., with lower [narrower] categories within higher [broader] categories). It can also be thought of as a system of subsets within sets (i.e., species within genera, genera within families, and so on). The Linnaean system contains seven basic ranks, from "kingdom" at the highest level (e.g., "Plantae" for all plants and "Animalia" for all animals), to "species" at the lowest levels (e.g., *Canis lupus* for dogs, or *Homo sapiens* for humans) (fig. 17). The seven basic Linnaean ranks, in order, are as follows:

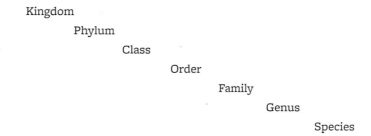

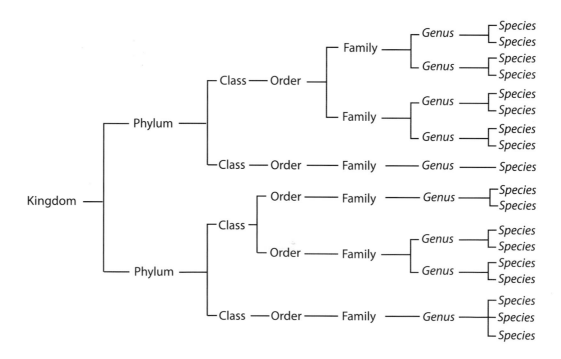

Representative model of the Linnaean hierarchy

Prefixes can also be applied to increase the number of rank levels (e.g., superfamilies and subfamilies added above and below the Linnaean family category, and infraclass added between subclass and superorder). Species is the terminal rank (or lowest rank) for most TAXONOMISTS, although some also use the term "subspecies" for subdivisions of a species. An enormous amount of progress in categorizing the diversity of life has been made thanks to the Linnaean system.

The HIERARCHICAL ARRANGEMENT of the Linnaean system can also express some basic evolutionary relationships among taxa. The Linnaean classifications of plants and animals are not detailed, perfect reflections of phylogenetic patterns. This is because our knowledge of biodiversity and hypotheses of evolutionary relationships are constantly changing. Also, phylogenetic trees can be extremely complex, and there are not enough standardized rank levels to easily express the phylogeny of all life in a single tree. Nevertheless, the Linnaean system still reflects basic elements of phylogeny, revises its groups sometimes to reflect new knowledge about evolutionary patterns, and usually avoids obviously unnatural groups. For example, many taxa originally described and named by Linnaeus (1758) were revised or eliminated because they were recognized as unnatural, such as *Simia* for a group containing all apes but excluding humans. (Even Linnaeus thought humans should be classified with the apes but avoided doing so mainly to avoid conflicts with religious authorities at the time.) If we look at a functional group such as "animals that fly," this is an unnatural group that is grossly out of sync with phylogeny, and it is not given a formal group name in the Linnaean system. Butterflies are classified within the insects (class Insecta) rather than with birds (superorder Aves here), and bats are classified within mammals (class Mammalia) rather than with birds or butterflies, which better follows the natural relationships. Today birds are frequently classified with reptiles as they are here (class Reptilia) because they share a closer relationship to crocodiles and dinosaurs than to mammals, amphibians, or invertebrate animals. This pattern of relationship became more and more clear with scientific progress (e.g., discovery of new species such as †*Archaeopteryx* and "feathered dinosaurs," development of new techniques such as DNA analysis, better fossil preparation, and the creation of better analytical programs for scientific data). Sometimes scientific progress eventually trumps long tradition (e.g., no longer classifying birds as separate from reptiles). The basic correlation of phylogeny with classification makes the system much more informative on many levels, ranging from understanding biodiversity and anatomy to interpreting the evolutionary web of life on this planet.

With regard to rank levels in the Linnaean system, there are no absolute levels for higher taxa. In my classification here, Aves is a "superorder," but in many other classifications it is a "class" or "infraclass." The rules of nomenclature do not cover any taxa above the family-level taxa. And rank levels have phylogenetic significance only within the context of particular groups or studies. Ranking taxa

The Linnaean system of classification with the seven basic taxonomic ranks diagrammed as a sideways tree in order to show the hierarchical nature of the system. Although it is rare that phylogenetic tree branches correspond exactly to the Linnaean rankings, the idea of hierarchical organization of groups within groups is a concept common to both classification and phylogenetic research. Actual phylogenetic trees are usually in a constant state of revision as new information about characters and new taxa are discovered. But the similarity of philosophy is important. The group-within-group concept is critical to both classification and phylogenetic research.

should be thought of as a tool that can be applied in whatever way expresses the organizational needs of a particular scientific project or researcher.

Evolutionary study is basically a retrospective endeavor. We normally cannot directly observe the complete process of evolutionary SPECIATION, which could take thousands or even millions of years of time. Normally we can only see its end product. Consequently, modern evolutionary research is not a search for specific "ancestors." It is instead a search for patterns of evidence indicating relative relationship. Evidence showing two species as being more closely related

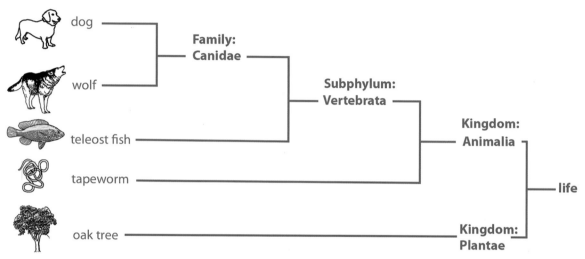

Heirarchical classification of species

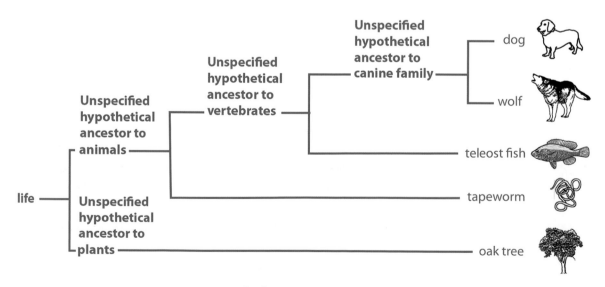

Phylogenetic tree

to each other than to any other species allows us to reconstruct an evolutionary tree. Such evidence of relationship consists of unique similarities or "derived characters" that are uniquely shared between species and larger groups (e.g., hair and mammary glands for mammals, feathers for a group of dinosaurs including birds). We explain the fact that many vertebrate species have hair and mammary glands (along with many other congruent characters) because these characteristics were all inherited from a common ancestor, making mammal species more closely related to each other than to any non-mammalian species. This is extrapolated from our being able to observe heritable change in real time, such as inherited characteristics in our own family members. For example, my sister and I share some peculiar characteristics that can be traced back to our mother or father. We also can observe heritable change by raising generations of fruit flies, guppies, or peas in scientific experiments. In these ways we can observe a process of generational inheritance and evolution. It is then intuitively easier to conceptualize evolutionary process as an explanation for the interconnectedness of all taxa over the billions of years that life has existed on this planet.

Just as a branching diagram can illustrate a hierarchical species classification, it can also often diagram sequential patterns of evolutionary diversification, resulting in what we call a **PHYLOGENETIC TREE**. The dog and the wolf in figure 18 share the characters of fur, four legs modified for walking on land, mammary glands, and many other derived features not shared by the teleost fish or the tapeworm. At a broader group level, the dog, the wolf, and the teleost fish share the characters of bone, vertebrae, specialized eyes, and many other derived characters not shared by the tapeworm. And at an even broader group level, the dog, wolf, teleost fish, and tapeworm all share many derived characters of animals not shared by plants, just as plants share many derived characters not shared by animals. Some patterns of relative relationship are very strong and clearly the result of a non-random process (such as the dog and wolf sharing many more derived characters with each other than with the tapeworm). Evolutionary biologists explain the origin of these strongly congruent patterns of shared characteristics as the result of evolutionary process. The shared ancestor between the dog and the tapeworm is much further back in time than the shared ancestor between the dog and the wolf. It all fits together very well, and in the words of the famous twentieth-century geneticist Theodosius Dobzhansky, "Nothing in biology makes sense except in the light of evolution." Biologists are continually making progress in tying all species together into a single tree or web of life based on their relative degree of relationship to each other. Each new species inserted into that web is a type of "missing link" in the overall network of life on this planet.

This book uses branching diagrams (phylogenetic trees) to illustrate relationships among species and unite all the FBM organisms illustrated here into a single web of life. The detailed character evidence supporting those diagrams will not be given here, although references will be given in the text for those who wish to dig deeper into the subject.

← FIGURE 18

Four animal species and a plant species grouped within a hierarchical arrangement. When read from left to right starting with species (*top*), the diagram represents a hierarchical classification based on special similarities. When read left to right from the "root" end (*bottom*), the diagram can be interpreted as representing a hypothesis of evolutionary divergence (a "phylogenetic tree"). The phylogenetic tree specifies relative degree of relatedness (and the hypothesized order of lineage splitting). This branching diagram indicates that a dog is more closely related to a wolf than it is to a teleost fish, that the teleost fish is more closely related to the group containing the wolf and dog than it is to the tapeworm, and so on. Blue-colored branching diagrams of this type will be used throughout this book to connect all of the known FBM organisms into a single network of life.

Official Scientific Names and Rules of Nomenclature

Frequently used in this book and elsewhere in scientific literature is the term "DESCRIBED." When used by a scientist, it also means "named." When a new species is officially "described," it is given its unique scientific name under official international codes of nomenclature that have been in place for many decades (summarized in Ride et al. 1999, for animals; and McNeill et al. 2006, for plants). In order for a scientific species name to be scientifically VALID, it must satisfy the rules of AVAILABILITY and PRIORITY. A name satisfies the rule of availability if it follows the rules of the International Code of Nomenclature. It satisfies the rule of priority if it is the first available name used to describe this species. The naming of species requires strict rules, policies, and oversight committees because order and consistency of method are necessary for the scientific naming process to be useful in biodiversity studies.

A species name is italicized and is in two parts (called a BINOMIAL). The first part is the genus name, which is capitalized, and the second part is the specific EPITHET, which is not capitalized. For example, *Homo sapiens*, the species name for humans, contains the genus name *Homo* and the epithet name *sapiens*.

Designation of a HOLOTYPE specimen, sometimes simply referred to as the "type specimen," is one of the requirements for a new species name to be valid. The holotype is a single specimen chosen as a typical example and the reference specimen to represent a new species name when it is first described. These are particularly important specimens in museum collections. Some of the specimens illustrated here are type specimens and are indicated as such in the figure captions.

When used in print, a species name is often followed by the last name of the scientist who first validly named and described the species. If the author's name is shown not in parentheses, then the species is still in the genus it was originally placed in by the author, such as "†*Priscacara liops* Cope, 1877." If the species name was later moved to another genus by a subsequent author, the original author's name for the species still follows the name, but it is now placed in parentheses, such as "†*Cockerellites liops* (Cope, 1877)." There are many other rules and recommendations covered by the official rules of nomenclature, and for further details see Ride et al. (1999) and McNeill et al. (2006).

Organization and Order of Presentation of FBM Taxa in This Book

In the next several sections of this book presenting a field guide of FBM diversity, species will be organized within the Linnaean system and in phylogenetic order (grouped according to ancestral relationship) wherever possible, and only the valid scientific names will be recognized. The hypotheses of evolutionary relationships are the result of hundreds of studies (reference citations will be given in most sections for further studies of these groups, and sources of phylogenies followed in this book are given in appendix F). A blue **PHYLOGENETIC TREE** illustrating the evolutionary relationships among the FBM species presented here precedes many of the taxonomic sections in this book. Names for extinct groups are preceded by a dagger symbol (†). The 16 different phylogenetic trees used in this book all overlap with one another, enabling them all to be connected into a single phylogenetic tree. That composite tree represents what is known about the overall web of life for Fossil Lake 52 million years ago.

The most basic phylogeny of FBM organisms can be summarized by the following tree:

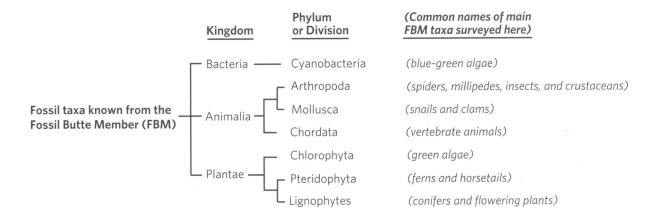

Kingdom	Phylum or Division	*(Common names of main FBM taxa surveyed here)*
Bacteria	Cyanobacteria	*(blue-green algae)*
Animalia	Arthropoda	*(spiders, millipedes, insects, and crustaceans)*
	Mollusca	*(snails and clams)*
	Chordata	*(vertebrate animals)*
Plantae	Chlorophyta	*(green algae)*
	Pteridophyta	*(ferns and horsetails)*
	Lignophytes	*(conifers and flowering plants)*

Fossil taxa known from the Fossil Butte Member (FBM)

This tree also diagrams the basic order of sections in the field guide portion of this book.

Bacteria

Preservation of fossils in the FBM occurs even at the microscopic level, although there has been very little work done on FBM MICRO-FOSSILS so far. Fossil bacteria known as cyanobacteria have been indirectly reported from Fossil Lake sediments by the presence of structures they build called STROMATOLITES, particularly in the latter stages of Fossil Lake's history (Loewen and Buchheim 1998). Although more common in the overlying Angelo Member, stromato-lites have also been reported from the FBM (Buchheim, Cushman, and Biaggi 2012). Although cyanobacteria are sometimes called "blue-green algae," they should not be confused with green algae, a plant. Blue-green algae are actually PHOTOSYNTHETIC bacteria. Green algae have also been reported from the FBM (Cushman 1983), and these fossils are discussed on page 283 (fig. 158).

No isolated bacteria cells have yet been reported as fossils from the FBM. However, Wilmot H. Bradley (1931) described and illustrated highly magnified photographs of isolated bacteria and blue-green algae cells as well as other microfossils from neighboring

Lake Uinta of the Green River Formation. Some of the rich organic layers of the FBM should contain similar specimens of fossil bacteria, should someone decide to look for them in the future. But for the rest of this book, it is the MACROFOSSILS that will be emphasized, including arthropods, mollusks, vertebrates, and plants.

Arthropods

(Phylum Arthropoda)

Arthropods are invertebrate animals with an external armor-like skeleton (EXOSKELETON), a segmented body, and jointed appendages. Their hard parts are made primarily of a cellulose-like substance called CHITIN. They are one of the two major animal groups very well represented in the FBM, the other being vertebrates. With well over a million living species, arthropods account for more than 80 percent of all living animal species, and they are represented in the FBM mostly by insects. The main groups of arthropods discovered so far from the FBM include the following:

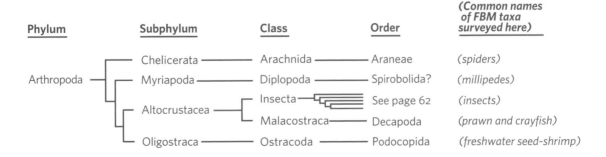

Phylum	Subphylum	Class	Order	*(Common names of FBM taxa surveyed here)*
Arthropoda	Chelicerata	Arachnida	Araneae	*(spiders)*
	Myriapoda	Diplopoda	Spirobolida?	*(millipedes)*
	Altocrustacea	Insecta	See page 62	*(insects)*
		Malacostraca	Decapoda	*(prawn and crayfish)*
	Oligostraca	Ostracoda	Podocopida	*(freshwater seed-shrimp)*

Spiders (Subphylum Chelicerata, Class Arachnida)

Spiders, being mostly non-aquatic and without wings, are exceptionally rare in the FBM deposits. Occasionally they probably blew into the lake in the wind or fell into it from overhanging vegetation. Another reason for their scarcity in the FBM is probably TAPHONOMIC. Although spiders have a chitinous EXOSKELETON, they are relatively soft-bodied and decay more rapidly after death than insects. (This is why modern museum collections of spiders are kept in alcohol instead of on pins like insects.) I am aware of only three spiders from the FBM (fig. 19). They remain UNDESCRIBED, but each represents a distinct variety. One appears to be

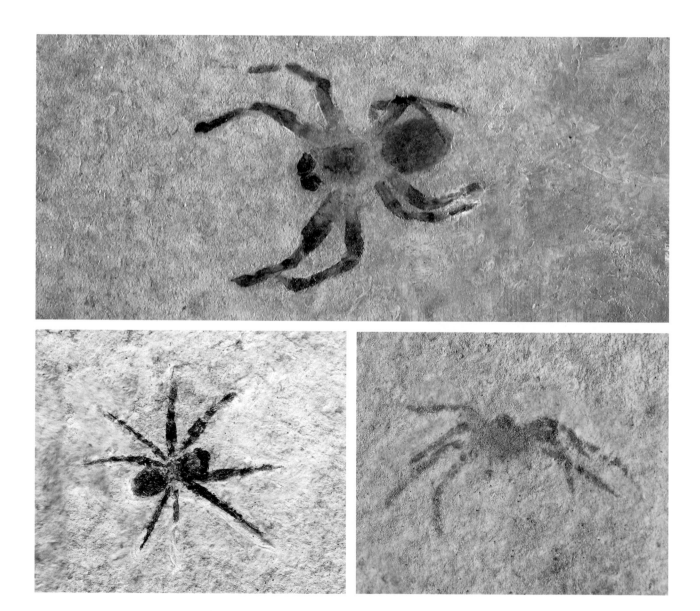

a crab spider (family Thomisidae) (fig. 19, *top*). The Thomisidae are called crab spiders because their flattened bodies and stout legs resemble crabs. Their webs are not made to trap prey like other spiders. Instead they are hunters who sit on leaves, other vegetation, or tree bark and ambush their prey. The specimen shown here might have washed in with leaves that fell into the lake's tributaries from the surrounding highlands. Crab spiders are represented today by more than 3,000 described living species.

Jumping spiders (family Salticidae) are also represented in the FBM (fig. 19, *bottom left*). This family has over 5,000 described living species today, most of which are found in the tropics, and the FBM fossil is one of the oldest known examples of this family. Like the crab spiders, the jumping spiders do not rely on a web to catch food but are instead active hunters who stalk their prey. Some jumping spiders eat plant nectar and pollen in addition to insect prey.

A third type of spider known from the FBM is a poorly preserved specimen of **CURSORIAL** spider, possibly a wolf spider (family Lycosidae, fig. 19, *bottom right*). Again, like the other two FBM spiders, wolf spiders actively hunt insects and other spiders rather than relying on their web for catching food. Some even burrow underground. There are over 2,000 living species of wolf spiders today.

The spiders discussed and illustrated here are the first reported from the FBM.

Millipedes (Subphylum Myriapoda, Class Diplopoda)

Also extremely rare in the FBM are millipedes (fig. 20). The millipede illustrated here is the first reported from the FBM. Millipedes, like spiders, are not aquatic animals and do not travel through the air, so their rarity in lake deposits is not

←··· FIGURE 19
Spider fossils from the FBM. *Top:* A crab spider, family Thomisidae, with a body length of 10 millimeters (0.4 inch), from FBM Locality A (FMNH PE60320). *Bottom left:* A jumping spider, family Salticidae, with a body length of 8 millimeters (0.3 inch) from FBM Locality A. Private collection; photo courtesy of Bill Rieger. *Bottom right:* A lightly preserved cursorial spider resembling a wolf spider (Lycosidae?) from FBM Locality H. Width from right to left is 15 millimeters (0.6 inch) (FOBU 13456).

···⋗ FIGURE 20
Millipede fossil from the sandwich beds of FBM Locality H. Specimen is 65 millimeters (2.6 inches) long (FOBU 13457). This is the first reported specimen of a millipede from the FBM.

ARTHROPODS

surprising. Millipedes are easily identified by the possession of two pairs of legs per body segment, and the FBM specimen has about 44 body segments. It probably belongs to the order Spirobolida, but family determination has not been made. Of millipedes living today, there are about 10,000 described species in 145 families and 16 orders, but many more species than that remain to be described by scientists. The Spirobolida has about 1,000 living species today, most of which live in the tropics. Millipedes are an extremely old group. The class Diplopoda is thought to be among the first major animal groups to have colonized land, during the Silurian, over 400 million years ago. In the Pennsylvanian geologic period, the millipede relative *Arthropleura* grew to a length of 2.6 meters (8.5 feet!), making it the largest known land invertebrate.

Most millipedes like the one from the FBM are DETRITIVORES that feed on decomposing vegetation or other organic matter in the soil. They burrow in the ground and coil up into a ball if threatened, leaving them encased in a protective EXOSKELETON. There is much work yet to be done with the description and classification of living millipedes in order to allow definitive integration of fossil millipedes into that classification.

Insects (Subphylum Altocrustacea, Class Insecta)

Some scientists believe that insects, like so many other organisms, suffered a mass extinction at the end of the Cretaceous (Labandeira, Johnson, and Wilf 2002; Wilf et al. 2006), although this is controversial. Some species may have been extinguished directly from the impact of the CHICXULUB ASTEROID, but many more probably become extinct as the species of trees that they fed on died out in the following decades. But by the early Eocene, insect speciation was on the rise again. Insects, plants, and other organisms rebounded together through CO-EVOLUTIONARY PROCESSES, throughout Paleocene and Eocene time. Pollinating insects such as bees, wasps, and flies diversified in parallel with flowering plants. Plant-eating insects diversified in parallel with the trees, shrubs, and other angiosperms that they fed on. And mammals, birds, and other vertebrates diversified in parallel with the insects that they fed on. Many insect families present in the FBM have diversified spectacularly and are today represented by thousands, or even tens of thousands, of living species covering much of the world.

To date, very little TAXONOMIC work has been done on the FBM insect fossils, and until ENTOMOLOGISTS study and describe this material, its contribution to the story of Fossil Lake will be far from complete. Here I illustrate many families of insects previously unknown from the FBM, but nearly all of these extinct species remain undescribed. Insects are the dominant and most species-rich group of animals on the planet today, and there are still millions of living insect species in need of scientific description by entomologists. Consequently, there are not enough professional entomologists to understand the full range of living insect diversity, let alone the extinct fossil forms. The priority for professional entomologists has therefore been to focus on the more than 1 million described living species, and relatively few experts have ever focused their attention on the FBM insects. It is always an intimidating task to classify and record Earth's enormous diversity of species, both living and fossil, and insects are the most formidable of all animal groups. Of the 25 million objects housed in the Field Museum, where I work, about 13 million of those are insects.

Work on fossil insects is further complicated by lack of preserved detail. Many living insect species are identified by delicate features not preserved in the FBM, such as reproductive anatomy, patterns of fine hairs on the body, color patterns, soft-part anatomy, or even DNA. Some FBM fossils appear indistinguishable from living species due to lack of detailed preservation of parts that could possibly separate them. It is unlikely that any of the FBM species are still living today, given their great age, but we may never know for some of them.

Within the Green River Formation, most of the limited scientific work on insects has been done on the younger middle Eocene deposits of Lake Uinta (e.g., Douglas Pass and near Rifle, Colorado, and Bonanza, Utah) and Lake Gosiute (near Green River), where more than 300 species of fossil insects have been described (see Grande 1984, 242–45; and Dayvault et al. 1995 for references). In contrast, only a few insect species have been described from the FBM deposits of Fossil Lake. There is also an insect layer in the overlying Angelo Member of Fossil Lake, but only the Fossil Butte insects will be discussed and illustrated here.

Insect fossils are tissue-thin stains on the rock, so they must split out in the field to be visible (e.g., fig. 31, *bottom*). They do not show from beneath the rock like fossil vertebrates do. Even as thin film-like remains, the detail can be stunning, with fine details of wing venation, hairs on legs and body, and compound eyes. Most FBM insect fossils that have been discovered are from the mid-lake 18-inch layer. In the nearshore sandwich beds, insects are most often colorless impressions, and thus extremely hard to see. Many examples of 18-inch layer insects are illustrated and discussed here, presented in taxonomic order.

Although the FBM insect fossils are not as abundant as those from the Lake Gosiute and Lake Uinta deposits, they nevertheless represent a diverse sample of insects that lived in and around Fossil Lake. The sample is heavily biased toward flying and aquatic insects because it consists exclusively of species that either fell into the lake or were living in it to begin with. Insects that were neither winged nor aquatic (e.g., non-flying ants) are very rare in the FBM. More than 85 percent of the insects I have found over the years in the FBM are all of a single species, the March fly †*Plecia pealei* Scudder, 1890 (fig. 34). This species, described over 120 years ago by Samuel Hubbard Scudder, is one of the very few FBM insects to have been described and named. It is known by thousands of specimens from the FBM and occurs mostly at the base of the 18-inch layer in the bottom "capping" layer. While finding †*Plecia* can be somewhat monotonous to the fossil insect enthusiast looking for variety, the remaining 15 percent of the FBM insect fauna contains a great deal of diversity belonging to many taxonomic orders. A brief sample of this material (most of which still awaits scientific description) is illustrated here. Most of these are reported here for the first time from the FBM. The FBM insect fossils presented below are as follows:

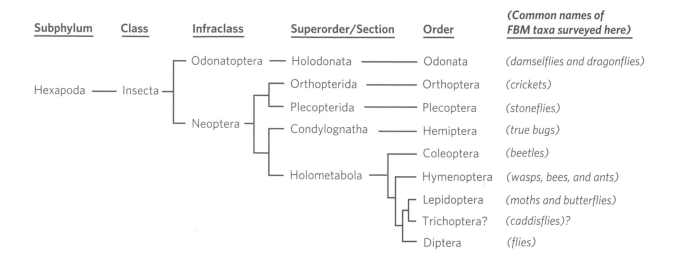

Subphylum	Class	Infraclass	Superorder/Section	Order	(Common names of FBM taxa surveyed here)
Hexapoda	Insecta	Odonatoptera	Holodonata	Odonata	(damselflies and dragonflies)
		Neoptera	Orthopterida	Orthoptera	(crickets)
			Plecopterida	Plecoptera	(stoneflies)
			Condylognatha	Hemiptera	(true bugs)
			Holometabola	Coleoptera	(beetles)
				Hymenoptera	(wasps, bees, and ants)
				Lepidoptera	(moths and butterflies)
				Trichoptera?	(caddisflies)?
				Diptera	(flies)

Dragonflies and Damselflies (Order Odonata)

These four-winged insects are some of the largest and most beautiful insects in the FBM, where they are represented by several undescribed species. Dragonflies and damselflies are day-flying predaceous insects that feed mostly on smaller insects. Immature stages (nymphs) are freshwater aquatic, and adults spend much of their lives on the wing near water, which explains how so many of them ended up in Fossil Lake. Dragonfly nymphs are aquatic predators, feeding on other insects and even small fishes. Present-day Odonata range from 20 to 130 millimeters (0.75 to 5 inches) in length, but 240 million years ago giant species reached wingspans of 640 millimeters (2.5 feet)! The pattern of veins in the wings helps identify the family of Odonata to which a species belongs. The vein patterns in some FBM specimens are striking (e.g., fig. 21, *top right*). Like the butterfly that emerges from the skin of a caterpillar, dragonflies and damselflies emerge from the final molt of the nymph, which lives most of its life underwater with functional gills.

Damselflies (suborder Zygoptera) are represented in the FBM by several families including the "pond damselflies" (family Coenagrionidae, with 1,100 living species) and the "broad-winged damselflies" or "demoiselles" (family Calopterygidae, with 150 living species). The Field Museum's collection includes two beautiful undescribed species of pond damselflies (fig. 21, *top*). The Fossil Butte National Monument museum has an extremely fine broad-winged damselfly in its collection with the color pattern preserved in its wings (fig. 21, *bottom*). The earliest known zygopteran fossils are from the Triassic of South America, Australia, and Central Asia (Grimaldi and Engel 2005).

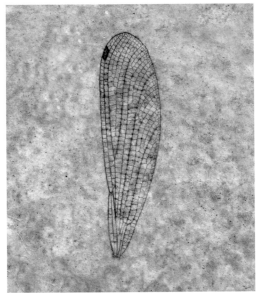

FIGURE 21

Damselflies (suborder Zygoptera) from the FBM 18-inch layer. *Top left:* Undescribed species of the family Coenagrionidae (the "pond damselflies") with body length of 39 millimeters (1.5 inches) from FBM Locality A (FMNH PE51414). *Top right:* Well-preserved wing of an undescribed transparent-winged species of Coenagrionidae, 27 millimeters (1.1 inches) long (FMNH PE51415). *Bottom:* Undescribed species of the family Calopterygidae (the "demoiselles") with a wing span of 60 millimeters (2.4 inches) from FBM Locality A (FOBU 448).

⚲ FIGURE 22

Dragonflies (suborder Epiprocta) from the 18-inch layer of the FBM.
Top: Undescribed species of the family Libellulidae (the "skimmers" and
"perchers") with darkly pigmented wings and a wing span of 71 millimeters (2.8 inches) from FBM Locality A (FMNH PE51413). *Bottom left:*
Undescribed species of the family Libellulidae with transparent wings
containing dark vein patterns and a wing span of 59 millimeters (2.3
inches). Private collection; photo courtesy of William Rieger. *Bottom
right:* Dragonfly nymph from FBM Locality B, measuring 8 millimeters
(0.3 inch) long (FOBU 13518).

⤏ FIGURE 23

Top: Cricket (family Gryllidae) from the sandwich beds
of FBM Locality L, with a body length of 22 millimeters
(0.9 inch). Private collection; photo courtesy of Arvid
Aase and FOBU. *Bottom left:* Cricket (family Gryllidae),
with a body length of 11 millimeters (0.4 inch), from the
18-inch layer of FBM Locality B (FMNH PE60993). *Bottom
right:* Stone fly nymph (order Plecoptera, family unknown) from the 18-inch layer of FBM Locality E, with a
body length of 7 millimeters (0.3 inch) (FMNH PE60988).

Dragonflies (suborder Epiprocta) are closely related to damselflies and are represented in the FBM by the family Libellulidae (skimmers and perchers). This family today contains over 1,000 living species. There have been some spectacular specimens found in the FBM; some have transparent wings and sharply defined venation patterns, and others have darkly pigmented wings (fig. 22). None are yet scientifically described or named as species. Dragonfly nymphs also are found occasionally (fig. 22, *bottom right*), and they are a further indication of the freshwater conditions in the Eocene lake.

Crickets (Order Orthoptera)

Although the order Orthoptera includes crickets, grasshoppers, and katydids, in the FBM this order is so far represented only by crickets (fig. 23, *top* and *bottom*

left). There are two cricket specimens known so far from the FBM, and they probably fell into the lake while on the fly. The FBM crickets are undescribed, but both closely resemble living species in the family Gryllidae ("true crickets").

There are about 600 living species of gryllids today, most of which are nocturnal. They are also found in Early Cretaceous deposits of Brazil. As anyone but the staunchest urbanite knows, crickets "chirp" by rubbing their wings together. The FBM crickets were probably a major part of the night sounds around Eocene Fossil Lake. Crickets are omnivores that feed on decaying plant material, fungi, very small plants, and even dead crickets. They also serve as an important food source for amphibians, lizards, and spiders today, and likely did so around Fossil Lake.

Stoneflies (Order Plecoptera)

Another group of insects found in the FBM are the stoneflies, known by several specimens. The stonefly NYMPH in figure 23 (*bottom right*) could not be identified to family. Stonefly nymphs are aquatic herbivores that live in well-oxygenated freshwater. Plecopterans remain as nymphs from one to four years depending on the species. When they finally transform into adults, they generally only survive for a few weeks. Today there are over 3,500 living species of stoneflies, and they are found worldwide, on all continents except Antarctica. They are indicators of excellent water quality and are highly intolerant of pollution. The oldest known stoneflies (literally "stone" flies in this case) are from Late Permian deposits.

True Bugs (Order Hemiptera)

The order Hemiptera contains the "true bugs." Although the term "bug" is often used for many insects and other animals, it is actually properly used for only a

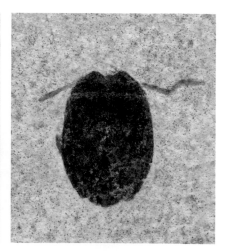

select group of insects diagnosed by a number of specialized characteristics of the mouthparts, wings, and legs. This is one of the most diversely represented groups in the FBM, which is not surprising given that today there are well over 60,000 living species in this order. Their great diversity is thought to be related to the diversification of flowering plants beginning in the Cretaceous (Grimaldi and Engel 2005).

One of the true bug families represented in the FBM is comprised of the shield-backed bugs, or jewel bugs, family Scutelleridae (fig. 24, *left*). Today the family includes about 450 living species and has a worldwide distribution. Species in this family feed on plant juices. They inject enzyme-rich saliva into plants. The enzymes digest plant matter into a liquid that they then suck up. Although jewel bugs superficially resemble beetles, they are true bugs.

Another bug family in the FBM is the stinkbug and shield bug family, Pentatomidae. A number of specimens from this family have been found (e.g., fig. 24, *middle*). The name "shield bug" comes from the shield-shaped body of these insects. Living stinkbugs emit a rancid almond scent when disturbed, as a defense mechanism. Some species feed on other insects, while most feed on plants. The family is represented today by about 5,000 living species.

Another group of true bugs found in the FBM are cicadas, both as adults and as nymphs (e.g., fig. 25). Cicadas (family Cicadidae) spend from 2 to 17 years underground in a nymph stage, after which they emerge and metamorphose into large flying insects, living only a few additional weeks as adults in trees and shrubs. The distinctive songs of the adult cicadas must have periodically filled the day and night skies around Fossil Lake during mating season. Cicadas feed on tree sap by way of a long, sharp proboscis under their head. Today they live in tropical to temperate climates and are represented by about 1,500 living species. The earliest known fossil cicada is from Early Jurassic deposits of England.

FIGURE 25
Cicadas (family Cicadidae), from the 18-inch layer of FBM Locality A, showing two different life stages. *Left:* Part and counterpart of a nymph with a body length of 20 millimeters (0.8 inch) (FMNH PE51409). *Right:* Unidentified large bug, possibly Cicadidae or Fulgoroidea, with a body length of 31 millimeters (1.2 inches) (FMNH PE60997).

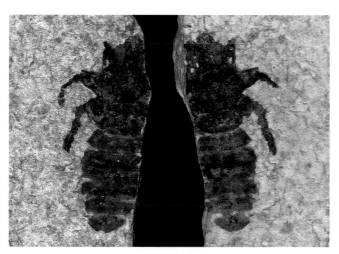

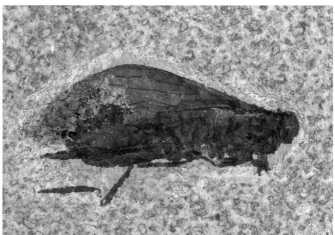

There is one aquatic true bug that is fairly abundant in the FBM, the water striders (family Gerridae) (fig. 26). Hundreds of these have been found over the years in the 18-inch layer quarries and in the layers near the K-spar tuff bed. Water striders have lightweight bodies and long legs that allow them to live on the surface of the water, running or "skating" on the surface. They feed on smaller insects that fall onto the surface or small insects just below the surface of the water. The four rear legs are long, for skimming over the water's surface, and the front pair of legs is short and used to capture food. There are about 500 known

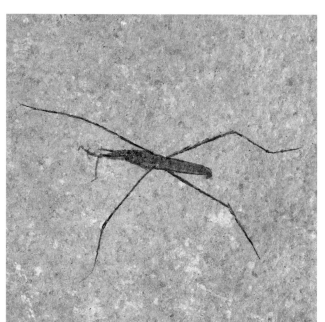
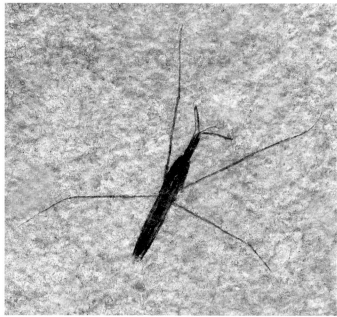

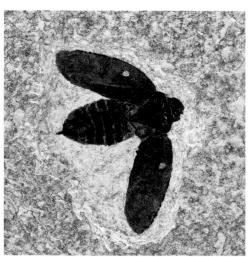

living species of water strider. The family Gerridae is thought to have originated in the Cretaceous (Grimaldi and Engel 2005).

Plant hoppers (superfamily Fulgoroidea) are a major group of true bugs represented by several different families in the FBM, including the Cixiidae, the Achilidae, and the Fulgoridae (fig. 27). Cixiidae are considered to be "stem" or "primitive" plant hoppers and generally have transparent wings with dark-colored veins running through them (fig. 27, *left*). This ancient group, one of the largest families of plant hoppers, is at least 150 million years old and has 2,000 described living species and a worldwide distribution. The NYMPHS of this family today live underground, feeding on plant roots or rotten wood, while the adults mainly eat leaves of trees and shrubs.

The achilid leafhoppers (family Achilidae) are also found in the FBM (fig. 27, *middle*). Today species of this family feed mainly on fungus, are tropical to subtropical, and are represented by about 500 living species.

The fulgorid leafhoppers (family Fulgoridae) are also present (fig. 27, *right*). This plant-eating family today contains at least 550 species and is especially diverse in the tropics.

Beetles (Order Coleoptera)

Beetles are the largest order of insects living today, with nearly 400,000 described species and many more awaiting scientific description. They were already stunningly diverse by the Jurassic, some 160 million years ago (Grimaldi and Engel 2005). This amazing diversity is one reason that beetles in the FBM remain undescribed and unnamed. There is so much work yet to do with living beetles that entomologists have still largely ignored most fossil beetles. Other than the aquatic forms, beetles either were washed into the lake in tributary streams or from along the shoreline, or fell into it while on the fly.

Based on my survey, the most abundant beetles in FBM are the jewel beetles (family Buprestidae), also called the metallic wood-boring beetles. The common names are derived from the colorful metallic-looking dorsal surface of their body in life. Buprestids are some of the largest beetles in the FBM, reaching a length of nearly 50 millimeters (2 inches), but some smaller species also exist there. Several species have been found in the FBM, but none have been formally described (figs. 28 and 29, *middle*). Both adults and their larvae are plant eaters, and the eggs are usually laid in the bark of trees. When larvae hatch, they tunnel under the bark of the tree. Today this family is represented by 15,000 living species.

Ground beetles (family Carabidae) are another group of beetles represented in the FBM (fig. 29, *bottom right*). This is an extremely large family with more than 40,000 living species today worldwide. They are often predaceous, feeding on other insects, but as might be expected with a group this large, they occupy a variety of ecological habitats. Several distinct species have been found in the FBM, although they remain undescribed.

← FIGURE 26
Water striders (family Gerridae). *Left:* Specimen with a body length of 25 millimeters (1 inch), from the 18-inch layer of FBM Locality E (FMNH PE60996). *Right:* Specimen with a body length of 24 millimeters (0.9 inch), from near the K-spar tuff layer in FBM Locality A (FMNH PE52419).

← FIGURE 27
Undescribed species of plant hoppers (superfamily Fulgoroidea) from the 18-inch layer. *Left:* A representative species of the family Cixiidae, with a body length of 10 millimeters (0.4 inch), from FBM Locality A (FMNH PE60943). *Middle:* A representative species of the family Achilidae, with a body length of 16 millimeters (0.6 inch), from FBM Locality A (FMNH PE60322). *Right:* A representative of the family Fulgoridae with a body length of 10 millimeters (0.4 inch), from FBM Locality E (BMNH In. 64612).

↑ FIGURE 28

Large, undescribed "metallic wood-boring beetles," family Buprestidae, from the FBM 18-inch layer. *Left:* Large-headed species from FBM Locality B with a body length of 35 millimeters (1.4 inches) (FOBU 11699). *Middle:* Small-headed species from FBM Locality B with wing covers partly extended, body length 45 millimeters (1.8 inches) (FOBU 13384); photo courtesy of Arvid Aase and the Fossil Butte National Monument. *Right:* Small-headed species from FBM Locality A with a body length of 35 millimeters (1.4 inches) (FMNH PE60930).

⋯⋙ FIGURE 29

Various beetles from the FBM. *Top left:* An unidentified beetle of the superfamily Tenebrionoidea, with a body length of 9 millimeters (0.4 inch), from FBM Locality G (BMNH In. 64613). *Top middle:* A weevil, or "snout-beetle," of the family Curculionidae, with a wing span of 12 millimeters (0.5 inch), from FBM Locality A (FMNH PE60944). *Top right:* Another weevil from FBM Locality A, measuring 12 millimeters (0.5 inch) long (FMNH PE60945). *Center left:* A reticulated beetle of the family Cupedidae measuring 17 millimeters (0.7 inch) in length, with the color pattern preserved on the elytra, from FBM Locality B (FOBU 13518). *Center middle:* An unidentified beetle (Buprestidae?) from FBM Locality A with a body length of 15 millimeters (0.6 inch) (FMNH PE60947). *Center right:* A small species of buprestid beetle from FBM Locality L with a body length of 10 millimeters (0.4 inch) (FOBU 13458). *Bottom left:* Possible buprestid beetle with a body length of 15 millimeters (0.6 inch), from FBM Locality B (FMNH PE60940). *Bottom middle:* A tiny beetle from the 18-inch layer of FBM Locality B, measuring only 3 millimeters (0.1 inch) in length of a yet-undetermined family (FMNH PE60094). *Bottom right:* A species of ground beetle of the family Carabidae, with a body length of 13 millimeters (0.5 inch), from the 18-inch layer of FBM Locality G (FMNH PE61048).

The weevils or snout beetles (family Curculionidae) are also represented in the FBM. Examples of undescribed FBM weevil species are illustrated in figure 29 (*top middle* and *top right*). This group has been highly successful over time and has diversified to over 50,000 living species, making it one of the largest animal families today. Weevils feed on plant tissues and in some instances aid in pollination.

The reticulated beetles (family Cupedidae) are also present in the FBM. Today this family consists of about 30 living species distributed worldwide. It is an ancient group, with fossil representatives going back to the Early Triassic (about

250 million years before present). Larvae of these beetles are wood borers, typically living in damp, fungus-infected wood. Adults eat foliage or fungus under the bark of trees. One species in the FBM is known with a pigmentation pattern preserved on its ELYTRA (fig. 29, *middle left*).

Wasps, Bees, and Ants (Order Hymenoptera)

These closely related groups include SOCIAL INSECTS with a high level of social behavior. Social insect groups have reproductive queens and sterile workers. Wasps and bees are quite abundant in the FBM deposits, but ants, particularly wingless ants, are rare. This is again a factor of ecology and preservation. Those terrestrial animals preserved in the FBM are primarily those that were flying over or washed into Fossil Lake. Like beetles and flies, the Hymenoptera is a hyper-diverse group today, with about 125,000 described and named species, but with estimates of up to 1 million new species still awaiting description. The earliest

known fossil Hymenoptera are from Triassic deposits of Australia, Asia, and Africa (Grimaldi and Engel 2005).

Many bees and wasps today have a close association with flowers, and their diversity in the FBM, where there is also a diversity of flowering plants, is probably another reflection of co-evolving speciation in the Early Cenozoic. Wasps (superfamilies Vespoidea, Ichneumonoidea, and Sphecoidea) are well represented in the FBM. Many wasps are predaceous, feeding on other insects and spiders. Some wasps feed on flower nectar as adults, but also feed insects or spiders to their young.

One family of wasps reported here are the Vespidae, which sometimes reach large size in the FBM (fig. 30, *left*). This is a very large family with nearly 5,000 living species today. This family contains most of the highly social species of wasps (called **EUSOCIAL**). They build nests out of a paper-like material made from chewed wood and their saliva, or they nest opportunistically in hollow spaces. This family includes some of today's most dangerous stinging insects, such as the paper wasps, hornets, and yellow jackets.

Another family possibly represented in the FBM is that of the scorpion wasps, family Ichneumonidae (fig. 30, *right top* and *bottom*). Ichneumonid wasps are **PARASITOIDS**. Common hosts are the **LARVAE** and caterpillars of beetles, bees, wasps, moths, and butterflies. Ichneumonid wasps have very long ovipositors (the long "stinger-like" structure pointing posteriorly from the body). There are over 60,000 living species in this family today, and its geographic distribution is worldwide. It is possible, too, that this could be a member of the extinct family †Ephialtitidae, also with long ovipositors and a similar body shape. This family was previously known only from the Early Jurassic through the Early Cretaceous. Further study of this specimen is needed to resolve this definitively.

Another family that can be reported here from the FBM is that of the "flower wasps," family Scoliidae (fig. 31, *top left*). Scoliid wasps are represented today by about 300 living species. They feed on beetle larvae, and the adults are important pollinators of flowers.

All the FBM wasps remain largely unstudied, so the identifications of many specimens are yet inconclusive (e.g., fig. 31, *top right* and *bottom*). There are well over 100,000 living species of wasps today.

Bees (superfamily Apoidea) are represented in the FBM by several undescribed specimens (e.g., fig. 32, *top*). Today bees include approximately 20,000 known living species, and the group is probably the most beneficial to humans of all insect families in its role as pollinators for flowering plants. Bees feed primarily on flower nectar and pollen.

Ants (family Formicidae) are also present in the FBM. Ants appear to have evolved from a wasp-like ancestor hundreds of millions of years ago. They have been extremely successful ever since then, and today they are represented by more than 12,500 described living species. Within most ant species, the colony is organized as a caste society with different morphological types, including very

FIGURE 31

Wasps from the 18-inch layer of FBM Locality A. *Top left:* A "flower wasp," family Scoliidae, with a body length of 35 millimeters (1.4 inches) (FMNH PE60851). *Top right:* An unknown variety of wasp, with a body length of 25 millimeters (1 inch) (FMNH PE60849). *Bottom:* Part and counterpart split of an unknown variety of wasp with a body length of 20 millimeters (0.8 inch) (FMNH PE60938a and b).

FIGURE 32

Bees (superfamily Apoidea) and ants (family Formicidae) from the 18-inch layer. *Top left:* A bee with a body length of 20 millimeters (0.8 inch), from FBM Locality A (FMNH PE60931). *Top right:* A bee with a body length of 16 millimeters (0.6 inch) from FBM Locality E (FMNH PE60992). *Bottom left:* A worker or soldier ant with a body length of 4 millimeters (0.2 inch), from FBM Locality A (FMNH PE52593). *Bottom right:* A queen ant with a body length of 20 millimeters (0.8 inch), from FBM Locality A (FMNH PE57071).

large fertile females called "queens" and smaller "workers" and "soldiers." The strict organization and interconnectedness within colonies of some species make them behave like a single organism. Both workers and queens are reported here from the FBM for the first time (fig. 32, *bottom*). Non-flying ants are extremely rare in the FBM, not well preserved, and probably washed into the lake from tributaries.

Moths and Butterflies (Order Lepidoptera)

The order Lepidoptera contains moths and butterflies. Moths are first known in the Jurassic period about 170 million years ago, but the earliest known butterfly fossil is much younger. The FBM butterfly reported here for the first time is one of the earliest known, with the earliest known butterfly from the late Paleocene Fur Formation of Denmark. All of the FBM moths and butterflies remain unde-

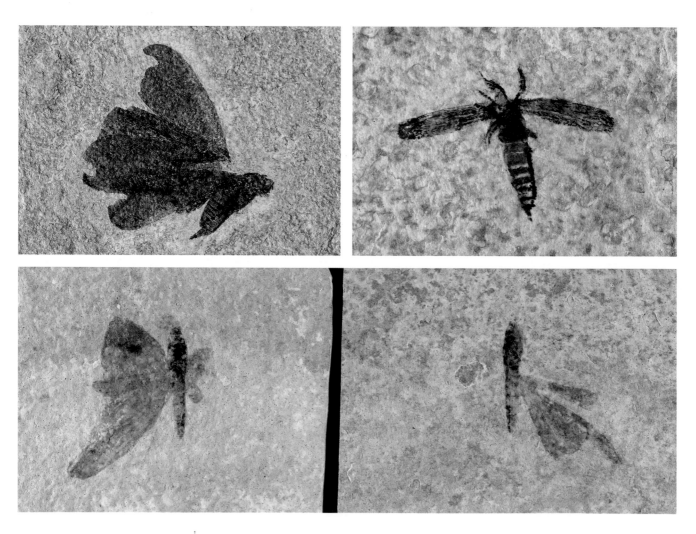

scribed (e.g., fig. 33). The earliest known moths had mandibles, somewhat like grasshoppers do today, and they probably fed on pollen, like some living moths do today. Butterflies and some moths no longer have chewing mandibles, but instead have a long siphoning tongue, called a "proboscis," used to feed on liquid nectar from flowers. Preliminary identifications of moths reported here from the FBM include two different families. One large specimen appears to be from the royal moth and giant silk moth family, Saturniidae (fig. 33, *top left*). This family today includes some of the largest of all moth species, with wing spans up to 300 millimeters (nearly 12 inches). Caterpillars feed on the leaves of plants where the adults lay their eggs (e.g., fig. 36, *bottom right*). The caterpillars are an important food source for small mammals and birds, and also serve as a host for the larvae of parasitic wasps and flies.

Another family of moths that is represented in the FBM is that of the plume moths, family Pterophoridae (fig. 33, *top right*). Modern plume moths have caterpillars that either eat leaves or bore into the roots and stems of plants. The family is today represented by nearly 1,000 living species.

True butterflies (superfamily Papilionoidea) are also present in the FBM. The fossil in figure 33 (*bottom*) appears to be one of the earliest known papilionoids. This specimen awaits full description. Another well-preserved butterfly wing from FBM Locality B exists in a private collection. A species was previously reported from the Upper Parachute Creek Member of the Green River Formation in Colorado, but that locality is much younger than the FBM. Butterflies are day-flying insects that are important pollinators for flowering plants. Some living butterflies have evolved SYMBIOTIC or PARASITIC relationships with social insects like ants or wasps. For example, some caterpillars secrete a sugar from the plants they eat. Ants feed on this secretion and, in turn, protect the caterpillar from many types of predators. Today there are about 25,000 species of living butterflies, and the relatively large number of described species is partly an artifact of the great interest that entomologists and hobbyists have historically had in this group.

In summary, moths and butterflies were already well represented in the early Eocene and were developing in parallel with the deciduous tree and plant flora, which probably served as a food source for them. Moths themselves probably were an important food source for bats, some species of birds, and other insectivorous species. Butterflies are much rarer in the fossil record and show up first in the FBM.

Caddisflies (Order Trichoptera)

Supposed caddisfly larval cases have been reported from the FBM by Buchheim, Cushman, and Biaggi (2012). This report is somewhat tenuous given that only the larval cases have been found and no actual insect bodies have yet been discovered. Buchheim, Cushman, and Biaggi (2012) also do not illustrate or reference any specific caddisfly specimens from the FBM in their paper. Caddisflies are

Moths and a butterfly from the FBM. *Top left:* A large "royal moth," family Saturniidae, from FBM Locality F, with a wing height of about 60 millimeters (2.4 inches). Private collection of Mr. Dick Dayvault, photo courtesy of Dick Dayvault. *Top right:* A "plume-moth," family Pterophoridae, with a wing span of 25 millimeters (1 inch) from FBM Locality A (FMNH PE60846). *Bottom:* An unidentified butterfly, superfamily Papilionidea, part and counterpart from FBM Locality L. This is one of the earliest known fossil butterfly fossils. Specimen has a body length of 20 millimeters (0.8 inch) (FMNH PE60942). Top pieces from the 18-inch layer; bottom pieces from the sandwich beds.

thought to indicate clean freshwater. There are about 12,000 species in this order living today, occupying a variety of ecological niches ranging from predator to algal grazers.

Flies (Order Diptera)

Flies rule among the FBM insects. There are many species of flies represented in the FBM, and they make up over 90 percent of the insect fossils there. And of those fossil flies, more than 80 percent belong to a single species, the March fly †*Plecia pealei* (fig. 34). This is one of the few FBM insect species that has been studied and named, and this was done by Samuel Hubbard Scudder over 120 years ago. The family Bibionidae, to which this species belongs, is also sometimes referred to as the "love bug" family, because living bibionids are very frequently found in copulation (evidently making mad passionate insect love). One outcome

of that activity is their extreme abundance. Modern-day bibionids sometimes occur in enormous swarms that lead to hazardous driving conditions. The FBM bibionids are often found in mass-mortality layers, indicating similar swarms in the early Eocene (fig. 34, *bottom*). Modern adult bibionids feed primarily on flowers, and the group extends at least as far back as the Cretaceous and is possibly another group that co-evolved with flowering plants.

Soldier flies (family Stratiomyidae) are large flies that sometimes mimic wasps (e.g., an undescribed fossil soldier fly from the FBM illustrated in fig. 35A, *top left*). Several of these have been found in the FBM. Today there are about 1,500 living species of soldier flies, most of which live in damp environments of the neotropics. Their larvae live in tree bark, soil, or decaying organic matter.

Crane flies (family Tipulidae) are known by a number of specimens from the FBM (e.g., fig. 35A, *bottom*). Adults of many crane fly species feed on nectar from flowering plants. Adults of other species feed on nothing at all, living most of their life as flightless larvae and metamorphosing to adults only to mate and die. Most of the larvae are aquatic and feed on detritus or plant tissue. Some are even carnivorous. Today there are about 15,000 living species of crane fly, making Tipulidae the largest family of flies in the world.

Meniscus midges are a variety of gnat (family Dixidae) that are occasionally found in the FBM (fig. 35B, *top left*). They are tiny, stilt-legged forms whose adult stages are very poor fliers. The larvae are aquatic and live in unpolluted standing or slow-moving freshwater, usually among nearshore vegetation. They filter-feed fine particles of algae from just below the water's surface film. Today there are about 180 known living species in the family.

Dark-winged fungus gnats (family Sciaridae) comprise another family of gnats present in the FBM (fig. 35B, *top middle*). Scairid species live most of their lives in their larval form, and they live only about five days in their adult stage; just long enough to mate and produce eggs. There are about 1,700 described living species today, and it is estimated that there are another 20,000 or more living species awaiting discovery, mainly in the tropics. Most species live in forests, swamps, or moist vegetation. This family has a worldwide distribution, although it is most common in tropical regions.

Dagger flies (family Empididae) are also present in the FBM (fig. 35B, *top right*). This family has over 3,000 described species living today, mostly on continents of the Northern Hemisphere. These small, long-legged flies are mostly predatory, feeding on other insects, and the name "dagger-fly" refers to the sharp, piercing mouthparts of some species. They occupy a wide range of habitats, both aquatic and terrestrial. The larvae live in mud and detritus, or in algal mats and mosses floating in slow-moving freshwater.

There are many other small mosquito or gnat-size flies in the FBM that remain unidentified (e.g., fig. 35B, *bottom left*). These are often missed by the quarriers because they are so small and difficult to see when excavating the larger specimens.

A "love bug" or "march fly," | *Plecia pealei* Scudder, 1890 (family Bibbionidae), from the 18-inch layer. This is one of the few species of FBM insects that has been described and named. *Top left:* An individual from FBM Locality G, with a body length of 11 millimeters (0.4 inch) (FMNH PE60929). *Top right:* An individual in ventral view with a body length of 13 millimeters (0.5 inch) from FBM Locality A (FMNH PE60948). *Bottom:* Part of a mass mortality from the base of the 18-inch layer of FBM Locality A, with body lengths ranging between 8 and 10 millimeters (0.3 and 0.4 inch) (FMNH PF51407).

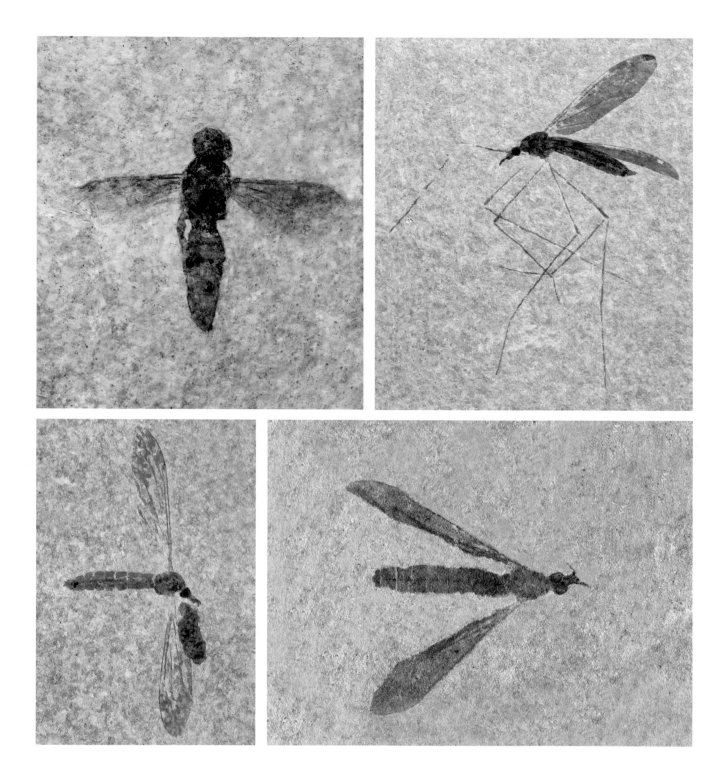

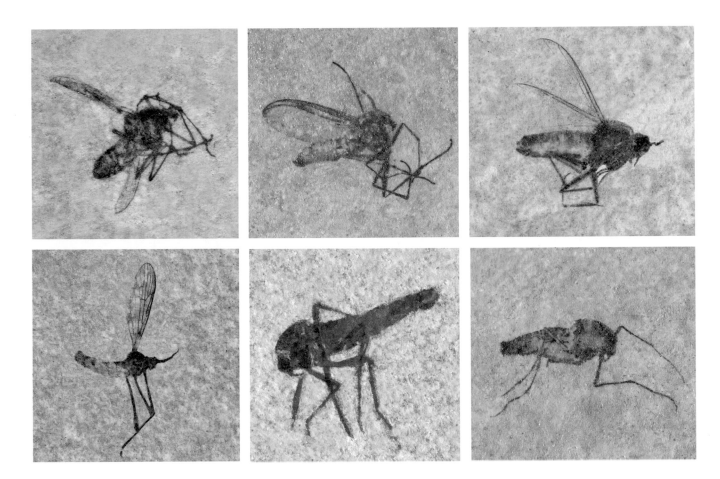

↑ FIGURE 35B

Various tiny flies and gnats from the 18-inch layer of the FBM. *Top left:* A meniscus midge (family Dixidae) with a body length of 5 millimeters (0.2 inch), from FBM Locality A (FMNH PE60949). *Top middle:* A dark-winged fungus gnat (family Sciaridae) with a body length of 8 millimeters (0.3 inch), from FBM Locality A (FMNH PE60872). *Top right:* A dagger fly (family Empididae) from FBM Locality A, with a body length of 6 millimeters (0.2 inch) (FMNH PE61050). *Bottom left:* A mosquito or other small biting fly from FBM Locality A, with a body length of 7 millimeters (0.3 inch) (FOBU 808). *Bottom middle:* Non-biting midge (family Chironomidae) from FBM Locality E, with a body length of 7 millimeters (0.3 inch) (FMNH PE60990). *Bottom right:* Non-biting midge (family Chironomidae) from FBM Locality A, with a body length of 6 millimeters (0.2 inch) (FMNH PE61018).

←⋯ FIGURE 35A

Soldier fly and crane flies (families Stratiomyidae and Tipulidae) from the 18-inch layer of the FBM. *Top left:* A soldier fly with a body length of 23 millimeters (0.9 inch) from FBM Locality A (FMNH PE52594). *Top right, bottom left, and bottom right:* Crane flies. *Top right:* Undescribed species A (FMNH PE61135), a large-bodied species from FBM Locality A with a body length of 24 millimeters. *Bottom left:* Another specimen of undescribed species A (FMNH PE60946) from FBM Locality A with a body length of 23 millimeters. *Bottom right:* Undescribed species B (FMNH PE60995), a smaller-bodied species from FBM Locality E with a body length of 14 millimeters.

Chironomid flies, also called non-biting midges (family Chironomidae), are fairly abundant and are the most abundant of the tiny, non-bibionid flies from the FBM (fig. 35B, *bottom middle* and *bottom right*). Some species resemble mosquitoes, but they lack the elongate mouthparts of the mosquito family Culicidae. Today the family Chironomidae is a large one with a global distribution and over 10,000 living species. There appear to be several different species present in the FBM, but all remain undescribed. The larval stages are aquatic, and they are so abundant that they form a large and important part of the aquatic biomass and ecosystem. The larvae serve as an important food source for fishes and were probably an important part of the diet of adult and juvenile †*Knightia*, †*Cockerellites*, †*Priscacara*, and juveniles of many other species. Chironomid larvae are also capable of inhabiting highly polluted waters with very low oxygen content. In fact, large numbers of them are sometimes used to indicate poor water quality.

Other flies are also present in the FBM, but their family relationships are still mostly unknown. Of an estimated 500,000 or more living species of flies, the vast majority of them await description by entomologists, which makes precise identification of many of the FBM fossil flies very difficult.

Insect Damage to Plants in the FBM

Leaves and other plants showing insect predation are abundant in the FBM, which might be expected based on recent research. According to some scientists, there is a strong correlation between temperature and insect predation on leaves. When temperatures increase, so does insect-feeding damage to plant species. Insects in the tropics (i.e., warmer climates) consume significantly more plants than do insects in temperate regions. And according to scientists, the time in which the FBM was deposited was the warmest that Earth has been in the last 100 million years (Zachos, Dickens, and Zeebe 2008). It has been proposed that the global warming trend in the late Paleocene and early Eocene allowed many new species of herbivorous insects from the PALEOGENE tropics to migrate north, leading to increased insect predation on plants in regions such as the one around Fossil Lake (Currano et al. 2008). There are lessons to be learned from the fossil record. Increased insect predation due to global warming is a great concern today for many regions of the world. In the words of Winston Churchill, "The farther back you can look, the farther forward you are likely to see."

Some of the different types of insect predation we see on leaves from the FBM include leaf-blister GALLS, leaf-mining trails, and a variety of different chew-mark patterns. Leaf-blister galls are abnormal outgrowths of plant tissue often formed after herbivorous insects lay their eggs in early developing plant tissue. The leaf responds by forming a blister of thick tissue around the egg or larva, which leaves a round blister or scar on the leaf (fig. 36, *top*). The developing insect larva uses the surrounding tissue for both protection and as a food source. Some of the most common insects to provoke the formation of galls with their

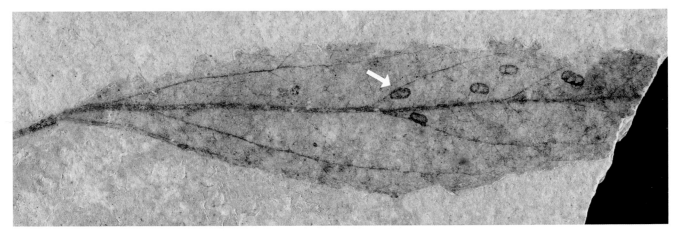

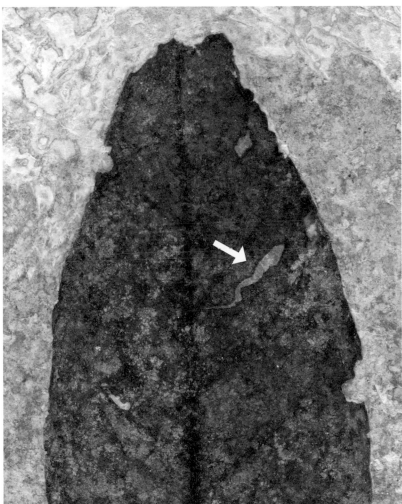

FIGURE 36

Examples of insect predation on leaves from the 18-inch layer of the FBM. *Top:* A leaf with five blister-galls (arrow pointing to one). Largest gall is 2 millimeters (0.1 inch) long, and leaf is specimen FMNH PP45984. *Bottom left:* A leaf showing damage from a leaf-mining insect (arrow). Leaf is 24 millimeters (0.9 inch) wide (FMNH PP54793). *Bottom right:* A leaf with insect chew marks around the edges (FMNH PP54792). Leaf measures 88 millimeters (3.5 inches) from stem tip to leaf tip.

young are gall wasps and gall midges. Today there are over 1,000 living species of gall wasps, and some of the early forms were no doubt present in the time of Fossil Lake.

Leaf mining is another specialized type of insect predation on plants. It is used today primarily by moths, sawflies, and various families of true flies, and by a few odd species of beetles and wasps. Leaf mining involves the penetration of the hard, CELLULOSE-rich surface of the leaf to reach the thin inner layer that has the least amount of cellulose and is therefore easiest to digest. This leaves a peculiar "trail"-like scar on the leaf (fig. 36, *bottom left*).

The most common signs of insect herbivory in the FBM are the many leaves with obvious chew marks on them (e.g., fig. 36, *bottom right*). Such chew marks are characteristic of caterpillars and a variety of other herbivorous insects. Sometimes these marks are "nibbles" around the edge of the leaf, while other times they are holes that occur within the leaf. The chew-mark type of predation leaves a wide variety of damage patterns on the leaves.

Most of the FBM leaves show some kind of insect damage; and like today many Eocene insect species probably depended on particular plant species for food. The insect-plant relationship is yet another community aspect of the early Eocene that is beautifully illustrated by the FBM fossils. As the diversity of plants and insects continued to expand to the present day, so did the complex ecological relationships between them. The diversity of the plants will be discussed in a later chapter of this book.

Other Arthropods

Other than insects, millipedes, and spiders, there are a number of aquatic arthropod fossils known from the FBM. These include crayfish and prawns (class Malacostraca, order Decapoda) and freshwater seed-shrimp (class Ostracoda, order Podocopida). The restricted presence of these aquatic species within specific regions of the FBM is a reflection of the different ecological zones within Fossil Lake. The crayfish and prawns (together called "decapods") occur mainly in the nearshore deposits. Formerly, decapods and ostracods were grouped together in the subphylum "Crustacea," but that was later deemed to be an unnatural group because decapods were found to be more closely related to insects than to ostracods (Edgecombe 2010; Regier et al. 2010).

Prawn and Crayfish (Subphylum Altocrustacea, Class Malacostraca)

In the FBM, there are two species of described decapods: the crayfish †*Procambarus primaevus* and the freshwater shrimp or prawn †*Bechleja rostrata* (figs. 37, 38). The decapods from the FBM are primarily in the nearshore sandwich bed quarries. I have not seen a decapod from the mid-lake 18-inch layer localities during my 30 years of fieldwork in this layer, but I have seen more than 200 crayfish and prawn from the nearshore sandwich beds, mostly from FBM Localities H, L, and M. As bottom-living organisms that require well-circulated waters, decapods probably could not tolerate the stagnant deep zones of the mid-lake regions of Fossil Lake that deposited the 18-inch layer sediments. They may also have come from nearby tributary streams that were connected to Fossil Lake.

The crayfish †*Procambarus primaevus* (Packard, 1880) is one of the first invertebrate fossils ever described from the FBM. In the late 1870s, Joseph Leidy borrowed two specimens of this species from a Dr. J. Van A. Carter of Evanston, Wyoming, and sent them to the **PALEOENTOMOLOGIST** Alpheus Spring Packard

FIGURE 38

†*Bechleja rostrata* Feldman et al., 1981. A freshwater shrimp or prawn from the sandwich beds of FBM Locality H. Body length (excluding antennae) is 86 millimeters (3.4 inches) (FMNH PE14047).

for description. After Packard described the specimens in 1880, he returned the pieces to the owner, and they have not been seen since. These may have been the first recorded fossils from any of the nearshore sandwich bed localities, but unfortunately without the original TYPE SPECIMENS, we will never be able to verify that. A NEOTYPE was designated for the species by Feldmann et al. (1981).

Procambarus is an EXTANT genus in the family Cambaridae whose species mostly prefer warm, fresh, slow-moving rivers or streams, or nearshore marsh habitat. The genus today has about 160 living species ranging through North and Central America. Feldmann et al. (1981) found the extinct FBM †Procambarus primaevus to be most closely related to species in the subgenus Austrocambarus. Several specimens of †P. primaevus have been found with ostracods in their ABDOMINAL REGIONS (Feldmann et al. 1981), indicating that they may have been feeding on these.

The freshwater shrimp †Bechleja rostrata (Feldmann et al. 1981) wasn't described and named until more than 100 years after the crayfish. It is SYMPATRIC with the crayfish but slightly less common. †Bechleja is an extinct genus in the living family Palaemonidae. Today palaemonids are ecologically diverse, with some species in freshwater and some in marine environments. They are true shrimp, but like other families of freshwater shrimp, they are sometimes called "prawns." Palaemonids are mainly carnivores that eat small invertebrates such as insect larvae and ostracods (both of which occurred in abundance in Fossil Lake). †Bechleja rostrata was probably an important food source for several of the fish species in Fossil Lake that possessed dentitions suited for crushing arthropods (e.g., †Heliobatis, †Asterotrygon, Amia, †Cyclurus, and †Masillosteus).

Freshwater Seed-Shrimp (Subphylum Oligostraca, Class Ostracoda)

Ostracods are a class of tiny aquatic arthropods sometimes known as seed-shrimp because of their appearance. They are the most common of all freshwater arthropods in the fossil record. The ostracods in the FBM are only about 1 millimeter long, and usually only the oval-shaped, bivalved "shell" is preserved. They are particularly abundant at the Warfield Springs FBM Locality K (fig. 39) and also in the small fish layers near the K-spar tuff layer in the mid-lake deposits. Although there are several taxonomic orders of ostracods, all freshwater ostracods are currently classified in the order Podocopida. The FBM species appear to be mainly in the extinct genus †Hemicyprinotus, which is in the family Cyprididae. This genus is commonly found associated with plant material. Like living cypridid ostracods, the FBM cyprids were probably an important food source for aquatic animals in Fossil Lake that fed on ZOOPLANKTON. Although mostly tiny organisms, living cyprids have a wide range of diets. Some species are filter feeders, while others are carnivores, herbivores, or scavengers. They are capable of reproducing in vast numbers, even parthogenetically (requiring only one sex to reproduce). The

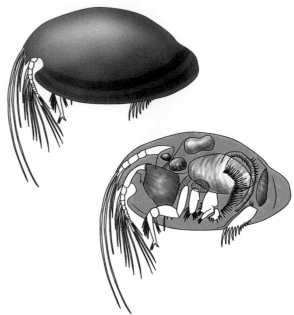

FIGURE 39
Ostracods; tiny crustaceans that occur by the thousands in particular layers of the FBM. *Left:* Several individual ostracod VALVES (or "shells") from the genus †*Hemicyprinotus* (family Cyprididae) on a slab from the sandwich beds of FBM Locality K (USNM 276504). The ostracod shells are the tiny white oval-shaped objects on the slab, each averaging about 1 millimeter (0.04 inch) in length. Shown with a U.S. dime for scale. Also on slab are leaves of *Rhus* (sumac). *Right:* Drawing of a living ostracod to show what the entire animal looked like (*above*) and with the left valve removed showing internal structure (*below*).

presence of freshwater ostracods is thought to indicate relatively shallow, freshwater conditions, and this fits with Fossil Lake's estimated maximum depth of 30 meters and that Locality K represents a shallow deltaic nearshore region of the lake (Grande and Buchheim 1994). Today there are about 1,000 described living species in the family Cyprididae. The family ranges from the tropics to the polar regions and today occupies a wide range of habitats, including freshwater lakes, ponds, rivers, streams, and even moist leaf litter on the ground of rain forests.

Mollusks

(Phylum Mollusca)

The FBM contains a number of fossil mollusks (snails and clams). As with the crayfish and shrimp, the restricted presence of these particular mollusks within specific geographic regions of the FBM is evidence of the different ecological zones within Eocene Fossil Lake. Although mollusks can give us some information about Fossil Lake's **PALEOECOLOGY**, their value as ecological indicators is somewhat limited due to uncertain classification and **PHYLOGENETIC RELATIONSHIPS** of the FBM species. Living mollusk species are often distinguished by features of soft-part anatomy, but fossil mollusks are usually preserved only by their hard shell. Shell shape is often unreliable as a diagnostic species character because very similar shell types are present among different families of living mollusks. Another challenge to accurate identification of FBM mollusks is that most studies of living and fossil mollusks have so far been regional in scope rather than global. The external shape and appearance of a fossil shell might resemble only a single North American genus and therefore seem to be easy to classify when studied only within a

North American context. But within a global context, the shell can resemble more than one genus or family of gastropod with different ecological parameters. Such is often the challenge of paleontology and of SYSTEMATICS in general. Many of the early Eocene plants, insects, and vertebrates of the FBM show strong Asian or European affinities. Therefore, the FBM mollusks must be compared to freshwater snails in North America, Europe, Asia, and elsewhere in order to fully understand the possible relationships and classification of these species.

Mollusks from the FBM include the following taxa:

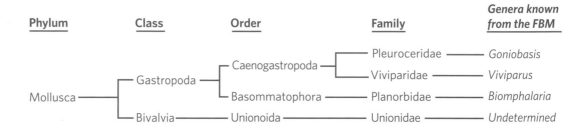

Phylum	Class	Order	Family	Genera known from the FBM
Mollusca	Gastropoda	Caenogastropoda	Pleuroceridae	*Goniobasis*
			Viviparidae	*Viviparus*
		Basommatophora	Planorbidae	*Biomphalaria*
	Bivalvia	Unionoida	Unionidae	*Undetermined*

Snail Fossils from the FBM (Class Gastropoda)

In the FBM, there are primarily three varieties: a low-spired species that occurs in mid-lake localities ("*Viviparus*" sp.), a high-spired species that occurs in the nearshore localities ("*Goniobasis*" sp.), and a disc-shaped snail ("*Biomphalaria*" sp.). A globally comprehensive study is needed to better establish the validity of these generic assignments, which is why I have put the names in quotation marks.

Pleurocerid Snails (Family Pleuroceridae)

The common FBM snail species with a high-spired shell is usually classified in the living genus *Goniobasis* (fig. 40, *top*). Living *Goniobasis* (sometimes called *Elimia* instead) belong in the family Pleuroceridae, a family of gilled snails that are widespread today in tropical to temperate climates of Asia, Africa, North America, and Central America. Most species of the family today live in rivers and streams, often with gravel bottoms. Fossil Lake would probably not have been a suitable environment for them, and therefore I suggest that they lived in the tributary streams and rivers and died when they were washed into Fossil Lake. This is probably why the shells are known only from the nearshore sandwich bed location on Thompson Ranch, where several other fossils also indicate there may have been a nearby tributary to the lake.

Some of the *Goniobasis* sp. specimens, although flattened, show a significant number of morphological details. One specimen has whorl spines and some of its rasping teeth (radular teeth) preserved next to it (fig. 40, *top middle* and *top right*). Another specimen appears to have its color pattern preserved (fig. 40, *top left*). When they are found in the beds near the K-spar tuff layer, they are often

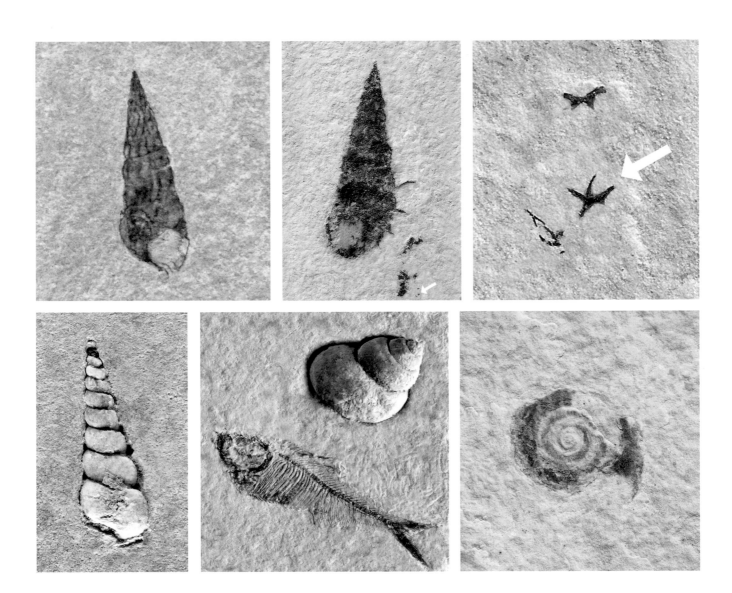

FIGURE 40

Snails (Gastropoda) from the FBM. *Top left:* Specimen of *Goniobasis* sp.
(family Pleuroceridae) from the sandwich beds of FBM Locality H with color
pattern preserved, with shell height of 33 millimeters (1.3 inches) (FMNH
PE60939). *Top middle and top right:* Specimen of *Goniobasis* sp. from the
sandwich beds of FBM locality H with spines and RADULAR teeth preserved
(teeth enlarged at right), with shell height of 44 millimeters (1.7 inches)
(BMNH GG. 21510). Photo courtesy of the Natural History Museum in London;
photographer P. Crabb. *Bottom left:* STEINKERN of *Goniobasis* sp. with a shell
height of 16 millimeters (0.6 inch), from bottom capping directly beneath the
18-inch layer (FMNH PE60848). *Bottom middle:* Steinkern of *Viviparus* sp. (family Viviparidae) with a shell height of 20 millimeters (0.8 inch), on slab with
juvenile †*Diplomystus dentatus* (FMNH PF13506). *Bottom right:* A PLANISPIRAL
snail (*Biomphalaria* sp., family Biomphalaria) from the sandwich beds of FBM
Locality H; shell measures 14 millimeters (0.6 inch) in diameter (FOBU 703).

only STEINKERNS (e.g., fig. 40, *bottom left*). "*Goniobasis*" sp. from the FBM might actually be misclassified as a pleurocerid and instead belong to a different family, the Thiaridae—a family of tropical freshwater snails (Robert Dillon, pers. comm., 2010). When the globally comprehensive systematics of fossil and living gastropods is better resolved, the FBM gastropods will be able to tell us more about the conditions in Fossil Lake.

Mystery Snails (Family Viviparidae)

"*Viviparus*" sp. is the snail species associated with the mid-lake deposits of the FBM (fig. 40, *bottom middle*). It is, by far, the most common snail species in the FBM, where they are preserved mostly as internal molds (steinkerns). They occur primarily near the base of the 18-inch layer, but at FBM Locality A they also occur commonly near the K-spar tuff layer in a zone associated with a mass mortality of small †*Mioplosus*, †*Knightia*, and †*Diplomystus* (e.g., see frontispiece figure). It is possible that this species may belong to one of the other viviparid genera, such as *Bellamya* (Robert Dillon, pers. comm., 2010). Viviparid snails are sometimes known as "river snails" or "mystery snails." The family and genus name is derived from the word "viviparous," meaning "bear alive," because they give birth to fully developed snails (other types of snails are instead oviparous, meaning they lay eggs).

Most viviparids are freshwater browsers who feed on detritus or micro algae and prefer tropical to temperate climate. Living species are nearly worldwide in distribution (although absent from South America) with about 20 different genera.

Ramshorn Snails (Family Planorbidae)

A rare snail in the FBM is the ramshorn snail *Biomphalaria* sp., from the family Planorbidae (fig. 40, *bottom right*). This family has a planispiral shell (coiled flat like a disk rather than high spired). Its living members are air breathing and normally inhabit freshwater. There are several hundred living species in the family, and they have a widespread geographic distribution. They are extremely rare in the FBM deposits, where they are known mainly from Locality H. Due to this scarcity and presence near a suspected tributary to Fossil Lake, I suspect that they may not have been normal inhabitants of the lake but were instead washed in from the tributary, much like the unionid clams.

Clam Fossils from the FBM (Class Bivalvia)

There is only one family of clams known from the FBM so far, the unionids. They occur in the nearshore sandwich beds at and near FBM Locality H, where they are rare.

River Mussels (Family Unionidae)

Unionid clams (also called mussels or bivalves) are very rare items in the FBM, but they occasionally occur in the sandwich beds near suspected tributaries, such as Locality H (fig. 41). They are represented in the FBM only by their shells, which consist of two oval-shaped valves that are hinged together. In life, these two valves enclose and protect the soft body of the clam. Today the family Unionidae contains about 800 living species that mostly live on the bottom of streams or rivers, the mouths of lake tributaries, or other bodies of moving freshwater. They burrow into the bottom sediments and leave their posterior end exposed. The exposed end pumps water through their bodies for both oxygen and food. Like all mussels, unionid clams are filter feeders that filter microscopic creatures such as plankton out of the water circulating through their bodies. The young are parasitic on the gills of fishes for several weeks. Once they reach a certain stage, they break free from the host and drop to the water bottom to begin an independent life.

The geographic range of unionids is worldwide, and they are most diverse in North America and Asia. Their absence from the mid-lake deposits is most likely due to inhospitable bottom waters.

FIGURE 41

Clam (bivalve; family Unionidae), from the sandwich beds of FBM Locality H preserved as a STEINKERN. Shell is open showing both valves. Height of each shell is 40 millimeters (1.6 inches) (FOBU 6615). Bivalves are very rare in the FBM localities, suggesting that this specimen may have washed in from a connecting river or stream.

Vertebrates

Among the fossils of the FBM, the vertebrate animals have been the main target of both commercial and scientific interests for more than a century. In contrast to the invertebrate animals and plants of the FBM, most of the known vertebrate animal species of the FBM have been studied, scientifically described, and named. Vertebrates preserved in the FBM include "fishes" (completely aquatic vertebrates with fins) and **TETRAPODS** (vertebrates with four functional feet except in rare cases where they have been secondarily lost such as in snakes, or transformed such as in birds). The term "fishes" is a common (vernacular) name for an unnatural assemblage of two completely independent lineages of vertebrate animals: the cartilaginous fishes (Chondrichthys) and the ray-finned fishes (Actinopterygii). The ray-finned fishes are more closely related to four-footed land vertebrates—such as frogs, alligators, birds, and mammals—than they are to cartilaginous fishes (sharks and rays) (see branching diagram on page 96). Vertebrate animals from the FBM are organized below in the following way:

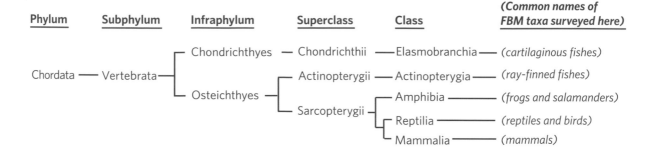

Phylum	Subphylum	Infraphylum	Superclass	Class	(Common names of FBM taxa surveyed here)
Chordata	Vertebrata	Chondrichthyes	Chondrichthii	Elasmobranchia	(cartilaginous fishes)
		Osteichthyes	Actinopterygii	Actinopterygia	(ray-finned fishes)
			Sarcopterygii	Amphibia	(frogs and salamanders)
				Reptilia	(reptiles and birds)
				Mammalia	(mammals)

Within the subphylum Vertebrata, there are three classes that are discussed here: Chondrichthyii (cartilaginous fishes), Actinopterygii (ray-finned fishes), and Sarcopterygii (including all the "four-footed" vertebrates or amphibians, reptiles, and mammals). Because birds have a reptilian ancestry, they are included here in Reptilia.

My professional career has been spent primarily as an ichthyologist, and I have included many FBM species in my research throughout the years. Consequently, the FBM "fishes" (Chondrichthii and Actinopterygii) are better known than many of the other groups of FBM fossils discussed in this book. It is mainly the fishes that the FBM is most famous for and that have generated the intense paleontological quarrying activity for the area. Within the sections on fishes, I will briefly describe a new family name (†Asterotrygonidae) and a new genus name (†Hypsiprisca gen. nov.), in part to help illustrate some of the concepts introduced in the chapter on classification.

Cartilaginous Fishes

(Superclass Chondrichthii)

The superclass Chondrichthii (infraphylum Chondrichthyes) includes all sharks and rays, and today has about 1,000 living species. In the FBM, this group is represented only by stingrays. The FBM Chondrichthii presented here are organized below in the following way:

Superclass	Class	Subclass	Order	Family	*(Common names of FBM taxa surveyed here)*
Chondrichthii —	Elasmobranchia —	Elasmobranchii —	Myliobatiformes —	†Heliobatidae	*(whip-tailed stingray)*
				†Asterotrygonidae	*(fat-tailed stingray)*

Stingrays (Order Myliobatiformes, Families †Heliobatidae and †Asterotrygonidae n. fam.)

Stingrays are some of the most beautiful of the FBM fishes. Although sometimes mistakenly referred to as "skates" in the popular literature, these are true stingrays (order Myliobatiformes) not skates (order Rajiformes). The most obvious stingray character is the presence of tail barbs or "stings." The stings are sharply pointed, with barbed edges so they could not have been easily dislodged by their victims, and in living stingrays these barbs are often venomous. Stingrays use the barbs as an effective defensive mechanism. Both FBM species appear to have had three to four stings when fully armed, but many specimens appear to have expended one or more of their stings prior to their death and eventual fossilization.

Fragmentary stingray fossils consisting of isolated teeth, skin denticles, and stings (caudal barbs) are widespread in the fossil record. Such material is known as far back as the Early Cretaceous and from dozens of localities in North America, Europe, Africa, Asia, South America, and Cuba (Carvalho, Maisey, and Grande 2004). However, there are presently only two localities in the world that produce quantities of complete articulated fossil stingrays. One is the Eocene marine fossil locality of Monte Bolca, in Italy, and the other is the FBM. Not even the other deposits within the Green River Formation (from Eocene Lake Uinta and Lake Gosiute) have produced complete articulated stingray skeletons so far.

There are two distinct genera and species of stingray in the Fossil Butte Member: †*Heliobatis radians* in the family †Heliobatidae (figs. 42–44) and †*Asterotrygon maloneyi* in the new family †Asterotrygonidae (figs. 45–47). †*Heliobatis* is a whip-tail stingray that was described over 130 years ago. The valid genus name is †*Heliobatis*, predating the JUNIOR SYNONYMS †*Xiphotrygon* and †*Paleodasybatis*. †*Asterotrygon* is a recently described primitive "fat-tailed" stingray genus near the base of the myliobatiform evolutionary tree.

Stingrays are basically flattened sharks and are classified with sharks in the subclass Elasmobranchii. Stingrays have a disc-shaped head with their eyes located on top of the disc and their mouths located on the underside of the disc. Most living species of stingrays spend the majority of their time on the bottom of whatever body of water they inhabit, often burying themselves in sediments when not in search of food. They feed mostly on crayfishes, shrimps, mollusks, and other invertebrates that live on the bottom, which helps explain why stingrays are more common in the nearshore sandwich beds than in the mid-lake 18-inch layer. The nearshore sandwich beds have more abundant crayfish, shrimp, and *Goniobasis* mollusk fossils and the bottom waters there were more oxygenated than the mid-lake 18-inch layer. The middle regions of the early Eocene lake probably had a frequent buildup of toxins in the bottom waters due to poor circulation, which would have been inhospitable for bottom-dwelling fishes most of the time.

Stingrays are a diverse group comprising an entire order (Myliobatiformes) whose living members include at least 8 families, 27 genera, and 185 species. They are globally widespread, living in all temperate, subtropical, and tropical seas. Although many stingray species (including †*Heliobatis*) have at one time or another been referred to as being in the family "Dasyatidae," this name has generally been used merely as a **WASTEBASKET GROUP** of convenience for various stingray species that cannot be accurately classified in another family. Carvalho, Maisey, and Grande (2004) placed †*Heliobatis* outside of Dasyatidae in its own family, †Heliobatidae. They also classified †*Asterotrygon* outside of Dasyatidae as "family uncertain" within the suborder Myliobatoidei, because there was no evidence to include it in any existing family. Here I create the family name †Asterotrygonidae for it (see below). Most living species of stingrays are marine, but one neotropical family in the Amazon Basin of South America with more than 20 species (Potamotrygonidae) is exclusively freshwater and cannot tolerate saltwater. Certain species from other stingray families living in warm temperate to tropical regions of Asia, Africa, New Guinea, and Australia also inhabit freshwater. One such species is the giant freshwater whip ray, *Himantura chaophraya*, which inhabits the Mae Klong River of Thailand. This freshwater giant reaches lengths of over 4 meters (14 feet), can weigh nearly a half ton, and has a sting that can be 38 centimeters (15 inches) or more in length! Based on comprehensive morphological study of all living and fossil stingrays (e.g., Carvalho, Maisey, and Grande 2004), the FBM stingrays are not potamotrygonids, nor are they known to share a close relationship with each other or any particular living freshwater species. Instead they represent two independent lineages of stingrays that radiated into freshwater in North America and eventually became extinct, possibly by the middle or late Eocene. The FBM species are unique in that they appear to have been endemic to a freshwater lake. Today stingrays that inhabit freshwater live in river systems rather than lakes. It may also be that the presence of these stingray species in Fossil Lake was due to a connection to a major river system.

The world's best-preserved fossil stingrays occur in the FBM deposits, and they are beautiful when properly prepared. This makes them a pleasure to study by ichthyologists, favorite exhibit pieces for museums, and treasured acquisitions by collectors. They are also a prime example of just how much the FBM fossils can visually reveal to us about life in the Eocene. Their blunt teeth, robust jaws, and proximity to rock containing mollusk, crayfish, and shrimp fossils give us strong clues to their diet. Living rays with similar crushing-type jaw and tooth morphology feed on mollusks and decapods. Like living stingrays, the FBM species were sexually dimorphic, with the males bearing reproductive organs called claspers (figs. 42, 43, *top right*, and 45). It doesn't take much of an imagination to envision many parts of the life cycle of these long extinct species.

†*Heliobatis* is known only from the FBM deposits by a single described species, †*H. radians*. †*Heliobatis* ranges from about 8 centimeters (3 inches) to nearly 90 centimeters (3 feet) in length and is the more common of the two stingray genera in the FBM deposits. It is most easily distinguished from †*Asterotrygon* by its thin, whip-like tail, like modern whip-tail rays, and is often found with the tail in tightly curved positions (e.g., fig. 44, *top*). For a detailed description of †*Heliobatis radians*, see Carvalho, Maisey, and Grande (2004, 65–72). Although

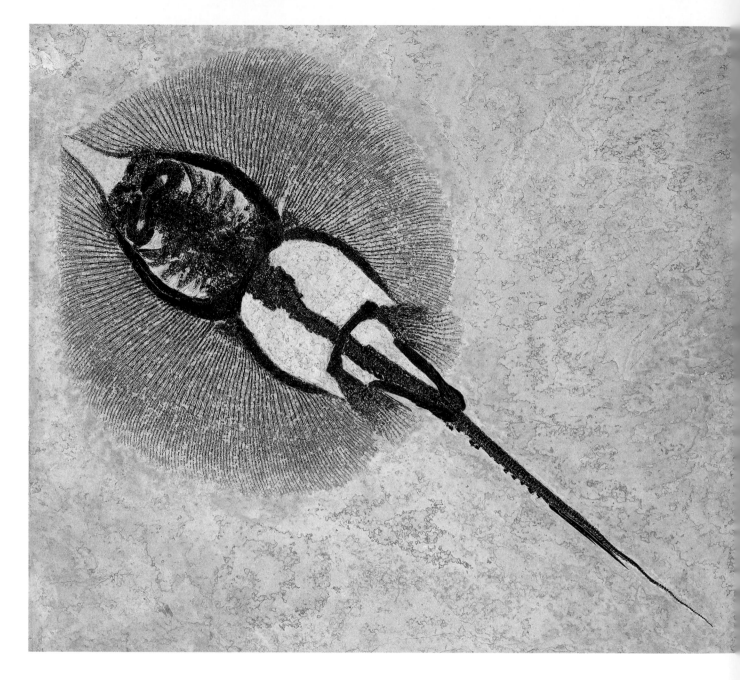

FIGURE 42
The Eocene whip-tail stingray †*Heliobatis radians* Marsh, 1877 (family †Heliobatidae), from the 18-inch layer deposits of the FBM (probably FBM Locality B). Complete male specimen in ventral view with well-preserved jaws and claspers, with a length of 510 millimeters (1.7 feet) (FMNH P25009). See fig. 43 for close-up of jaws, claspers, and tail stings.

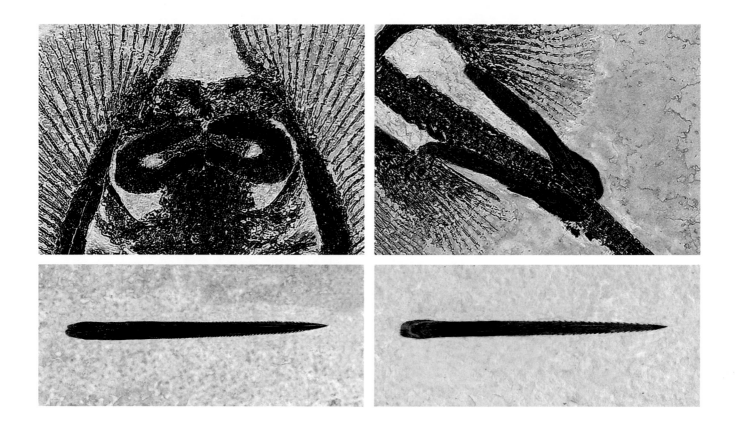

FIGURE 43

†*Heliobatis radians* Marsh, 1877, from the 18-inch layer deposits of the FBM.

Top: Close-up of the jaws (*left*) and claspers (*right*) from the specimen in fig. 42.

Bottom left: Tail sting in dorsal view, 55 millimeters (2.2 inches) long (from FMNH PF15365). *Bottom right:* Tail sting in ventral view, 47 millimeters (1.9 inches) long (from FMNH PF15364). Note the sharp point and barbed edges of the stings, which makes them formidable protective weapons.

relatively rare as a percentage of the fishes found in the FBM (see table 1, pages 176–77), the massive numbers of fishes excavated from the FBM over the last several decades have resulted in hundreds of †*H. radians* specimens. Although most specimens today reside in private collections, there are also many in museum collections (e.g., 10 specimens in the Field Museum collection at the time I wrote this book). Some commercially prepared specimens are heavily restored. Prepared specimens acquired from commercial vendors should be examined under magnification to verify originality. A black light can sometimes be used to show areas of restoration, because the natural parts will usually fluoresce differently than the restored parts. Some commercial specimens are composites (specimens reassembled from more than one individual). An X-ray of the slab will reveal such reconstructions.

†*Asterotrygon* currently contains a single species, †*Asterotrygon maloneyi*, although there is some morphological variation in the denticle pattern indicating that there may be more †*Asterotrygon* species that are yet undescribed. A number of specimens indicate that Fossil Lake was a breeding ground for †*Asterotrygon*. Although there are only a few dozen specimens of this species known, the known sample includes one apparent mating pair (fig. 45), a female specimen with an embryo inside (fig. 46), and a specimen with newly born babies beside it

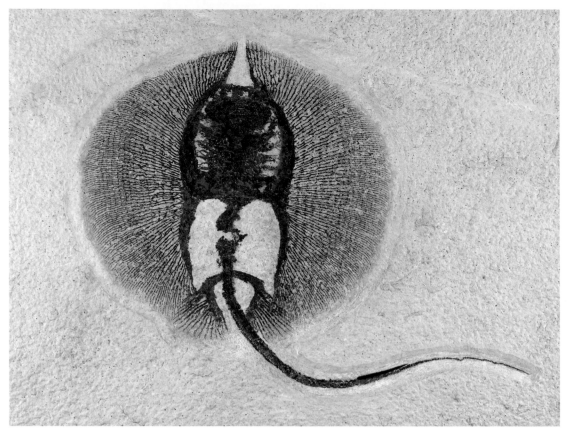

⇡ FIGURE 44

†*Heliobatis radians* Marsh, 1877, from the sandwich beds of the FBM. *Top:* Complete female specimen in dorsal view showing the flexibility of the whip-like tail in this species. The lack of claspers indicates this to be a female, with a length of 255 millimeters (10 inches), from FBM Locality H (AMNH P 19665). *Bottom:* Baby individual only 64 millimeters (2.5 inches) long, from FBM Locality H (FMNH PF15367).

⇢ FIGURE 45

The Eocene fat-tailed stingray †*Asterotrygon maloneyi* De Carvalho et al., 2004 (family †Asterotrygonidae). This remarkable plate shows a probable mated pair from the 18-inch-layer deposits of FBM locality G. †*Asterotrygon* is extremely rare in the FBM and not a schooling fish. The occurrence of two breeding-size individuals, one male (lower specimen) and one female (upper specimen), on the same slab suggests that they may have been either mating or about to mate just before they were killed. The male specimen measures 480 millimeters (1.6 feet) long. Photo by Mark Mauthner; courtesy of Heritage Auctions.

(fig. 47, *bottom*). The specimen with the embryo inside indicates that this species gave live birth (i.e., they were viviparous).

†*Asterotrygon* is a relatively primitive type of stingray whose family relationships have not yet been clearly resolved. The study by Carvalho, Maisey, and Grande (2004) classified this genus outside of all known families of stingrays, so here I place it in the new **MONOTYPIC** family †Asterotrygonidae for convenience, with †*Asterotrygon maloneyi* as the **TYPE SPECIES**. It is diagnosed from other myliobatoid families by the combination of characters given in Carvalho, Maisey, and Grande (2004). This is the species previously referred to as the "undescribed ray" (Grande 1980, 1984) or the "fat tailed ray," and was not officially described and named until 2004. It is easily distinguished from †*Heliobatis* in its having a

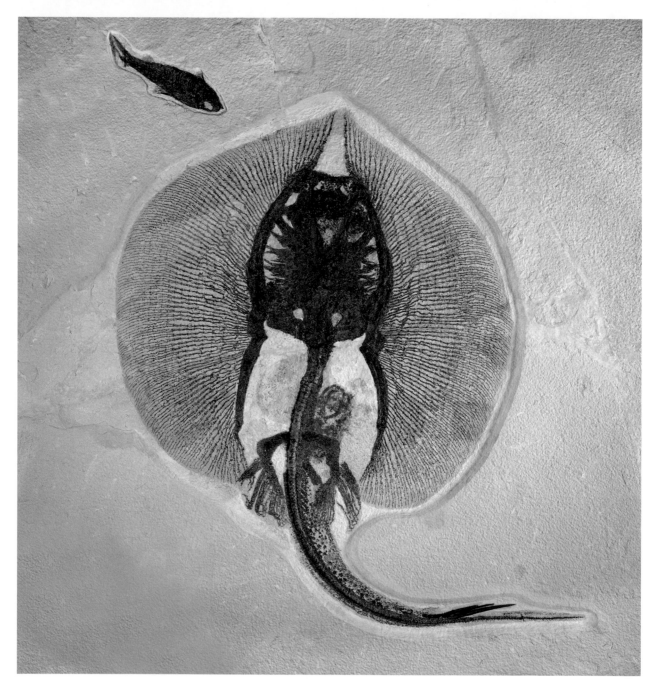

FIGURE 46

The Eocene fat-tail stingray †*Asterotrygon maloneyi* De Carvalho et al., 2004, from the sandwich beds of FBM Locality H, in dorsal view. This remarkable specimen is a pregnant female with a small baby ray coiled up inside her (enlarged in fig. 47, *bottom left*). This beautifully preserved and finely prepared specimen is 625 millimeters (2.1 feet) long and is the HOLOTYPE for the species (FMNH PF15166). Teleost fish on slab with the ray is the herring †*Knightia eocaena*.

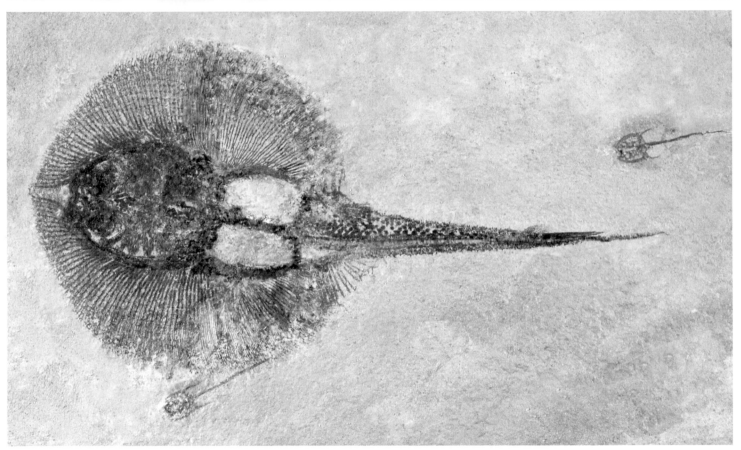

FIGURE 47

†*Asterotrygon maloneyi* De Carvalho et al., 2004, specimens from the sandwich beds of FBM Locality H. *Top left:* Close-up of the embryo from fig. 46 showing its coiled vertebral column and the tiny tail sting already developed (arrow). *Top right:* Close-up of the end of a tail from a specimen with an estimated length of 250 millimeters (9.8 inches), showing the heavy denticle covering of the "fat-tail" and the well-developed dorsal fin preceding the tail stings; two characteristics of this genus and species (FMNH PF15181). *Bottom:* A female ray with two aborted late-term fetuses next to her (AMNH P 11557); specimen is the PARATYPE for the species. The mother ray is 378 millimeters (1.2 feet) long. Specimen was poorly prepared, and the disc of the adult is badly damaged (AMNH P 11557).

dense covering of skin denticles, some of which bear curved, hook-like spines. It also has a fat tail covered with small bony plates ("dermal denticles"). The tail of †*Asterotrygon* was probably not quite as flexible as the "whip tail" of †*Heliobatis*, and a dorsal fin just anterior to the stings (fig. 47, *top right*). †*Asterotrygon* is known only from the FBM deposits, where it is very rare. I estimate that it is about 40 times rarer than †*Heliobatis* (page 176). At the time this book was written, there were 11 specimens of this species in the collection of the Field Museum. When my colleagues and I described this species, we named it *maloneyi*, after the late fossil collector Thomas Maloney, who donated one of the PARATYPES (the female specimen in fig. 47, *bottom*, with the two babies beside it). For a detailed description of †*Asterotrygon maloneyi*, see Carvalho, Maisey, and Grande (2004, 12–64).

Ray-Finned Fishes

(Superclass Actinopterygii)

The vast majority of fossils that have been mined from the FBM over the last century and a half have been fossil ray-finned fishes, or **ACTINOPTERYGIANS**. Literally millions of complete fossil ray-finned fish skeletons have been excavated from the FBM, the majority of which have been recovered in the last 30 years because of a post-1970s boom in the number of commercial fossil operations. Almost all vertebrate fossils in the FBM are actinopterygian fishes, with perhaps 1 out of 2,500 being a stingray and 1 out of every 5,000 to 10,000 being a **TETRAPOD**.

Some actinopterygian groups are still poorly understood because of their great diversity. One such group is the spiny-rayed suborder Percoidei with over 3,200 living species (including perch, bass, sunfishes, and thousands of other species with pointed spines in their fins). Until the living percoid species are better known, accurate classification of the FBM percoids (†*Mioplosus*, †*Priscacara*, †*Hypsiprisca*, and undescribed percoid genera) will be unsatisfactory.

Length measurements given here for actinopterygians were made from the tip of the snout to the very end of the tail fin (= total length). The FBM actinopterygian fishes presented below are as follows:

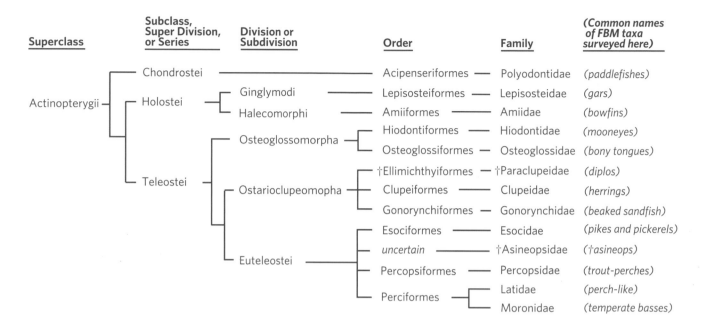

Superclass	Subclass, Super Division, or Series	Division or Subdivision	Order	Family	(Common names of FBM taxa surveyed here)
Actinopterygii	Chondrostei		Acipenseriformes	Polyodontidae	(paddlefishes)
	Holostei	Ginglymodi	Lepisosteiformes	Lepisosteidae	(gars)
		Halecomorphi	Amiiformes	Amiidae	(bowfins)
	Teleostei	Osteoglossomorpha	Hiodontiformes	Hiodontidae	(mooneyes)
			Osteoglossiformes	Osteoglossidae	(bony tongues)
		Ostarioclupeomopha	†Ellimichthyiformes	†Paraclupeidae	(diplos)
			Clupeiformes	Clupeidae	(herrings)
			Gonorynchiformes	Gonorynchidae	(beaked sandfish)
		Euteleostei	Esociformes	Esocidae	(pikes and pickerels)
			uncertain	†Asineopsidae	(†asineops)
			Percopsiformes	Percopsidae	(trout-perches)
			Perciformes	Latidae	(perch-like)
				Moronidae	(temperate basses)

Paddlefishes (Order Acipenseriformes, Family Polyodontidae)

Paddlefishes are relatively rare in the FBM, represented by the species †*Crossopholis magnicaudatus* (fig. 48). †*Crossopholis* has a very long snout region, or "paddle." Living paddlefishes are sometimes called "spoonbills," "spoonies," or even "spoonbill catfish." The last of those common names is misleading because paddlefishes are not closely related to catfishes and are instead close relatives of sturgeons. Living paddlefishes have tiny eyes, and the paddle is a sensory device lined with thousands of electroreceptor organs that allow paddlefishes to navigate and locate food, even in dark or cloudy waters. Grande and Bemis (1991, 73) found evidence that the FBM †*Crossopholis* also had such electro-sensory organs.

Paddlefishes, like their close relatives or SISTER GROUP the sturgeons, lack robust, solid vertebrae of the type found in other bony fishes of the FBM. The axial skeleton ("vertebral column") of paddlefishes is composed largely of soft tissues that do not preserve in the fossils. When first described from the FBM by Cope in 1883, †*Crossopholis magnicaudatus* was known only by a partial body and tail section lacking a head. In 1886 Cope described a second specimen consisting of a partial skull. It was not until almost a century later that the first nearly complete skeletons were reported (Grande 1980). One reason it took so long to discover a complete skeleton of †*Crossopholis* is that some of the commercial quarries did not recognize them, thinking them to be poorly preserved gars or some other badly preserved fish not worth collecting. The first complete paddlefish I reported in 1980 (Grande 1980, fig. II.8a) was assembled from discarded pieces I found in a scrap pile in a commercial quarry. Once *Bulletin* 63 (Grande 1980) was published and adopted as a field guide, many paddlefishes from the FBM began to show up, and today there are dozens of complete skeletons known (e.g., figs. 48, 49). Most of the paddlefishes come either from the Thompson Ranch sandwich bed quarries (e.g., FBM Locality H) or from the layers above the 18-inch layer near the K-spar tuff bed.

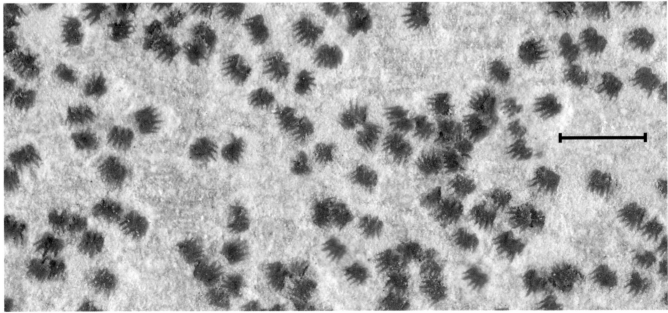

FIGURE 49

†*Crossopholis magnicaudatus* Cope, 1883, two individuals from the sandwich beds of FBM Locality H with large fishes preserved within their stomach regions. Many such individuals have been found, indicating that †*Crossopholis* was a voracious predator that fed on other fishes. This mode of feeding is in contrast to that of the living North American paddlefish, *Polyodon*, which is a filter feeder that feeds primarily on plankton. *Top:* †*Crossopholis* specimen preserved on slab with some of the fishes it was evidently feeding on (†*Diplomystus dentatus* and †*Knightia eocaena*); length of paddlefish 825 millimeters (2.7 feet) (FOBU 407). *Bottom:* †*Crossopholis* specimen preserved with a large †*Mioplosus labracoides* in its stomach; length of paddlefish is 855 millimeters (2.8 feet) (FMNH PF13552).

The paddlefish family Polyodontidae is today represented by only two modern-day freshwater species: *Polyodon spathula* from the Mississippi River drainage and contiguous areas of eastern and central North America, and *Psephurus gladius* from the Yangtze River drainage of China. These two species are sometimes referred to as "living fossils" because of their relatively primitive anatomy and the fact that they are some of the few surviving members of Chondrostei, a group that has a very old fossil record. Paddlefishes have a somewhat shark-like body profile and were, in fact, first described as sharks in 1792. The living North American *Polyodon* has a large toothless, thinly boned mouth and highly specialized, net-like gill rakers in its throat. It swims through the water column with its mouth wide open, straining water through its mouth and out its gill slits, trapping tiny ZOOPLANKTON in its gill rakers. Because it lives primarily on these microorganisms filtered out of the water, it is called a FILTER FEEDER. It reaches a documented length of about 1.6 meters (5 feet). In contrast, the living Chinese *Psephurus* is a carnivorous predator with a small, heavily boned mouth and PROTRUSIBLE JAWS and short tooth-like gill rakers. It feeds primarily on other fishes and decapods and reaches a documented length of about 3 meters (just over 9 feet), although there are undocumented reports of it reaching lengths of 6 meters (20 feet). There have been no sightings of *Psephurus gladius* in the wild since 2003, and this species is now thought to probably be extinct.

The FBM †*Crossopholis* is somewhat of an enigma. Its closest relative is the North American *Polyodon* (Grande and Bemis 1991), but *Crossopholis* was not equipped to be a filter feeder. Instead, it had small heavily boned protrusible jaws like *Psephurus* and fed largely on other fishes, as evidenced by the large percentage of specimens that have been found with large fishes preserved in their stomachs (fig. 49). This points to PISCIVORY as being the primitive mode of feeding in the family Polyodontidae and the highly specialized filter-feeding of *Polyodon* being uniquely evolved within the family for that genus. The known species of fossil and living paddlefishes are described and presented in detail in a monograph by Grande and Bemis (1991) and Grande, Jin, Yabumoto, and Bemis (2002).

Paddlefishes make up less than 0.02 percent of the total fishes excavated from the FBM. There are numerous specimens in the collection of the Field Museum in Chicago. The FBM species is briefly summarized below. †*Crossopholis* is so far known only from the FBM within the Green River Formation. It has a maximum known size of about 1,500 millimeters (59 inches) total length (FMNH PF13551 is 1,480 millimeters). †*Crossopholis* has thousands of very tiny fringed scales on much of its body, although much of the body, particularly just behind the head, is naked (fig. 48). The scales are the source of the name †*Crossopholis*, which means "fringed scale."

The fact that †*Crossopholis* is more common in the nearshore Thompson Ranch sandwich bed deposits to the northwest (e.g., FBM Locality K) than in the mid-lake 18-inch layer deposits indicates that they may have migrated into the lake from connecting rivers such as the one thought to have existed to the north. In the mid-lake regions (e.g., FBM Locality A), they are much more abundant in the overlying fish layers near the K-spar tuff beds than in the 18-inch layer, possibly reflecting a changing environment in the lake over time. Another line of evidence that †*Crossopholis* spent much of its life in connecting rivers to Fossil Lake is the absence of larval and young juvenile-size specimens of †*Crossopholis* in the FBM. The smallest †*Crossopholis* specimen I am aware of is about 260 millimeters in total length (Grande and Bemis 1991, fig. 58G) out of about 100 specimens so far discovered. There are 16 specimens of †*Crossopholis magnicaudatus* in the collection of the Field Museum. †*Crossopholis* is described and illustrated in detail in Grande and Bemis (1991, 64–85).

Gars (Order Lepisosteiformes, Family Lepisosteidae)

Gars have long been of interest to ichthyologists and paleontologists alike. Their primitive look is a thing of both beauty and scientific appeal. There are currently four species of gars in the FBM, and all of them are very rare components of the fish fauna there. Fewer than 1 in every 5,000 fishes excavated from the FBM are gars. Gars have long cylindrical bodies covered with thick, shiny, diamond-shaped scales. These thick, armor-like scales make the large adults virtually immune from being preyed upon by any other animals. Historically in human culture, scales from living gars have been used as arrow heads, and the scale-covered hides of large gars were used for protective breastplates for warriors. Loose gar scales are extremely common fossils in Cretaceous through Eocene freshwater deposits of North America, but complete fossil skeletons are very rare. There are seven living gar species today, restricted to freshwaters of central and eastern North America, Central America, and Cuba (four species of *Lepisosteus* and three of *Atractosteus*). The largest living gar is the alligator gar, which reaches a length of just over 3 meters (about 10 feet). Large alligator gars have even been seen to strike and swallow waterbirds from the surface of the water, although this is rare.

Fossil gars have a much broader range, further extending it to western North America, Europe, Africa, Madagascar, India, and South America. The earliest gars are Early Cretaceous in age (about 110 million years before present). Living gars are sometimes referred to as "living fossils," and except for †*Masillosteus*, the FBM gars look similar to living gars.

Living gars prey mostly on other fishes and invertebrates such as shrimps and crayfishes for food. They are slow moving except for the short periods when they strike at prey; then they become voracious predators capable of extremely fast movement. Their swim bladders can function as lungs, allowing them to survive in stagnant or warm water where the oxygen content is low. Their preferred habitat, though, is shallow weedy areas near shore. The scarcity of gars in

FIGURE 50

The Eocene needle-nose gar †*Lepisosteus bemisi* Grande, 2010 (family Lepisosteidae), from the 18-inch layer of the FBM. Note the elongate, extremely narrow snout and the 14 rays in the tail fin. Both beautifully preserved specimens from FBM Locality B. *Top:* Specimen is in lateral view and is 720 millimeters (2.4 feet) long (PU 14585). *Bottom:* Large specimen in oblique, ventro-lateral view on slab with small †*Diplomystus dentatus*. Gar is 968 millimeters (3.2 feet) long and is the PARATYPE for the species (FMNH PF15169).

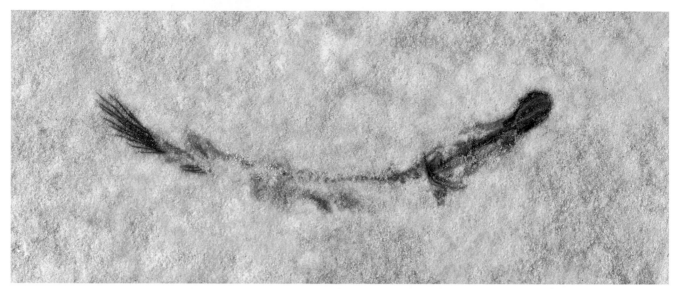

the FBM suggests that the Fossil Lake species may have preferred the tributary streams and rivers, only rarely entering Fossil Lake itself. Gars usually migrate up streams and rivers to spawn, which explains the extreme scarcity of baby gars in the FBM. The 20-millimeter hatchling in figure 51 (*top*) is so far unique within the FBM, and the next smallest FBM gar specimen I am aware of is over 10 times that size.

Of the four FBM gar species, three appear to be similar in morphology and probable habitat to living gars: †*Lepisosteus bemisi*, †*Atractosteus simplex*, and †*Atractosteus atrox*. But one species of FBM gar in the extinct genus †*Masillos-*

teus is unlike any modern gar. †*Masillosteus janei* lacks the long jaws lined with sharply pointed needle-like teeth suited for preying on other fishes. Instead, it has very short, stout, muscular jaws with only a few pointed teeth but many rounded or semi-flattened teeth more suited for crushing small invertebrates such as snails, shrimps, or crayfishes. All known fossil and living gars (including four FBM species) are described in detail in Grande (2010).

†*Lepisosteus bemisi* is so far known only from the FBM, where it is represented by about 100 specimens. This species, also called the Green River longnose or needle-nose, has an extremely long snout consisting of upper and lower jaws lined with needle-like teeth. This species can also be distinguished from all other FBM gars in having 13 to 14 tail fin rays rather than the 12 present in other gar species. Other differentiating characteristics are described in Grande (2010, 249–84). The largest known †*L. bemisi* is 1,610 millimeters (63 inches) in total length. This species is represented in the Field Museum collection by 10 specimens, including one with a †*Diplomystus dentatus* in its stomach (FMNH PF15187). This species preyed upon other fishes for food.

†*Atractosteus simplex* is the most common of the four gar species in the FBM and is known by several hundred specimens. It is sometimes called the "simplex gar." It has 12 tail-fin rays, smooth-sided scales, a medium-length snout, and many sharply pointed teeth in its jaws. Maximum known size for this species is 705 millimeters (28 inches) in total length. There are 15 specimens of this species in the collection at the Field Museum. Several individuals of this species have been found with fishes in their jaws or stomachs (e.g., fig. 52), attesting to its voracious nature.

†*Atractosteus atrox* is a species that is only known by very large individuals ranging in size from 1,600 to 2,055 millimeters (63 to 81 inches) in total length. Perhaps the young never ventured out of the tributary rivers into Fossil Lake. Like †*A. simplex*, this species has a medium-length snout, jaws lined with sharply pointed teeth, and 12 tail-fin rays. Unlike †*A. simplex*, it has highly ornamented scales. †*Atractosteus atrox* also has more lateral-line scales than any other FBM gar (62 to 64). For a more complete description, see Grande (2010, 533–49). This is one of the rarer species of gar in the FBM. There are two specimens in the Field Museum collection.

The genus †*Masillosteus* was first described from middle Eocene deposits of Messel, Germany (Micklich and Klappert 2001), with the species †*Masillosteus kelleri*. I later described a second species from the FBM in 2010, †*Masillosteus janei*.

†*Masillosteus janei* is one of the rarest fishes in the FBM and also one of the most peculiar. It has extremely short, robust jaws and crushing molariform teeth rather than the long rows of pointed teeth that other species of gars possess. It has so far been found only in the zones of Fossil Lake where the snails, shrimps, and crayfishes are found. The proximity to these shelled invertebrates and the presence of crushing teeth and jaws strongly suggest that †*Masillosteus*

←·· FIGURE 51

The medium-snouted gar †*Atractosteus simplex* (Leidy, 1873) (family Lepisosteidae), from the FBM. *Top:* Baby gar measuring only 20 millimeters long (0.8 inch), from the sandwich beds of FBM Locality H. Note that the scales and many of the skeletal elements were not yet ossified in this tiny hatchling (FMNH PF15366). *Bottom:* Nicely preserved adult specimen in lateral view from the 18-inch layer of FBM Locality A, with a length of 674 millimeters (2.2 feet) (FMNH PF12597).

FIGURE 52

†*Atractosteus simplex* (Leidy, 1873); specimens from the 18-inch layer with †*Diplomystus dentatus* in their jaws, indicating that this fossil species was a piscivour (fish eater) like living gars today. *Top:* Specimen measuring 424 millimeters (1.4 feet) long, from FBM Locality B (PU 22911). *Bottom:* Specimen measuring 700 millimeters (2.3 feet) from FBM Locality E. Private collection; photo courtesy of Bruce Baganz.

FIGURE 53

The largest FBM species, †*Atractosteus atrox* (Leidy, 1873) (family Lepisosteidae), a species so far known from only very large specimens. *Top:* This crudely prepared specimen was collected by Lee Craig in the late nineteenth century and is probably from the 18-inch layer of FBM Locality B. Specimen is now in the collection of the Harvard Museum of Comparative Zoology and is 1,680 millimeters (5.5 feet) long (MCZ P.5168). *Bottom:* Scattered scales showing the typical shape of gar scales. From a disarticulated specimen with an estimated total length of 1,900 millimeters (6.2 feet), from the 18-inch layer of FBM Locality A (FMNH PF13056). Scale bar equals 20 millimeters (0.8 inch).

FIGURE 54

The ultrashort-nose gar †*Masillostues janeae* Grande, 2010 (family Lepisosteidae, subfamily †Masillo-steidae). Two specimens from the sandwich beds of FBM Locality H. This recently described species is unlike the other gars in the FBM, with its extremely short, muscular jaws and its semi-flattened teeth suited for crushing mollusks or arthropods. This is in contrast to all living gars, which have sharply pointed teeth for preying on other fishes. *Top:* HOLOTYPE specimen with a length of 945 millimeters (3.1 feet) (FMNH PF15196), on a slab with a †*Knightia eocaena*. *Bottom:* PARATYPE specimen with a length of 632 millimeters (2.1 feet) (FMNH PF15195), on a slab with a †*Knightia eocaena*.

janei fed on these invertebrates. †*Masillosteus* shared an ecological niche with the bowfin †*Cyclurus*, which also has crushing teeth and occurs in the same beds. †*Masillosteus* probably spawned in the tributary rivers and streams to Fossil Lake because the smallest known specimen from Fossil Lake is about 400 millimeters (16 inches) in total length. The largest known specimen is 1,060 millimeters (42 inches) total length. There are three specimens of †*M. janei* in the collection of the Field Museum and only about eight specimens known in total. For a detailed description of this species, see Grande (2010, 635–61).

Bowfins (Order Amiiformes, Family Amiidae)

Bowfins are represented in the FBM by two species and genera: †*Amia pattersoni* and †*Cyclurus gurleyi*. The name "bowfin" comes from the long arching dorsal fin present in these genera. *Amia* and †*Cyclurus* are rare components of the FBM fish fauna, with †*Cyclurus gurleyi* being slightly more scarce than †*Amia pattersoni*. The adults are easily distinguishable from each other in overall body shape. †*Cyclurus gurleyi* has a short body that is FUSIFORM in shape (tapered at both ends like an American football), while *Amia* has a more elongate body form and reaches a much larger size. They also differ from each other in their mouth structure and probable diet. *Amia* has long jaws and a mouth filled with sharply pointed teeth, whereas †*Cyclurus* also has large patches of MOLARIFORM crushing teeth in its mouth. The family Amiidae is survived today by only a single living species, *Amia calva*, which lives in freshwaters of eastern North America. The extinct genus †*Cyclurus* is not known to have survived past the late Eocene (about 35 million years ago). Both *Amia* and †*Cyclurus* have well-preserved fossil species in western North America, Europe, and Asia going back to Late Cretaceous times (about 95 million years before present). The fossil and living species of the family Amiidae are described in detail in Grande and Bemis (1998), and an updated key to the species is given in Grande (2001, table 2).

Amia calva today inhabits sluggish, clear, lowland freshwaters and is extremely hardy. It goes by the common names "bowfin," "dogfish," "grindle," "choupique," "mudfish," "blackfish," "swamp muskie," "lawyer," and a number of other colorful vernaculars. Like its cousins the gars (Lepisosteidae), it is capable of breathing air if it needs to because its swim bladder can act as a functional lung. *Amia calva* is a voracious predator that feeds primarily on other fishes, but also on crayfishes and prawns. The FBM †*Amia pattersoni* was also evidently a voracious predator, because many of the known specimens have large fishes preserved in their stomachs (e.g., fig. 57; and also see Grande 1984, fig. II.20a). *Amia calva* reaches a length of about 870 millimeters (34 inches). The FBM †*Amia pattersoni* gets much larger, with specimens known over 1,400 millimeters in total length. The two genera and species of FBM bowfins, *Amia* and †*Cyclurus*, are briefly reviewed below.

FIGURE 55

The long bowfin †*Amia pattersoni* Grande and Bemis, 1998 (family Amiidae). *Top:* A baby measuring 96 millimeters (3.8 inches), from the sandwich beds of FBM Locality H (FMNH PF15388). *Bottom:* A large adult measuring 1,092 millimeters (3.6 feet), from the 18-inch layer of FBM Locality A (FMNH PF14091). Specimen is the HOLOTYPE for the species.

†*Amia pattersoni* is a rare component of the FBM fish fauna. Of the two FBM amiids, †*A. pattersoni* is the more common (a revision of my earlier statement in Grande [1984] that it was the less common species). Adult specimens of †*A. pattersoni* are slightly more common in the mid-lake deposits than the near shore, based on a known sample size of about 30 specimens. There are six specimens in the collection of the Field Museum. Unlike adult †*Cyclurus*, †*Amia pattersoni* adults have an elongate body form. They can also be distinguished by the presence of longer jaws, the lack of molariform teeth in the mouth, more vertebral CENTRA (83 to 87 versus 70 to 79 in †*Cyclurus gurleyi*), more dorsal fin rays (usually 45 to 47 versus 36 to 39 in †*C. gurleyi*). The maximum known size for †*Amia pattersoni* is 1,404 millimeters (4.6 feet) in total length (fig. 56). For

FIGURE 56

†*Amia pattersoni* Grande and Bemis, 1998. *Top:* The largest known specimen, measuring about 1,400 millimeters (4.6 feet) long, from the 18-inch layer of FBM Locality G (FMNH PF10235). *Bottom:* Isolated scale from an individual estimated to be about 1,200 millimeters (3.9 feet) in total length, from the 18-inch layer of FBM Locality A (FMNH PF15369). Scale measures 31 millimeters (1.2 inches) in length.

†*Amia pattersoni* Grande and Bemis, 1998. Specimen from the 18-inch layer of the FBM Locality B with a large †*Mioplosus* within its stomach region, attesting to the voracity of this species. Several specimens of this species are known with large fishes in their stomachs. Specimen is 788 millimeters (2.6 feet) long. The back half of fish is preserved in a lateral view, but the skull and front half are twisted into a dorsal view, with both sides of skull showing. The stomach of this normally slender species is bulging with the large fish it swallowed. The other large fish on the slab is a †*Diplomystus dentatus*. Private collection; photo courtesy of Scott Wolter.

a more detailed description of †*Amia pattersoni*, see Grande and Bemis (1998, 185–207).

†*Cyclurus gurleyi* is the rarer of the two FBM amiids. The adults have a characteristic American football shape, and they do not get as large as †*Amia pattersoni* (maximum known total length of 789 millimeters [31 inches]). Adult †*Cyclurus gurleyi* is slightly more common in the mid-lake localities than in the nearshore localities, and the young are known only from the nearshore localities. There are only about 20 individuals of †*C. gurleyi* known so far, eight of which are in the collection of the Field Museum. It occurs in many of the same localities as the gar †*Masillosteus* and may have had a similar ecological niche in Fossil Lake (e.g., feeding on mollusks and arthropods). Because of its rarity in the lake, it may have also been a resident of surrounding tributaries. The earliest known occurrence of the genus †*Cyclurus* is in the Late Cretaceous about 65 million years ago, and the genus appears to have become extinct sometime around the early Oligocene about 30 million years ago. This genus probably included arthropods and gastropods in its diet, based on the molariform tooth patches in its mouth. All fossil species of †*Cyclurus* known through 1998 were reviewed in detail in Grande and Bemis (1998). More recently, a species of †*Cyclurus*, †*C. orientalis*, was described from the middle Eocene Xiawanpu Formation of Hunan Province, China (Chang, Wang, and Wu 2010). The Chinese species closely resembles the FBM species †*Cyclurus gurleyi*.

FIGURE 58

†*Cyclurus gurleyi* (Romer and Fryxell, 1928),
the short bowfin (family Amiidae). *Top:*
Baby, measuring 41 millimeters (1.6 inches)
in length, from the sandwich beds of FBM
Locality H (FMNH PF14072). *Bottom:* Large
specimen probably from the 18-inch layer of
FBM Locality B, measuring 683 millimeters
(2.2 feet) long, is the HOLOTYPE for the species (FMNH UC2201).

☝ FIGURE 59

†*Cyclurus gurleyi* (Romer and Fryxell, 1928). *Top:* Largest known specimen, from the 18-inch layer of FBM Locality E, measures 789 millimeters (2.6 feet) long (UMNH 5125). Photo courtesy of UMNH and Randy Imis. Note the FUSIFORM body shape characteristic of the adults of this species. *Bottom:* Isolated scales disarticulated from specimen FMNH PF14095, which measures 738 millimeters (2.4 feet) in total length, from the sandwich beds of FBM Locality H. Scale bar equals 20 millimeters (0.8 inch).

⇢ FIGURE 60

†*Hiodon falcatus* (Grande, 1979), the Eocene mooneye (family Hiodontidae). Two specimens from the sandwich beds of FBM Locality H. *Top:* Large adult specimen about 200 millimeters (7.9 inches) long (FMNH PF12516). *Bottom:* Scales disarticulated from specimen FMNH PF9897, a small individual 91 millimeters (3.6 inches) long. Scale bar equals 2 millimeters (0.08 inch).

Mooneyes (Order Hiodontiformes, Family Hiodontidae)

Mooneyes are relatively scarce in the FBM and were not discovered there until 1979. They are FUSIFORM-shaped fishes with a forked tail and are represented in the FBM by †*Hiodon falcatus* (figs. 60, 61). Mooneyes are today known by two living species of *Hiodon* in the freshwater of much of North America. They live in clean, clear waters of lakes, ponds, and rivers, and reach a length of about 520 millimeters (20 inches). They feed mostly on insects, but larger individuals also feed on crayfishes, shrimps, fishes, and frogs.

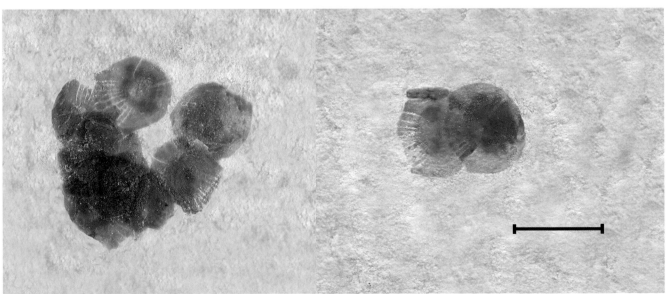

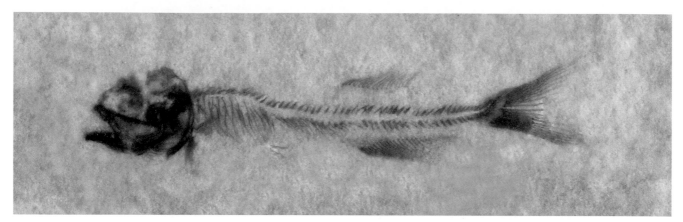

FIGURE 61

†*Hiodon falcatus* (Grande, 1979). Early growth series showing development of the skeleton. From the FBM sandwich beds. *Top:* Very young recently hatched individual of only 18 millimeters (0.7 inch) length, in which the vertebral discs (CENTRA) and scales have not yet formed; from FBM Locality M (FMNH PF15362). *Middle:* Slightly older individual of 39 millimeters (1.5 inches) long, in which the vertebral discs are partly formed as hollow rings and scales are only starting to form; from FBM Locality H (FMNH PF10955) *Bottom:* Specimen of 98 millimeters (3.9 inches) long with fully formed solid vertebral discs and scales completely covering body; from FBM Locality H (FMNH PF9878).

Fossil mooneye species, which are known from the early Eocene through the Oligocene, are known only from western North America and Asia. The geographic range of the fossils is most likely a vestige of the Cretaceous connection between Asia and the western North American subcontinent (see page xii).

The FBM species was originally described in the genus †*Eohiodon*. But a later detailed study of the family Hiodontidae (Hilton and Grande 2008) found that the characters used to differentiate the fossil genus †*Eohiodon* from the living genus *Hiodon* were incorrect, and so the two genera were synonymized and the older name, *Hiodon*, took precedence. †*Hiodon falcatus* is the best-preserved species of all fossil mooneyes and is even known by growth series from hatchlings to adults, which illustrate the ONTOGENETIC development of the skeleton (fig. 61).

In the FBM, †*Hiodon falcatus* occurs primarily in the northern nearshore sandwich beds of Thompson Ranch (e.g., FBM Locality H) from where I have seen over 200 specimens. A few specimens have also been found in the nearshore western quarries (FBM Locality L). I am only aware of one specimen ever being found in the mid-lake 18-inch layer quarries, and that was from the 18-inch layer beds of FBM Locality E. The distribution in Fossil Lake is indicative of a tributary to Fossil Lake in the vicinity of the Thompson Ranch quarries.

†*Hiodon falcatus* most closely resembles the osteoglossid †*Phareodus* but is easily distinguished from it by the lack of an enlarged pectoral fin, its smaller scales, and the presence of smaller dorsal and anal fins, smaller teeth, and a smaller head. Maximum known size for †*H. falcatus* is about 250 millimeters (10 inches) in length. There were 28 specimens of †*H. falcatus* in the collection of the Field Museum at the time this book was written. For a more complete description of †*H. falcatus*, see Grande (1979) and Hilton and Grande (2008).

Bony Tongues, or Arowanas
(Order Osteoglossiformes, Family Osteoglossidae)

The Eocene bony tongues are represented in the FBM by two species in the extinct genus †*Phareodus* (figs. 62, 66). These species are not particularly rare, making up about 0.5 percent of the mid-lake 18-inch layer fishes and about 2 percent of the nearshore sandwich bed fishes (listed at around 5 percent in Grande [1984, table II.10] because of a mass mortality of juveniles in that study sample). Juvenile †*Phareodus* are significantly more common in the nearshore localities than the mid-lake localities.

†*Phareodus* is one of the most popular large fishes with collectors because of their impressive teeth and interesting highly developed pectoral fin. This fish was predaceous on other fishes as indicated by several known specimens preserved with fishes in their jaw or stomach region (e.g., fig. 64). The main food fishes found in specimens of †*Phareodus* are †*Knightia*, †*Mioplosus*, and †*Diplomystus*, but a specimen with †*Priscacara* spines in its stomach region is also known.

⬆ FIGURE 62

The extinct bony tongue †*Phareodus encaustus* (Cope, 1871), on a slab with two †*Knightia eocaena*. All Osteoglossiformes are extinct in North America today but survive in tropical regions of South America, Asia, and Australia. This large fish is 630 millimeters (2.1 feet) long and is from the 18-inch layer of FBM Locality A (FMNH PF14040). Note the large scales (see also fig. 65, *right*).

⋯➡ FIGURE 63

†*Phareodus encaustus* (Cope, 1871) (family Osteoglossidae). Early growth series showing gradual formation of the skeleton from the FBM sandwich beds. *Top:* Very young recently hatched individual of 21 millimeters (0.8 inch) in length, from FBM Locality A (FMNH PF15591). *Middle:* Slightly older individual of 28 millimeters (1.1 inches) in length, from FBM Locality M (FMNH PF15372). *Bottom:* Slightly older specimen of 43 millimeters (1.7 inches) in length, from FBM Locality A (FMNH PF15590). Note that all three growth stages shown here are prior to the formation of the scales.

↑ FIGURE 64

†*Phareodus encaustus* caught in the act of swallowing other fishes. *Top:* Large specimen from the 18-inch layer of FBM Locality B, with a length of 603 millimeters (2 feet), swallowing a large †*Mioplosus labracoides*. Private collection; photo courtesy of Tom Lindgren. *Bottom:* Part and counterpart of a small specimen 150 millimeters (5.9 inches) long from the sandwich beds of FBM Locality H, swallowing a †*Knightia eocaena*. Private collection; photo courtesy of Bill Rieger.

┈┈▶ FIGURE 65

†*Phareodus encaustus* (Cope, 1871), various isolated parts often found in the FBM. *Left and middle:* Isolated braincase in dorsal and ventral view, from 18-inch layer deposits of FBM Locality A, measures 100 millimeters (3.9 inches) from top to bottom (FMNH PF14262). *Right:* An isolated scale (anterior facing right) from 18-inch layer deposits of FBM Locality A, showing the RETICULATE pattern of lines characteristic of scales from the family Osteoglossidae. Scale height is 31 millimeters (1.2 inches) (FMNH PF10968).

†*Phareodus* is easy to distinguish from other FBM fishes because of its highly developed pectoral fin, jaws full of large teeth, and huge large scales. The scales also have a web-like mosaic of cells, which is indicative of the family Osteoglossidae. These scales are often found as isolated elements in the FBM (e.g., fig. 65, *right*). Isolated braincases of large †*Phareodus* are also occasionally found in the FBM (fig. 65, *left* and *middle*). Because of the abundance of †*Phareodus* in the FBM, complete growth series can be found from hatchlings up to large adults illustrating the ONTOGENETIC development of the skeleton (e.g., fig. 63).

Although adult †*Phareodus* are solitary fishes not generally found preserved with other †*Phareodus* individuals, juvenile specimens are occasionally found preserved as mass mortalities indicating schooling behavior (e.g., fig. 67, *bottom*). This parallels modern-day osteoglossids, which are solitary fishes as adults, but as young juveniles they occur as large schools of siblings guarded by their parent. Fishes that provide such parental care are rare. In fact, living arowanas are mouth-brooders, with the parent fish keeping the eggs and later the hatchlings in their mouths until the juveniles are better able to fend for themselves. Unlike some species that were schooling fishes even as adults (e.g., †*Knightia eocaena* and †*Cockerellites liops*), †*Phareodus* appears to have schooled only in the juvenile stages of life.

Today the family Osteoglossidae is represented by 10 living species found exclusively in tropical freshwaters of South America, Africa, Asia, and Australia. The arowana *Osteoglossum* is frequently found in aquarium stores. Living osteo-

FIGURE 66

†*Phareodus testis* (Cope, 1877), another FBM species of bony
tongue (family Osteoglossidae). Two adult specimens (one
from the sandwich beds and one from the 18-inch layer)
showing the extreme short FUSIFORM body shape and the
short jaws characteristic of this species. *Top:* Specimen from
the sandwich beds of FBM Locality H, measuring 177 millimeters (7 inches) long (FMNH PF12411). *Bottom:* Specimen from
the 18-inch layer of FBM Locality B, measuring 260 millimeters
(10.2 inches) long (FMNH P25014).

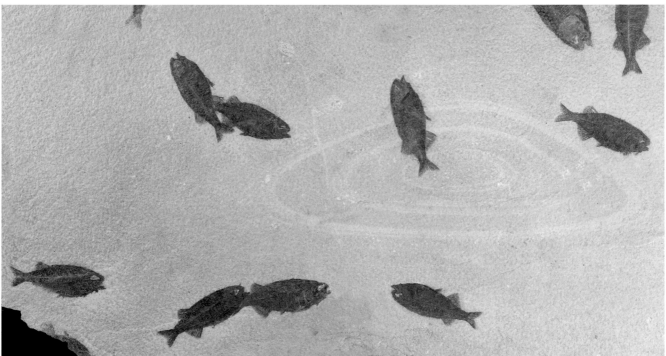

FIGURE 67

†*Phareodus testis* (Cope, 1877). Juveniles from the FBM sandwich beds. *Top:* A young juvenile with scales only thinly developed, clearly showing the internal skeleton. Specimen measures 49 millimeters (1.9 inches) long and is from FBM Locality M (FMNH PF15371). *Bottom:* A mass-mortality slab of 12 juveniles taken from a much larger slab (FMNH PF15368) of at least 20 individuals, from FBM Locality H. Individual fishes range from 55 to 78 millimeters (2.2 to 3.1 inches) in length. Adults are never found preserved in mass mortalities. This indicates that †*Phareodus* broods probably remained together until later in life when they eventually separated to become solitary individuals, similar to the social development of living osteoglossids.

glossids feed mostly near the surface on insects and fishes. Like gars and bow-fins, arowanas can breathe air at the surface using their blood vessel–lined swim bladder as a functional lung.

Although five species of †*Phareodus* have been described from the FBM, Grande (1984, 70–71) determined that only two of them are valid: †*Phareodus encaustus* and the smaller †*Phareodus testis*. The genus †*Phareodus* also occurs in Oligocene freshwater deposits of Queensland, Australia. See Grande (1984, 69–75) and Li, Grande, and Wilson (1997) for further discussion of the TAXONOMY and nomenclature of the genus.

†*Phareodus encaustus* is the larger of the two species, reaching a maximum known length of 750 millimeters (30 inches). It has a more elongate body profile, a greater number of dorsal-fin rays, and a smaller number of anal-fin rays than †*P. testis*. For a more precise key to distinguish between the two FBM species, see Grande (1984, 75). There were more than 80 specimens of †*Phareodus encaustus* in the collection of the Field Museum at the time this book was written. For further description of the morphology, see Taverne (2009, 175–84) and Li, Grande, and Wilson (1997, 494–505).

†*Phareodus testis* has a deeper belly, shorter body profile than †*P. encaustus* and has a maximum known size of about 380 millimeters (15 inches). Mass mortalities of the young of this species are occasionally found at FBM Locality H (e.g., fig. 67, *bottom*). Juveniles of this species are much more common in the near-shore localities (e.g., FBM Localities H and L) than in the mid-lake 18-inch layer localities. There were more than 70 specimens of this species in the collection of the Field Museum at the time this book was written. For further description of the morphology of †*Phareodus testis*, see Taverne (2009, 175–84) and Li, Grande, and Wilson (1997, 494–505).

Primitive Herring-like Fishes
(Order †Ellimmichthyiformes, Family †Paraclupeidae)

There is one valid species of the extinct family †Paraclupeidae in the FBM deposits, †*Diplomystus dentatus* Cope, 1877, commonly referred to as the "diplo" by the fossil quarriers. (Although the family name †Ellimmichthyidae is sometimes used for this group, the name †Paraclupeidae has PRIORITY and is therefore the valid name.) This species is one of the two most common fishes in the FBM, and it is also the most common large (over 305 millimeters [12 inches]) fossil fish species known from North America. It is commonly found in sizes ranging from embryonic specimens only 18 millimeters (0.7 inch) long up to full-size specimens of about 650 millimeters (26 inches) long (fig. 69A). The smallest larval-size individuals of 18 to 25 millimeters (0.7 to 1 inch) in length (e.g., fig. 69B, *top*) are much more common in the mid-lake localities than in the nearshore localities, suggesting that †*Diplomystus* spawned in the open waters of Fossil Lake, unlike most of

FIGURE 68

†*Diplomystus dentatus* Cope, 1877, the "diplo" (family †Paraclupeidae). *Top:* A large perfect specimen measuring 531 millimeters (1.7 feet) long, from the 18-inch layer of FBM Locality A (FMNH PF11935). *Bottom:* A young specimen of 100 millimeters (3.9 inches) long, from the sandwich beds of FBM Locality H, with a fully formed skeleton but very thinly developed scales clearly showing the internal skeleton (FMNH PF10628).

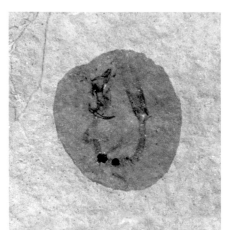
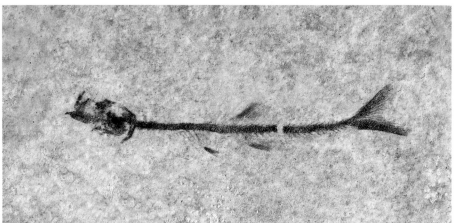
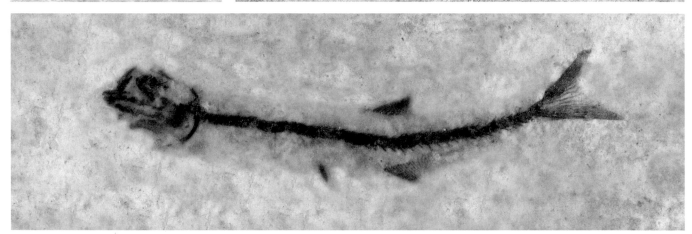

FIGURE 69A

†*Diplomystus dentatus* Cope, 1877. Early growth series from the FBM 18-inch layer, showing gradual formation of the skeleton. *Top left:* Prehatching specimen of about 18 millimeters (0.7 inch) in length, still coiled up in egg from FBM Locality E. Uncataloged uw specimen. *Top right:* Very young individual of 24 millimeters (0.9 inch) with much of the skeleton yet undeveloped, from FBM Locality A (FMNH PF12403) *Middle:* Specimen of 25 millimeters (1 inch) long, still without scutes and other skeletal features that form later in life, from FBM Locality A (FMNH PF10245). *Bottom:* Specimen from FBM Locality A of 34 millimeters (1.3 inches) long, which has a nearly completely developed internal skeleton, but the scales have yet to form (FMNH PF13575).

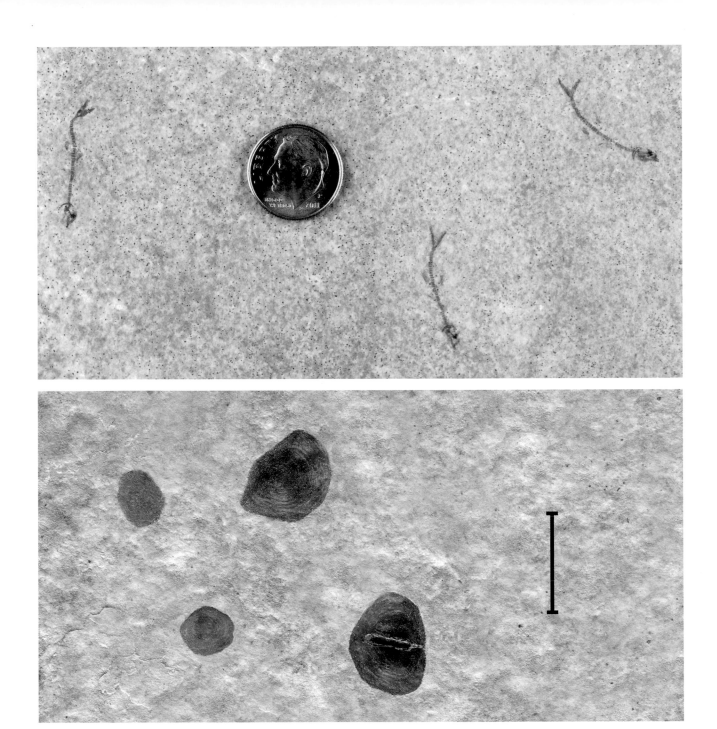

FIGURE 69B

†*Diplomystus dentatus* Cope, 1877, from the 18-inch layer. *Top:* Hatchling †*D. dentatus* occur in great numbers in the mid-lake localities but not so much in the nearshore deposits. This is in contrast to most other FBM fish species, for which the majority of baby fishes occur in the nearshore localities rather than the mid-lake. The three fishes are from FBM Locality G and range from 22 to 25 millimeters (0.9 to 1 inch) in length (FMNH PF10248). *Bottom:* Scales from a large adult with an estimated total length of about 380 millimeters (1.3 feet), from FBM Locality A (FMNH PF15562). Scale bar equals 5 millimeters (0.2 inch).

the other FBM fish species that spawned near shore. †*Diplomystus dentatus* is one of the truly classic species from the FBM, and it makes a beautiful specimen when it is well prepared (e.g., fig. 68). President Eisenhower purchased a large †*D. dentatus* from one of the commercial quarries that he presented as a gift to the emperor of Japan from the American people in 1960 (see page 28). †*Diplomystus dentatus* together with a smaller clupeomorph, †*Knightia eocaena*, make up well over half of all the fishes found in the FBM. In terms of sheer quantity, clupeomorph fishes (= †Ellimmichthyiformes plus Clupeiformes) definitely appear to have dominated the FBM fish community.

The †paraclupeids are herring-like fishes, but not true herrings as they are sometimes portrayed in the literature. They instead belong to an extinct order of primitive clupeomorph fishes, the †Ellimmichthyiformes, that survived from at least the Early Cretaceous until the middle Eocene (about 110 million to 48 million years before present). †*Diplomystus dentatus* was a living fossil in the Eocene because it was one of the last known species of the long †ellimmichthyiform lineage. The order appears to have died out completely by the middle Eocene. This extinct group had a very broad geographic distribution in their day, with fossil species known from North America, South America, Europe, Asia, the Middle East, and Africa. The closest known relative to the FBM †*Diplomystus dentatus* is †*Diplomystus shengliensis* (Zhang, Zhou, and Quin 1985), from Eocene freshwater deposits in China. This geographic distribution is another vestige of the earlier connection between Asia and the western North American subcontinent (fig. 1). †*Diplomystus dentatus* is known only from the Green River Formation, where it has been found in all three of the Eocene Green River lakes: Fossil Lake, Lake Gosiute, and Lake Uinta.

†*Diplomystus* may have been a mid-water to surface feeder given its sharply upturned jaws. Young †*D. dentatus* may have been primary consumers feeding on algae or zooplankton like many other clupeomorphs, but larger individuals were also piscivorous, including fishes in their diet. Many specimens of †*D. dentatus* are known with fishes preserved in their jaws and stomach regions. Usually these fishes are †*Knightia* or smaller †*Diplomystus* (e.g., fig. 70), but often it is found with †*Priscacara* or †*Cockerellites* preserved in its jaws or stomach region (fig. 71). These specimens were all swallowed headfirst, apparently to avoid getting stuck by the sharp percoid dorsal spines.

†*Diplomystus dentatus* is easy to distinguish from the other species of fishes in the FBM with its bony scutes along the belly margin and the upper body margin between the head and the dorsal fin (e.g., fig. 68). It reaches a much larger size than the other FBM clupeomorph genus †*Knightia* (next section) and had a maximum known length of about 650 millimeters (26 inches). There are hundreds of specimens of this species in the collection of the Field Museum. For more details about the anatomy and evolutionary relationships of this species, see Grande (1982a, 1985a).

FIGURE 70

†*Diplomystus dentatus* Cope, 1877. Dozens of †*D. dentatus* specimens are known with fishes preserved in their mouths or stomach regions, indicating that larger individuals included fishes in their diet. *Top:* Specimen caught in the act of cannibalism, swallowing a small †*Diplomystus dentatus*. Large fish measures 450 millimeters (1.5 feet) long and is from the 18-inch layer of FBM Locality A (FMNH PF11815). *Bottom:* Specimen that swallowed a †*Knightia eocaena*. (Note the small, slightly darker-colored †*Knightia* preserved between the ribs in the stomach region of the large †*Diplomystus*.) The larger fish is 460 millimeters (1.5 feet) long and is from the 18-inch layer of FBM Locality A (FMNH PF10384).

FIGURE 71

†*Diplomystus dentatus* Cope, 1877. This is one of the few species that has been well documented as having fed on †*Cockerellites liops*. †*Cockerellites* is a spiny-finned fish that must be swallowed headfirst to avoid getting caught in the throat by the sharply tipped, backward-pointing fin spines (e.g., see fig. 93). There are at least ten known specimens of †*Diplomystus dentatus* with †*Cockerellites* or †*Priscacara* individuals preserved in their mouths or within their stomach regions. These two specimens are from the 18-inch layer of FBM Locality B. Private collection; photos courtesy of Tom Lindgren. The two large †*Diplomystus* are 483 and 432 millimeters (1.6 and 1.4 feet) in length. The only other FBM species I have observed with †*Cockerellites* in its mouth or stomach is †*Phareodus encaustus*.

Fishes in the Herring Family (Order Clupeiformes, Family Clupeidae)

The herring family is represented in the FBM by two species in the extinct genus †*Knightia*. The more common of the two is the slender-bodied †*Knightia eocaena* (figs. 72, 73), which is the state fossil of Wyoming. Although the deeper-bodied †*Knightia alta* is the less common of the two species in the FBM, it is still abundant. Body depth in †*K. alta* ranges from moderately deep to very deep (fig. 75).

†*Knightia eocaena* was the most important fish species in Fossil Lake during the early Eocene. It was extremely common, forming immense schools, sometimes killed en masse, leaving fossilized mass-mortality layers (e.g., figs. 72, *bottom*; 74; and 214, *top*). Living clupeid species today lay as many as 200,000 eggs at once, giving them great capacity for multiplying quickly, even after mass die-offs, to which they are prone. They can die off suddenly because they are highly susceptible to sudden changes in water temperature or chemistry. †*Knightia eocaena* must have had a similar capacity and vulnerability, given the millions of them that have been excavated over the years. Some of the impressively large mass-mortality layers of †*Knightia eocaena* contain densities of up to a hundred fishes per square meter and extend for tens of thousands of square meters. The schools that produced these must have contained millions of fishes. Unlike some species that schooled only during the juvenile stages of life, †*Knightia* was apparently a schooling fish even as an adult (e.g., figs. 72, *bottom*, and 214, *top*).

†*Knightia eocaena* was key to the ecological food web of Eocene Fossil Lake and critical to the ecosystem in a very basic way. It was a **PRIMARY CONSUMER**, or herbivore, which like modern clupeids fed on abundant algae and zooplankton in the lake and quickly reproduced in great numbers. They in turn served as a critical food source for many other **SECONDARY CONSUMERS**. It has been found preserved in the mouth and/or stomach region of many large FBM fishes, including †*Crossopholis*, *Lepisosteus*, *Amia*, *Atractosteus*, †*Phareodus*, †*Diplomystus*, †*Priscacara*, and †*Mioplosus* (e.g., fig. 49, *top*; fig. 64, *bottom*; fig. 70, *bottom*; and fig. 85, *top*). They were also an important food source for frigatebirds, trionychid turtles, aquatic lizards, small crocodilians, the otter-like †pantolestid, and many other animals. †*Knightia eocaena* was a highly effective engine of the Fossil Lake ecosystem that transformed **AUTOTROPHS** such as microscopic plankton into a food source for many secondary consumers (predatory species).

Early growth stages of †*Knightia eocaena* are also well represented in the nearshore localities of the FBM. This material shows that the skeleton of †*Knightia* developed at a much earlier stage than the other FBM clupeomorph †*Diplomystus dentatus*. The internal skeleton of †*K. eocaena* was fully developed by the time they reached a total length of 20 millimeters (0.8 inch) (fig. 73, *top*), while the internal skeleton of †*D. dentatus*, even at 24 to 25 millimeters (1 inch), was still largely undeveloped (fig. 69A, *top right*, and 69B, *top*).

The family Clupeidae today includes about 200 living species and is one of the world's most important commercial fish species. The family includes true

FIGURE 72

†*Knightia eocaena* Jordan, 1907, in the herring family (Cluepeidae). This is one of the world's most common vertebrate fossils known by a complete skeleton. *Top:* A very large adult from the 18-inch layer of FBM Locality A, measuring 207 millimeters (8.1 inches) in length (FMNH PF15385). *Bottom:* Mass-mortality slab of individuals averaging about 130 millimeters (5.1 inches) long, from near the bottom of the 18-inch layer of FBM Locality E (FMNH PF10385).

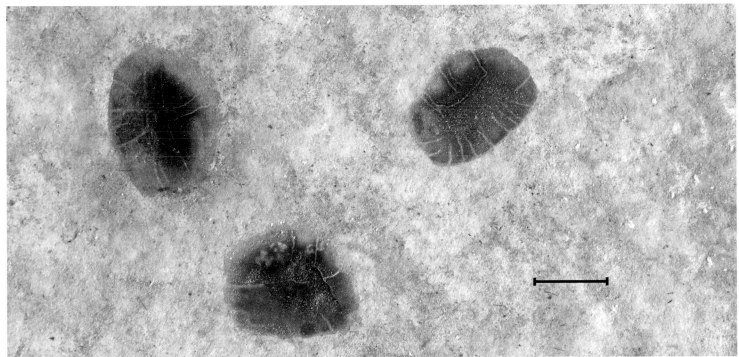

FIGURE 73

†*Knightia eocaena* Jordan, 1907, from the 18-inch layer deposits of FBM Locality A. *Top:* Juvenile individual of only 20 millimeters (0.8 inch) long (FMNH PF10950). Note that unlike †*Diplomystus*, the internal skeleton of †*Knightia* is fully formed by the time it reaches this size (compare to fig. 69A, *top right*). *Bottom:* Scales from an individual about 150 millimeters (5.9 inches) in length (FMNH PF15563). Scale bar equals 5 millimeters (0.2 inch).

FIGURE 74

†*Knightia eocaena* Jordan, 1907, mass-mortality slab of small specimens averaging between 50 and 60 millimeters (2 and 2.4 inches) in length. From the mass-mortality zone near the K-spar tuff bed at FBM Locality A.

FIGURE 75

†*Knightia alta* (Leidy, 1873) (family Clupeidae). Note the deeper body profile than in †*Knightia eocaena*. *Top:* Individual measuring 101 millimeters (4 inches) in length from the 18-inch layer of FBM Locality G (AMNH 10479). *Bottom:* A particularly deep-bodied individual measuring 124 millimeters (4.9 inches) long, possibly bloated or full of eggs. Specimen is from the sandwich beds of FBM Locality H (FMNH PF9671).

herrings, shads, gizzard shads, sardines, and round herrings, and most species are marine or **ANADROMOUS**. There are about 50 species of freshwater clupeids living today. The genus †*Knightia* also occurs in Eocene freshwater deposits of China and serves as another vestige of the Cretaceous connection between eastern Asia and the western North American subcontinent.

The name †*Knightia* was created by Jordan as part of a replacement name for a fossil species described from the Green River Formation by Leidy in 1856. Leidy described the fossil as "*Clupea humilis*" from the Laney Member of the Green River Formation, but at the time he did not realize that name had already been used in 1848 for a different species. Jordan (1907) subsequently created the name †*Knightia eocaena* to contain the Laney Member species. This species was later also discovered in the FBM, where it was found to be the most common fish species. For additional description and discussion of the FBM species of †*Knightia*, see Grande (1982b).

†*Knightia eocaena* is distinguished from †*K. alta* primarily by its more slender shape. Description and diagnostic characters are given in Grande (1982b, 7–8). The maximum known length of †*K. eocaena* is about 260 millimeters (10 inches), but this species rarely exceeds 150 millimeters (6 inches). Although †*K. eocaena* is the most common species of †*Knightia* by far in the FBM, †*K. alta* is the most common †*Knightia* species in the Laney Member Lake Gosiute deposits of the Green River Formation. There are hundreds of specimens of †*K. eocaena* from the FBM in the collection of the Field Museum. †*Knightia alta* is less common in the FBM, with a maximum known size of about 150 millimeters (6 inches). There are only about 10 FBM specimens in the collection.

Beaked Sandfishes (Order Gonorynchiformes, Family Gonorynchidae)

This family is represented in the FBM by a single species in the extinct genus and species †*Notogoneus osculus*. This species was the last surviving member of the order Gonorynchiformes in North America. It has an elongate body and a head that has ventrally located, toothless mouthparts suited for bottom feeding (fig. 76). The head, like all gonorynchids, is covered with peculiarly shaped scales (fig. 76, *bottom*). The head scales are usually prepared off of most fossils I have seen in museums. The shape of the scale is very diagnostic for this species (fig. 76, *bottom middle*). Young juvenile individuals of †*Notogoneus osculus* are abundant enough that the early development of the skeleton can be inferred from growth series (fig. 77).

†*Notogoneus* is somewhat scarcer than the clupeomorph and perciform fishes in the FBM, but not rare in the mid-lake 18-inch layer deposits, where it makes up between 0.3 and 2 percent of the fish fauna (Grande and Grande 2008). This species is effectively absent from the nearshore sandwich beds, the reasons for which are unclear. Only a single partial skeleton has ever been reported from the sandwich beds (not seen by me).

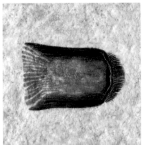

⋮ RAY-FINNED FISHES

FIGURE 76

†*Notogoneus osculus* Cope, 1885a, a beaked sandfish (family Gonorynchidae), from 18-inch layer deposits of FBM Locality A. Although abundant in the 18-inch layer deposits (see table 1, p. 177), †*Notogoneus* appears to be absent from the sandwich beds. *Top:* A well-preserved adult individual measuring 523 millimeters (1.7 feet) long, with the scales prepared off of the skull and other bones (FMNH PF11968). Normally, the head is covered with scales in this species. *Bottom left and right:* Two well-preserved adult-size skulls: one with scales still covering the head and the other with the scales removed from the black skull bones (FMNH PF15320 and FMNH PF11955). *Bottom middle:* An isolated scale measuring 6 millimeters (0.2 inch) long from an adult individual, anterior facing left (FMNH PF15323).

†*Notogoneus osculus* appears to have predominant, disjunct size classes in the FBM, consisting mostly of either very young juveniles from 29 to 40 millimeters (1 to 2 inches) total length, or large adults ranging from 250 millimeters to 900 millimeters (10 to 35 inches) in total length. The near lack of other size classes and the absence of †*Notogoneus osculus* around the shoreline deposits suggest that this species may have been migratory, living and perhaps spawning in the middle of Fossil Lake for part of its life, and moving to connecting streams and rivers at other parts of life. The abundant large individuals, averaging about 460 millimeters (18 inches) in total length, may even have been post-spawning deaths rather than permanent residents of the lake.

The family Gonorynchidae survives today with a single genus *Gonorynchus*, containing five described species. Living members of the family are confined to tropical marine regions of the Indo-Pacific, ranging from the Indian Ocean to the Hawaiian Islands, and very rarely in the southern Atlantic (St. Helena). The common name for *Gonorynchus* is the beaked sandfish, although it is also called the beaked salmon, the mouse fish, the sand fish, and the sand eel. *Gonorynchus* is a nocturnal fish that feeds on bottom-dwelling invertebrates and burrows into sand or mud during the day.

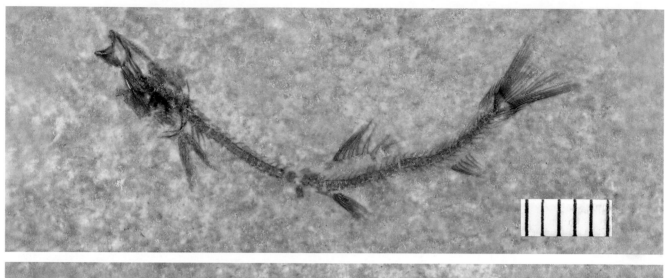

FIGURE 77

†*Notogoneus osculus* Cope, 1885a. Early growth series from the 18-inch layer of FBM Locality A, showing gradual formation of the skeleton (scale bar in millimeters). *Top:* A very young individual only recently hatched before fossilization, measuring 31 millimeters (1.2 inches) long (FMNH PF11958). *Middle:* Slightly larger individual with faint traces of scales starting to form near the tail and vertebral column, measuring 35 millimeters (1.4 inches) in length (FMNH PF15319). *Bottom:* Slightly larger individual measuring 47 millimeters (1.9 inches) long, showing scale covering almost completely formed except along the ABDOMINAL REGION (FMNH PF15232).

Given the ventrally located mouthparts of †*Notogoneus osculus* and similarities to the living genus *Gonorynchus*, it is likely that †*Notogoneus* also fed on bottom-dwelling invertebrates. I have seen no specimens of †*N. osculus* with fishes in its mouth or stomach. Martin, Vasquez-Prokopec, and Page (2010) described trace fossils in the mid-lake 18-inch layer deposits of the FBM, indicating a bottom-feeding fish that may have been †*Notogoneus* (fig. 210). The presence of abundant †*Notogoneus* in the 18-inch layer is noteworthy because other bottom-dwelling animals are rare there due to the apparent stratified water column (see page 358).

Although there are no living freshwater gonorynchids today, the extinct genus †*Notogoneus* is widespread in freshwater fossil deposits. Besides the FBM, the genus occurs in freshwater fossil deposits of Europe, Asia, and Australia. One species of †*Notogoneus* has been reported from both freshwater and brackish deposits (Gaudant 1981), possibly evidence of ANADROMY in the early history of the genus. This might explain the widespread geographic distribution of the genus in freshwater deposits.

The genus †*Protocatostomus* Whitfield, 1890, is a junior synonym of the genus †*Notogoneus*. The genus also occurs in the Cretaceous of Montana, the Paleocene of the Ukraine, the Oligocene of France and Australia, the Miocene of Germany, and possibly the Cenozoic of England and Alberta, Canada. There is only one species of †*Notogoneus* in the FBM.

†*Notogoneus osculus* was one of the last of the Gonorynchiformes in North America. It is by far the best-preserved fossil gonorynchid species and is described in detail in Grande and Grande (2008). There are over 100 specimens of †*Notogoneus osculus* in the collection of the Field Museum, all from the mid-lake 18-inch layer quarries of FBM.

Pikes and Pickerels (Order Esociformes, Family Esocidae)

This family is represented in the FBM by a single extinct species, †*Esox kronneri*, which is known by only a single specimen, making it the rarest fish species in the FBM by far. Fortunately, this specimen is a beautifully preserved complete skeleton (fig. 78). The discovery of this specimen from the FBM in 1996 in one of the commercial quarries was truly remarkable. In 1999 I described this species and named it after its discoverer Robert Kronner, who generously donated the specimen to the Field Museum for scientific description. After more than a century of excavating FBM fossils, this was the first example of an entire order of fishes (Esociformes) to be found anywhere in the Green River Formation. Out of more than a million fossil fishes collected from the FBM, this is the only known specimen of †*Esox kronneri*. The extreme rarity of this species is a graphic example of why paleontological sample sizes are never large enough to reveal a complete paleobiological community. Our knowledge is limited to those specimens that were preserved and are discovered. It usually takes decades of intense collecting and study to reconstruct past community histories with relative accuracy.

FIGURE 78

†*Esox kronneri* Grande, 1999, a pickerel (family Esocidae). This is the only specimen of the order Esociformes known so far from the FBM, out of the millions of fish specimens so far excavated. This is a good example of why large sample sizes are required to adequately understand extinct biological communities. Specimen is 118 millimeters (4.7 inches) long, from the sandwich beds of FBM Locality M (FMNH PF14918). Specimen is the HOLOTYPE for the species.

Why is †*Esox kronneri* so rare in the FBM? It was clearly not a normal resident of Fossil Lake; otherwise it would be more common. I believe †*E. kronneri* normally lived in the tributary systems to Fossil Lake, but not in the lake itself, and that the one anomalous specimen wandered into Fossil Lake by accident. The fact that the specimen comes from one of the Thompson Ranch nearshore quarries supports this hypothesis. These quarries (FBM Localities H and M) are the source of other unique or near-unique occurrences of aquatic species thought to have originated elsewhere (e.g., clams, certain snail species). The Thompson Ranch quarries also have a much heavier concentration of certain other species than elsewhere in the lake (e.g., decapods, paddlefishes, percopsids, †*Masillosteus*, hiodontids).

The pike and pickerel family (Esocidae) is a group of freshwater fishes with a rich fossil record ranging from Late Cretaceous to present. All living and known fossil esocids are restricted to the Northern Hemisphere. Living species belong to one genus (*Esox*) with five species and inhabit freshwaters in North America, Europe, and Asia. Modern esocids prefer slow-moving or still waters of rivers and lakes. They inhabit cooler temperate regions all the way into the Arctic regions and are not found in the tropics today, which may be another explanation for their scarcity in Fossil Lake. Pike and pickerel feed primarily on fishes and other aquatic vertebrates.

†*Esox kronneri* is easily identified from the other species in the FBM, with its elongate jaws lined with teeth and its numerous minute, thin scales, the posteriorly set dorsal and anal fins, and the deeply forked tail. The only known specimen of this species is 118 millimeters (4.6 inches) long and is in the collection of the Field Museum. Because there is only one specimen of this species, there are no data on size range. This species is described in detail in Grande (1999).

The Mystery Fish, †*Asineops* (Order Uncertain, Family †Asineopidae)

This fish is very well preserved, and the anatomy is well known, but after decades of study, it is uncertain even to which order this genus belongs. It is currently in its own family, †Asineopidae, as a family of uncertain relationship. The †Asineopidae appears not to have survived past the middle Eocene. There is one known valid species of †*Asineops* in the FBM, †*Asineops squamifrons*. It is easily distinguished from other FBM spiny-rayed fishes by its numerous short dorsal spines, rounded tail margin, and the morphology of its enormous mouth. This species is much more common in the Laney Member deposits of Lake Gosiute, where it is known by hundreds of specimens. In the 18-inch layer and sandwich bed deposits, this species is very rare. When it does occur in these beds, it is usually much larger than the typical specimens from the Laney Member. In the collection of the Field Museum, there are four individuals of FBM †*Asineops* specimens that are between 295 and 330 millimeters long (12 and 13 inches) (e.g., fig. 79). Young juveniles of this species are also rare in the 18-inch layer and sandwich beds, although they are abundant in the Laney Member Lake Gosiute deposits.

Part of the reason that the relationships of †Asineopidae are still so poorly known is that the evolutionary relationships among the more than 15,000 species of living spiny-rayed fishes (Acanthomorpha) are still insufficiently known, and †*Asineops* has no obvious peculiar features that can clearly tie it to another known family. It appears to be a relatively primitive acanthomorph fish. Much work is currently being done by ichthyologists on this group, and eventually the relationships of †Asineopidae will be better known.

It is unclear what the diet of †*Asineops* was. It had very tiny teeth in its jaws and an enormous gape when the mouth was open (e.g., fig. 79, *bottom*). I have not seen any specimens with fishes in their stomachs. Perhaps further understanding of the internal head morphology (e.g., the gill arches) might shed light on the dietary habits of this species.

Cope (1870) originally described three species of †*Asineops* from the Green River Formation, but later decided that there were only two: †*A. squamifrons* (represented by small specimens) and †*A. pauciradiatus* (represented by larger specimens). Rosen and Patterson (1969) suggest that these two species were syn-

FIGURE 79

†*Asineops squamifrons* Cope, 1870 (family †Asineopsidae), a mystery fish of uncertain affinity. Two large adults from FBM Locality A. †*Asineops* specimens from the mid-lake 18-inch layer deposits are very rare but also usually very large when they do occur. Also note the huge gape of the mouth in this species and the large jaw bones. *Top:* Specimen measuring 295 millimeters (11.6 inches) long (FMNH PF15168). *Bottom:* Specimen measuring 300 millimeters (11.8 inches) long, on slab with a tree of heaven seed (*Ailanthus* sp.) below the tail (FMNH PF14207).

FIGURE 80

†*Asineops squamifrons* Cope, 1870. Early growth series. *Top:* A very young juvenile from the sandwich beds of FBM Locality L, measuring 20 milli-meters (0.8 inch) in total length; in a private collection. *Bottom:* A slightly larger juvenile from the sandwich beds of FBM Locality H, measuring 40 millimeters (1.6 inches) long (FMNH PF12595).

onyms and that the differences were only different growth stages of the same species, †*Asineops squamifrons*.

†*Asineops squamifrons* is known from deposits of all three of the Green River lakes but is most common by far in the Laney Member Lake Gosiute deposits. The largest known FBM specimen is 330 millimeters (13 inches) in total length. There are seven specimens of this species from the FBM in the collection of the Field Museum. There are also 18 specimens of this species from the Laney Member Lake Gosiute deposits in the collection. This species is described in detail in Rosen and Patterson (1969), although the gill arches and other internal parts of the anatomy remain undescribed.

Trout-Perches (Order Percopsiformes, Family Percopsidae)

The trout-perch family is represented in the FBM by the species †*Amphiplaga brachyptera* (figs. 81 and 82). This is a small fish with a maximum size of about 140 millimeters (5.5 inches) in total length. It is closely related to the Laney Member Lake Gosiute species †*Erismatopterus levatus* (see Grande 1984). The name "trout-perch" is derived from a superficial resemblance to trout and perch, but percopsids are not closely related to either group. †*Amphiplaga* occurs almost exclusively in the Thompson Ranch nearshore sandwich beds (FBM Localities H and M). Although they make up less than 1 percent of the fish population, thousands of specimens are known. In the mid-lake 18-inch layer beds, this species is exceedingly rare. Fewer than a dozen examples from the mid-lake deposits are known. Larval-size individuals of †*Amphiplaga* are almost unheard of even in the Thompson Ranch quarries, suggesting that these fishes spawned upriver in the northern tributary of Fossil Lake. Today living percopsids also spawn in streams or gravel in lake shallows. Because there was no gravel bottom in Fossil Lake during deposition of the FBM, the hypothesis that †*Amphiplaga* spawned in the river tributary seems sound. This species is never found in mass mortality, suggesting that it was a solitary rather than a schooling fish.

Today the trout-perch family is survived by the genus *Percopsis* with two living species restricted to freshwaters of North America. †*Amphiplaga brachyptera* is the earliest known occurrence of the family in the fossil record. Today living percopsids inhabit streams and shallow lakes in temperate to cool climates. The subtropical climate of Fossil Lake may have been too warm for †*Amphiplaga* and the reason it is found almost exclusively in the Thompson Ranch localities within Fossil Lake. These quarries were near a tributary to Fossil Lake from the north. †*Amphiplaga*, like modern percopsids, probably fed primarily on insects, ostracods, and zooplankton.

†*Amphiplaga*, together with the genus †*Erismatopterus*, was placed in a new family, †Erismatopteridae, by Jordan (1905), but Rosen and Patterson (1969) found both genera to belong in the extant family Percopsidae. †*Amphiplaga* is very similar to the Laney Member †*Erismatopterus* except for some MERISTIC

FIGURE 81

†*Amphiplaga brachyptera* Cope, 1877, a primitive trout-perch
(family Percopsidae). Two exceptionally well-preserved adults,
from the sandwich beds of FBM Locality H. *Top:* Specimen
measuring 72 millimeters (2.8 inches) long (FMNH PF15311).
Bottom: Specimen measuring 96 millimeters (3.8 inches) long
(FMNH PF15376).

FIGURE 82

†*Amphiplaga brachyptera* Cope, 1877. *Top:* A juvenile from FBM Locality H, measuring 25 millimeters (1 inch) in length (FMNH PF15564). *Bottom:* An adult specimen on a slab with a young †*Asterotrygon maloneyi*. †*Amphiplaga* specimen is 88 millimeters (3.5 inches) long and is from the sandwich beds of FBM Locality H (FMNH PF15180).

features of the fin rays, the possession of a spiny gill cover, and a denser scale covering of the opercle (Rosen and Patterson 1969; Grande 1984). †*Amphiplaga brachyptera* is the only species in the genus †*Amphiplaga*, and it appears to be restricted to deposits from Fossil Lake. Although a relatively scarce component of the fish fauna, it is known by over 500 specimens. There are 40 specimens of †*A. brachyptera* in the collection of the Field Museum. This species is described in detail in Rosen and Patterson (1969).

Perches and Basses (Order Perciformes, Families Latidae and Moronidae)

Perciform fishes, including fishes commonly called perches and basses, comprise the most diversified of living fish orders existing today. In fact, it is the largest of all vertebrate animal orders. The perch order Perciformes is well represented in the FBM, although the "true perch" family Percidae is not present (contra earlier classifications by Grande [1984], and others, who placed †*Mioplosus* in Percidae). The FBM perciforms are all clearly members of the suborder Percoidei, but their family relationships have been more difficult to unambiguously identify (Grande 2001, 28). The known FBM perciforms are represented by four genera (including one new genus described here) that belong to at least two different families. These genera are †*Mioplosus* (family Latidae), †*Priscacara* (family Moronidae), †*Cockerellites* (family Moronidae), and †*Hypsiprisca* (new genus, probably belonging to Moronidae). One of the difficulties in confidently classifying these genera to family has been the lack of knowledge about categorizing the more than 10,000 living species of Perciformes. As with most all fossil species, it is necessary to understand the interrelationships among living relatives before we can begin to fully understand the fossil forms. †*Mioplosus labracoides*, †*Cockerellites liops*, and †*Priscacara serrata* are common species in the FBM. †*Hypsiprisca* n. gen. and the undescribed †*Priscacara* species are relatively rare. There is probably more diversity present than currently recognized (e.g., the two undescribed species of †*Priscacara* and the one undescribed species of †*Hypsiprisca* mentioned below). There are several hundred complete FBM percoid skeletons at the Field Museum collected by me and my field crews over the last several decades that are awaiting comprehensive study.

†*Mioplosus* is the largest of the FBM perciform. Although Cope (1877) originally described five species from the FBM in this genus, Grande (1984, 142) found there to be only a single valid species, †*M. labracoides*. †*Mioplosus* makes up about 2 to 4 percent of the total fish excavated from the 18-inch layer and sandwich bed layer of the FBM (Grande 1984, 167, 169). There is also a mass-mortality zone of small †*Mioplosus* that often occurs in the gastropod layer of the FBM near the K-spar tuff bed at FBM Locality A (fig. 83, *bottom*).

†*Mioplosus* is easy to distinguish from the other perciform genera in the FBM because of its very elongate FUSIFORM body shape, the presence of two distinct dorsal fins, and a moderately forked tail fin. (In the other FBM percoid

FIGURE 83

†*Mioplosus labracoides* Cope, 1877, a common bass-like fish (family Latidae) from the FBM. These specimens are from the small †*Mioplosus* mass-mortality zone near the K-spar tuff layer in FBM Locality A. *Top:* A beautifully prepared, perfect specimen measuring 140 millimeters (5.5 inches) long, from a slab with four individuals plus a bird and a snail (see another view of same slab on page ii) (FMNH PA783). *Bottom:* A mass-mortality slab (FMNH PF15374) with five individuals ranging from 126 to 143 millimeters (5 to 5.6 inches) in length. Unlike the specimen above, these specimens have only a very rough preparation.

genera, the separation between the spiny dorsal fin and the segmented ray dorsal fin is much less divided.)

†*Mioplosus* has been classified in a number of different families over the last century and a half. Cope (1877) originally thought it resembled the striped bass, *Morone saxatilis*, thus implying classification in what we today call Moronidae (the "temperate bass" family). Later authors (e.g., Jordan 1923; Grande 1984) referred †*Mioplosus* to the "true perch" family Percidae. Cavender (1986) and others thought †*Mioplosus* was in Percichthyidae (the "temperate perch" family), and later even put †*Mioplosus* in its own family, †Mioplosidae (Cavender 1998, 150). Most recent work by Whitlock (2010) places †*Mioplosus* together with the genus *Lates*, family Latidae. Latidae today live in fresh, brackish, or marine waters of the Indo-Pacific and Africa.

†*Mioplosus labracoides* is common in the FBM but rare in the other Green River deposits of Lake Gosiute (e.g., the Laney Member) and Lake Uinta. The maximum size for this species is about 500 millimeters (20 inches), although specimens rarely exceed 400 millimeters (16 inches). Larval and young juvenile specimens are abundant in the nearshore FBM deposits, particularly in the Thompson Ranch localities (fig. 84). Unlike †*Diplomystus* and †*Notogoneus*, the nursery grounds for this species was in nearshore waters. The larger specimens are most common in the mid-lake 18-inch layer deposits (e.g., FBM Localities A–G). †*Mioplosus* probably occupied middle and upper lake zones. It was evidently a very predaceous species, as indicated by dozens of known specimens with fishes in their jaws or stomachs (e.g., fig. 85). Even as very young recently hatched individuals, this species was already cannibalistic, feeding on other small †*Mioplosus* (probably siblings) (fig. 86). Adult †*Mioplosus* were apparently solitary fishes because they are usually found preserved that way, but juvenile and larval individuals often traveled in schools, as evidenced by abundant mass-mortality slabs preserving groups of similarly sized individuals (figs. 83, *bottom*; and 84, *top*). The morphology of †*Mioplosus labracoides* is described in more detail in Whitlock (2010).

Between 1877 and 1886, Cope described eight species of †*Priscacara* from the FBM (and a ninth species from the "Manti beds" of Utah). He made †*Priscacara serrata* the TYPE SPECIES for the genus. Grande (1984, 150) listed all of these nominal species and indicated that several of these species are invalid because they cannot be differentiated from either †*Priscacara serrata* or Cope's †*"Priscacara" liops* (here placed in the genus †*Cockerellites* Jordan, 1923). In addition to †*P. serrata* and †*C. liops*, Cope's (1886) †*"Priscacara" hypsacantha* is also considered valid here but probably belongs in a new genus (briefly described here as †*Hypsiprisca* n. gen.). Although †*Priscacara* is relatively common in the FBM, it is extremely rare from Lake Gosiute and Lake Uinta deposits. The reports of †*Priscacara* from Lake Gosiute deposits that I have examined are actually specimens of †*Asineops*. Even though Buchheim and Surdam (1981) report †*Priscacara*

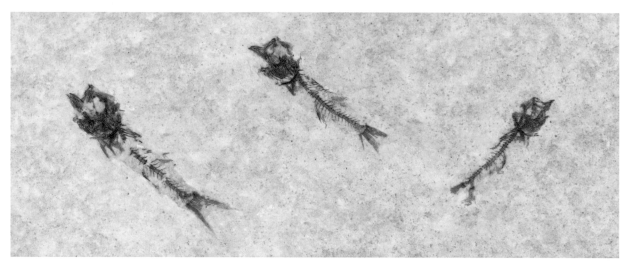

FIGURE 84

†*Mioplosus labracoides* Cope, 1877. Early growth series showing gradual formation of the internal skeleton. *Top:* Small slab with three hatchlings ranging from 12 to 16 millimeters (0.5 to 0.6 inch) long, from the sandwich beds of FBM Locality L (FMNH PF15592). *Middle:* Slightly larger individual measuring 17 millimeters (0.7 inch) long, showing more development in the internal skeleton; from the sandwich beds of FBM Locality H (FMNH PF10957). *Bottom:* Individual measuring 38 millimeters (1.5 inches) long, with internal skeleton fully formed but scales not yet fully developed; from the 18-inch layer of FBM Locality A (FMNH PF15363).

FIGURE 85

†*Mioplosus labracoides* Cope, 1877. This was a voracious species that fed largely on other fishes. ASPIRATION SPECIMENS (specimens preserved with fishes in their mouths or stomachs) are more common for this species than for any other fish species in the FBM. *Top:* †*Mioplosus labracoides* measuring 180 millimeters (7.1 inches) long, preserved with a †*Knightia eocaena* in its mouth; from the sandwich beds of FBM Locality H (FMNH PF10180). *Bottom:* Large †*Mioplosus labracoides* measuring 402 millimeters (1.3 feet) long, with a †*Diplomystus dentatus* preserved in its stomach region; from the 18-inch layer of FBM Locality G (FMNH PF15375).

FIGURE 86

†*Mioplosus labracoides* Cope, 1877. *Top:* This species was predaceous nearly from birth, as shown here by this very young cannibalistic individual only 33 millimeters (1.3 inches) long that has swallowed another small †*Mioplosus*; from the sandwich beds of FBM Locality M (FMNH PF15370). *Bottom:* Scales disarticulated from a specimen 192 millimeters (7.6 inches) in total length, from the sandwich beds of FBM Locality H (anterior facing left) (FMNH PF10624). Scale bar equals 2 millimeters (0.08 inch).

from the Laney Member Lake Gosiute deposits, the only example they provide (Buchheim and Surdam 1981, fig. 149B) is from the FBM, not Lake Gosiute.

Although previous authors have classified †*Priscacara* in its own family, †Priscacaridae (e.g., Grande 1984; Cavender 1998), it has also been suggested as a close relative of Pomacentridae (damselfishes), Labridae (wrasses), Percichthyidae (temperate perches), and Centrarchidae (sunfishes). Most recently, studies by Whitlock (2010) indicate this genus to be a member of the family Moronidae (the "temperate basses"). Today moronids live in fresh, brackish, or marine waters of North America, Europe, North Africa, and possibly Asia (if the genus *Lateolabrax* is included, as suggested by some authors. The genus †*Priscacara* also occurs in the middle Eocene freshwater deposits of Washington State and British Columbia, Canada. †*Priscacara* is not known to have survived past the middle Eocene.

†*Priscacara serrata* is relatively common, although proportionately it makes up less than 1 percent of the total fish fauna where it occurs. Young juveniles of †*P. serrata* commonly occur in the nearshore sandwich bed localities (e.g., FBM Localities H, L, and M), and early growth series of this species can be assembled to study the early development of its skeleton (fig. 88). This species is a deep-bodied with stout spines and a slightly rounded tail fin margin, with molar-like tooth pads in its throat suitable for crushing and grinding arthropods or even small mollusks (fig. 89, *bottom*). †*Priscacara serrata* can occasionally reach sizes of over 400 millimeters (16 inches). Occasionally †*P. serrata* is found with fish bones in its mouth or stomach, suggesting that it occasionally supplemented its diet with small fishes (fig. 90). These fishes are a favorite of fossil collectors, particularly those specimens with their dorsal spines pointing up. There are well over 100 specimens of †*P. serrata* in the collection at the Field Museum. For a more detailed description of this species, see Grande (1984) and Whitlock (2010).

There are two undescribed species that appear to belong in this genus, sharing the serrated preopercular margin and molariform teeth on the pharyngeal tooth plates. These are briefly mentioned here but not formally described. †*Priscacara* new species A is a very round-bodied form with weak dorsal spines that is still undescribed (fig. 91, *top*). This is a very rare species in the FBM, and its formal description is in progress. †*Priscacara* new species B is a variety with four anal spines, rather than the normal three in other species (fig. 91, *bottom*). It is possible that this is an abnormal variant of †*P. serrata*, and more study is warranted to verify it as a new species.

The so-called †*"Priscacara liops"* was placed in the genus †*Cockerellites* by Jordan (1923) because of the number of differences between this species and †*Priscacara serrata*. Unlike †*Priscacara serrata*, †*Cockerellites liops* has a larger number of dorsal fin rays, a larger number of anal fin rays, a less strongly serrated preopercular margin, and a thinner pelvic spine (see Grande 1984, table II.7). In addition, it is a smaller fish, seldom exceeding 120 millimeters (5 inches)

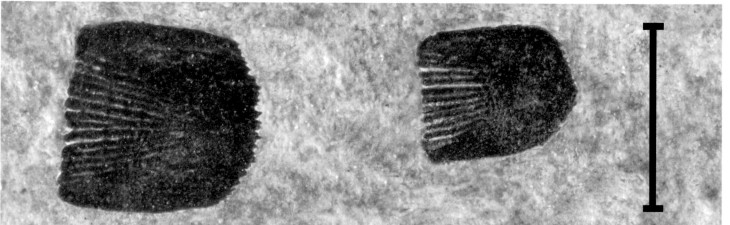

FIGURE 87

†*Priscacara serrata* Cope, 1877, a round-bodied bass-like fish (family Moronidae). *Top:* A well-preserved adult measuring 230 millimeters (9.1 inches) long, from the 18-inch layer of FBM Locality A (FMNH PF10296). *Bottom:* Two scales disarticulated from FMNH PF13016, a specimen that is 115 millimeters (4.5 inches) in total length, from the 18-inch layer of FBM Locality A (anterior facing left). Scale bar equals 2 millimeters (0.08 inch).

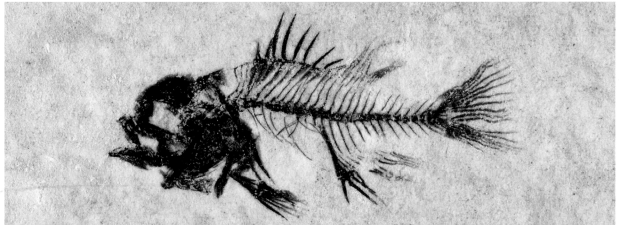

FIGURE 88

†*Priscacara serrata* Cope, 1877. Early growth series showing gradual formation of the skeleton. *Top:* Very young individual probably hatched only recently before fossilization, with a total length is 11 millimeters (0.4 inch); from the sandwich beds of FBM Locality A (FMNH PF10951). *Middle:* Slightly larger individual measuring 17 millimeters (0.7 inch) long, showing more development of internal skeleton but not scales forming yet; from the 18-inch layer of FBM Locality A (FMNH PF15587). *Bottom:* Slightly larger juvenile measuring 33 millimeters (1.3 inches) long with scales starting to develop over body; from the sandwich beds of FBM Locality H (FMNH PF14096).

FIGURE 89

†*Priscacara serrata* Cope, 1877. *Top:* A monstrously large individual of 414 millimeters (1.4 feet). *Bottom:* Close-up of gill arch region from above specimen showing the massive gill arch tooth plates covered with MOLARIFORM teeth; from the 18-inch layer of FBM Locality A (FMNH PF14929). These teeth were suited for crushing arthropods or mollusks.

FIGURE 90

†*Priscacara serrata* Cope, 1877. Although †*P. serrata* had the capacity to eat arthropods with its gill arch pads of crushing teeth (fig. 89, *bottom*), it also may have fed on smaller fishes, as evidenced by several individuals that have been found with fishes in their jaws. *Top:* A large specimen of 210 millimeters (8.3 inches) in length, with a †*Knightia eocaena* in its mouth. Specimen is from the 18-inch layer of FBM Locality G; in a private collection. Photo courtesy of Ron Mjos. *Bottom:* A young cannibalistic juvenile only 30 millimeters (1.2 inches) long, with a small †*Priscacara* in its jaws; from the 18-inch layer of FBM Locality A (FOBU 9445).

FIGURE 91

Two undescribed species of †*Priscacara. Top:* A yet-undescribed completely round-bodied specimen with eleven short dorsal spines, from the sandwich beds of FBM Locality H. Large individual of 160 millimeters (6.3 inches) in length (FMNH PF15310). *Bottom:* A second possible new species most easily recognized by the presence of four anal spines instead of three, and an elongate body. Specimen is from the 18-inch layer of FBM Locality B and is 129 millimeters (5.1 inches) in length (FMNH PF15383).

in length. Although I did not recognize this genus in my 1984 volume, I could find no unambiguous characteristics indicating close relationship to †*Priscacara serrata*, nor did Whitlock in his analysis of the FBM percoids together with extant species. Therefore, I resurrect the name †*Cockerellites liops* (Cope, 1877) here for this species discussed below.

†*Cockerellites liops* is much more common than †*Priscacara serrata*, particularly in the mid-lake quarries, where they are common and known by tens of thousands of specimens. Unlike †*P. serrata*, †*C. liops* must have been a schooling fish, based on the large number of mass-mortality plates that have been found of this species (e.g., fig. 93). Like †*Knightia eocaena*, †*C. liops* swam in schools even as adults. The maximum known size of †*C. liops* is about 150 millimeters (6 inches), although it rarely exceeds 120 millimeters (5 inches). This species probably fed on insects and insect larvae, and perhaps ZOOPLANKTON, because it had a very small mouth and has not been found with other fish bones preserved in its mouth or stomach. Although adults of this species are much more common than †*Priscacara serrata*, young juveniles of this species are much rarer than those of †*P. serrata*. †*Cockerellites liops* makes up less than 2.5 percent of the fish fauna in the nearshore quarries (e.g., Thompson Ranch), but ranges from 5 percent to about 20 percent in the mid-lake 18-inch layer quarries, although these numbers are largely influenced by the occurrence of mass-mortality layers of †*C. liops* in the mid-lake quarries. The large schools of this species must have preferred open water. There are several hundred specimens of †*C. liops* in the collection of the Field Museum.

†*Hypsiprisca* new genus is described here to contain †*Priscacara hypsacantha* (Cope, 1886) as the TYPE SPECIES and another new species previously referred to as "new percoid genus A" in Grande (2001). It is a genus of small percoid fishes with a maximum known size of 90 millimeters (3.5 inches) total length. Although it bears some similarity with the FBM moronids †*Priscacara* and †*Cockerellites*, it may not belong to the same family. Further study is required to determine the family to which this genus belongs.

†*Hypsiprisca* differs from moronid fish species other than †*Priscacara* and †*Cockerellites* in having an unforked, or barely forked, tail, and from †*Priscacara* and †*Cockerellites* in being more elongate and having a larger mouth gape and a less steep head profile. It differs from all Moronidae in the following unique combination of characters: two dorsal fins not separated, body margin under dorsal fin nearly parallel with skull roof, preopercular margin with long spines that point anteriorly along ventral edge of preopercle, nine dorsal spines, and rounded margins of the second dorsal and the anal fins.

†*Hypsiprisca hypsacantha* (Cope, 1886) is equivalent to what I have previously referred to as "new percoid genus A" (Grande 1984, 2001). The above section constitutes the initial description introducing the genus name †*Hypsiprisca*. It contains two species in the FBM; †*Hypsiprisca hypsacantha* and a new species that will be described elsewhere.

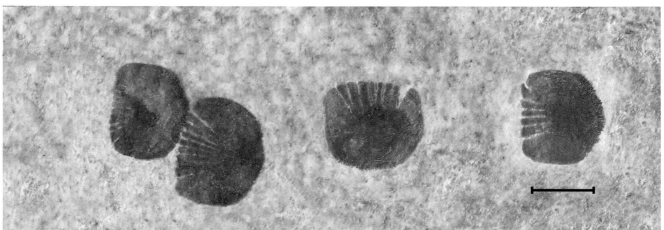

FIGURE 92

†*Cockerellites liops* (Cope, 1877), a bass-like fish (family Moronidae). *Top:* A well-prepared perfectly preserved adult skeleton measuring 125 millimeters (4.9 inches) long, from the 18-inch layer of FBM Locality A (FMNH PF13326). *Bottom:* Scales from FMNH PF12074 (see fig. 216), from an individual about 110 millimeters (4.3 inches) in length; from the 18-inch layer of FBM Locality G. Scale bar equals 2 millimeters (0.08 inch).

FIGURE 93
†*Cockerellites liops* (Cope, 1877). A slab of four adult individuals from a †*C. liops* mass-mortality zone. This zone occurs within a thin ash bed near the middle of the 18-inch layer at FBM Locality A. Specimens average about 130 millimeters (5.1 inches) long (FOBU 428).

†*Hypsiprisca hypsacantha* was originally described by Cope as †*Priscacara hypsacantha*. It is much rarer than either †*Priscacara serrata* or †*Cockerellites liops*. The holotype is **AMNH** 2453 (fig. 94, *top*). This species has a maximum known size of 61 millimeters (2.5 inches). There are very few known specimens of this species and only one at the Field Museum (fig. 94, *bottom*). The second species that remains undescribed (fig. 95) differs from †*H. hypsacantha* in being more slender-bodied and having a more convex posterior tail fin margin (†*H. hypsacantha* has a very slightly forked tail margin). It has a maximum known size of 90 millimeters (3.5 inches), is known mostly from the nearshore quarries, and its formal description will appear in a later publication.

←··· FIGURE 94

†*Hypsiprisca hypsacantha* (Cope, 1886), new genus of undetermined family. This is the TYPE SPECIES for a new genus (†*Hypsiprisca*) described here. Two adult specimens of
this small species; both have damaged skulls and appear
to have been collected from the upper part of the middle
unit of the FBM, near the K-spar tuff bed. *Top:* HOLOTYPE
specimen measuring 61 millimeters (2.4 inches) in length,
from Fossil Butte (AMNH 2453). *Bottom:* Specimen on slab
with the gastropod †*Goniobasis* sp. from FBM Locality A.
Fish is 79 millimeters (3.1 inches) long (FMNH PF15279).

⇡ FIGURE 95

†*Hypsiprisca* sp., a second yet-undescribed species (to be officially described at a later date).
Two beautifully preserved and finely prepared
adult-size specimens of a yet-undescribed species closely related to †*Hypsiprisca hypsacantha*,
from the FBM sandwich beds. This species is
less deep bodied and has a less forked tail-fin
margin than †*Hypsiprisca hypsacantha*. *Top:* An
individual 85 millimeters (3.4 inches) long,
from FBM Locality H (FMNH PF14941) (possible
HOLOTYPE). *Bottom:* An individual 80 millimeters (3.2 inches) long, from FBM Locality L
(FOBU 13413) (possible PARATYPE).

Abundance and Distribution of Fish Species

Having such large sample sizes of well-preserved fishes from different regions of Fossil Lake allows not only classification of these fishes, but also provides information for interpreting paleoecology of the lake's ecosystem.

From One in a Million to Common as Dirt: The Variation in Species Abundance within the Fossil Butte Member

The FBM has produced well over a million fossil fishes during the last century and a half that have been sold commercially or collected for museums. Today about 100,000 specimens are excavated annually, mostly by commercial fossil quarries. I estimate that less than half of those end up in the market because broken or incomplete common species are often not saved. All of this quarrying activity has produced amazingly good samples of even those species that are an extremely rare percentage of the fish assemblage. In the past, I have tried to estimate relative abundance of each genus in a number of quarries, relying on collections made by various quarriers (Grande 1984, tables II.8–II.10), but these previous surveys probably undercounted the most common species (†*Knightia eocaena*, †*Diplomystus dentatus*, †*Mioplosus labracoides*, and †*Cockerellites liops*) because quarriers are much more likely to discard damaged or incomplete specimens of these species than of rarer species. After 35 years of my own quarrying activity in the FBM in which I have overseen the excavation of more than 50,000 fish fossils, I have come up with a more detailed, conservative estimate of relative scarcity for each FBM fish species, summarized in table 1. Although I obviously did not examine every one of

Scarcity ranking (1 = rarest)	Species (*asterisk indicates FBM species not known from Eocene Lake Gosiute or Lake Uinta)	FBM specimens collected 1870–2010 (estimate)	Comments on scarcity and distribution (F-1 = mid-lake 18-inch layer deposits, F-2 = nearshore sandwich bed deposits)
1	†*Esox kronneri**	Only 1 known	Unique and rarest of all known fish species, known only from a nearshore F-2 deposit.
2	†*Masillosteus janei**	8 or more	Known mostly from nearshore deposits; young juveniles yet unknown.
3	†*Atractosteus atrox*	8 or more	May possibly be extremely large †*A. simplex*; smallest known specimen is > 1,500 mm total length.
4	†*Priscacara* n. sp.*	20 or more	Circular outline variety occurring in both F-1 and F-2; now that it is identified, more may be discovered.
5	†*Hypsiprisca hypsacantha**	20 or more	This deeper-bodied form is more rare than †*H.* n. sp. below.
6	†*Cyclurus gurleyi**	20 or more	Rarer than the other amiid, *Amia pattersoni*; juveniles are extremely rare.
7	†*Asineops squamifrons*	30 or more	Rare in the FBM, fairly common in Laney Member deposits from Lake Gosiute.
8	†*Asterotrygon maloneyi**	30 or more	More abundant in nearshore F-2 than in F-1; about 40 times rarer than the other FBM ray, †*Heliobatis*.
9	†*Amia pattersoni*	40 or more	Most are very large individuals over 1 m in length; juveniles are extremely rare.
10	†*Lepisosteus bemisi**	100 or more	Essentially equally abundant in F-1 and F-2; juveniles are unknown.
11	†*Hypsiprisca* n. sp.*	100 or more	More abundant in F-2 than in F-1; mostly very small individuals less than 60 mm long.
12	†*Crossopholis magnicaudatus**	100 or more	Most abundant in near shore and about 8–10 m above F-1.
13	†*Hiodon falcatus**	300 or more	Almost all specimens are from F-2; only 1 known from F-1.

the million-plus specimens listed in that table, I have probably seen more of the FBM material than any other professional scientist, and I have been studying the rates at which fossils are quarried in several of the commercial quarries. In the years since I published the first edition of *Bulletin 63* (Grande 1980), many of the quarriers have also continued to bring the most significant pieces they find to my attention. Therefore I present this table as a rough estimate of overall species abundance as of the year 2010. Some of the FBM species also occur in the Laney Member Lake Gosiute deposits, while most appear to have been present only in Fossil Lake (table 1, species with asterisk).

Scarcity ranking (1 = rarest)	Species (*asterisk indicates FBM species not known from Eocene Lake Gosiute or Lake Uinta)	FBM specimens collected 1870–2010 (estimate)	Comments on scarcity and distribution (F-1 = mid-lake 18-inch layer deposits, F-2 = nearshore sandwich bed deposits)
14	†*Atractosteus simplex*	300 or more	The most common gar species, equally abundant in F-1 and F-2; juveniles are extremely rare.
15	†*Amphiplaga brachyptera**	600 or more	Almost all specimens are from F-2; extremely rare in F-1.
16	†*Heliobatis radians**	1,000 or more	The most common stingray by far; more common in F-1 than F-2.
17	†*Notogoneus osculus**	4,000 or more	Unknown from F-2; occurs exclusively in F-1; known mostly by large individuals and tiny juveniles.
18	†*Phareodus testis*	6,000 or more	Juveniles more common in F-2 than F-1; extremely rare outside of the FBM.
19	†*Phareodus encaustus*	6,000 or more	Juveniles more common in F-2 than F-1; extremely rare outside of the FBM.
20	†*Priscacara serrata*	8,000 or more	Much less common than †*Cockerellites liops*; extremely rare outside of the FBM.
21	†*Knightia alta*	8,000 or more	Uncommon in the FBM, extremely common in the Laney Member deposits from Lake Gosiute.
22	†*Mioplosus labracoides**	30,000 or more	Extremely common; large individuals more common in mid-lake F-1 than nearshore F-2.
23	†*Cockerellites liops**	30,000 or more	Extremely common; much more common in F-1 than F-2. Extremely rare outside of the FBM.
24	†*Diplomystus dentatus*	over 350,000	One of the two most common FBM species by far, especially in F-1.
25	†*Knightia eocaena*	over 600,000	The most common of all FBM species; proportionately more common in F-2 than F-1.

Table 1. Relative abundance or scarcity of fish species in the FBM. Not shown here is the recently noticed "four-anal spine" variety of †*Priscaacara* because its abundance has yet to be determined. The proportionate numbers of †*Knightia* have greatly increased since quarriers have started mining the beds near the K-spar tuff bed in the mid-lake deposits, and since my counts in the FBM in Grande (1984).

Among fossil freshwater fish localities dominated by teleost species, the FBM has the highest known species diversity in the world. But species diversity seems low compared to today's modern, long-lived lakes of comparable size to Eocene Fossil Lake. In terms of fish family diversity (15 families), it ranks average to high, compared to modern lakes, but in terms of number of species, it ranks lower than many of today's living lake systems, such as the great African Rift lakes. This is partly an artifact of the incomplete preservation of fossils affecting the way we recognize species in fossil communities. In living fish communities, we can often differentiate species with only differences in color, color pattern, muscles, other soft tissue characters, or even molecular data. These types of characters are not preserved in fossils, and so we will never know the complete species diversity of Fossil Lake, even if we have a million well-preserved fossil skeletons. Some fish groups in the FBM are also in need of further study to determine the number of species in those groups (e.g., the perch-like fishes). Thus, the identification of 26 species here (appendix B) is an extremely conservative number, and probably underrepresents the true species diversity of the FBM fish community (Grande 2001, 33–34).

Some FBM species are much scarcer than others. What can we make of such drastic variation in species abundance in the FBM? Clearly, extremely rare species like the pickerel †*Esox kronneri* were not normal members of the main Fossil Lake community and probably wandered in accidentally from connected bodies of water. Other scarce species in the FBM that were possible inhabitants of river tributaries that occasionally wandered into Fossil Lake include †*Massilosteus*, †*Asterotrygon*, †*Crossopholis*, *Hiodon*, and †*Amphiplaga*. Some other examples of scarce species in the FBM are the gars *Lepisosteus* and *Atractosteus*, and the bowfins *Amia* and †*Cyclurus*. The gar and bowfin species in the FBM may be rare in part because gars and bowfin are today exceptionally hardy species that can survive in conditions that most other species cannot. Most of the FBM gar and bowfin species may have survived the mass-mortality events that killed the large numbers of individuals represented in the mass-mortality layers of other fishes.

Some of the variation in abundance was also probably a factor of environmental sensitivity and **FECUNDIDITY** providing enormous numbers of bodies for some species. The best example of this is †*Knightia eocaena*, a member of the herring family (family Clupeidae), which is the most common fish species in the FBM by far. Herrings are subject to mass die-offs due to even small changes in water temperature or chemistry. The shores along Lake Michigan near the museum where I work are occasionally littered with large piles of dead clupeids due to summer temperature changes and **UPWELLING** (fig. 214, *bottom*). Clupeids living in freshwater lake environments such as the alewife *Alosa pseudoharengus* are particularly sensitive. Clupeids are also prolific breeders, and a single female can lay up to 100,000 eggs. The combination of reproduction on a massive scale and massive mortalities of adults made the clupeid †*Knightia eocaena* the most common species of fish fossilized within the FBM.

In comparison to the other parts of the Green River Great Lake Complex (*sensu* Grande 2001), the FBM diversity is much higher than that known for Lake Gosiute or Lake Uinta. Lake Uinta had the least diverse fish fauna of the three Green River lakes, based on current paleontological data. The earlier Paleocene phase of Lake Uinta had two species and one genus of gar (†*Cuneatus cuneatus* and †*Cuneatus wileyi*) not yet known from the FBM (Grande 2010). There are no other fishes reported from Lake Uinta other than †*Knightia*, †*Amia*, †*Diplomystus*, †*Phareodus*, †*Erismatopterus*, †*Asineops*, and very rarely percoids probably belonging to †*Priscacara* and †*Mioplosus*. Lake Gosiute had a more diverse fish fauna than Lake Uinta, but it was also not as diverse as Fossil Lake. Like Fossil Lake, Lake Gosiute was dominated by the herring family but different species. Where †*Knightia eocaena* was the dominant clupeid species in the FBM, two other clupeids dominate the Laney Member deposits of Lake Gosiute: †*Knightia alta* and †*Gosiutichthyes parvus*. †*Knightia alta* is much less common in the FBM, and although †*Gosiutichthyes* is known by hundreds of thousands of specimens from Lake Gosiute, this genus is unknown from the FBM (Grande 2001). Lake Gosiute also had catfishes and suckers, two major groups absent from the FBM. (This is discussed further on page 180.) The genus †*Asineops* is much more common in Lake Gosiute than in Fossil Lake, although it is generally of smaller size in Lake Gosiute deposits. Fossil Lake also shared a number of taxa with Lake Gosiute (†*Knightia eocaena*, †*Knightia alta*, †*Atractosteus simplex*, †*Amia pattersoni*, †*Phareodus encaustus*, †*Diplomystus dentatus*, and |*Asineops squamifrons*). Lake Gosiute also has a species of trout perch (†*Erismatopterus levatus*), which is unknown from Fossil Lake. And Fossil Lake had numerous fish species found in neither Lake Gosiute nor Lake Uinta (see table 1, page 176). The fish faunas of Lake Gosiute and Lake Uinta are still in need of much study to understand the biodiversity of their fish communities. The fish fauna of Fossil Lake was clearly distinct from those known from Lakes Gosiute and Uinta.

Distribution Patterns among the FBM Fish Species

Within the FBM there are some significant differences between the mid-lake 18-inch layer beds and the nearshore sandwich-layer beds (see pages 176–77). In general, the hatchling to young juvenile growth stages of most genera are more common in the nearshore sandwich beds than in the mid-lake 18-inch layer (†*Asterotrygon*, †*Heliobatis*, †*Amia*, †*Hiodon*, †*Phareodus*, †*Knightia*, †*Amphiplaga*, †*Mioplosus*, †*Priscacara*, †*Cockerellites*, and †*Hypsiprisca* n. gen.), indicating that those areas were nursery grounds for most of the FBM fish species. Notable exceptions are the youngest larval growth stages of †*Notogoneus* and †*Diplomystus*, which are much more common in the mid-lake 18-inch layer beds than in the nearshore deposits. In fact, larval as well as adult †*Notogoneus* are completely absent from the nearshore sandwich bed deposits. Larval stages for some genera are unknown anywhere in the FBM (†*Crossopholis*, †*Masillosteus*, †*Lepisosteus*, *Esox*) or unique (*Atractosteus*, †*Asineops*), indicating that these fishes were spawning and spending their earliest lives in the tributary streams, rivers, or other connected bodies of water.

The adult stages of some genera are much rarer in the mid-lake 18-inch layer deposits than in the nearshore sandwich beds (†*Heliobatis*, †*Asterotrygon*, †*Crossopholis*, †*Masillosteus*, †*Hiodon*, †*Esox*, †*Amphiplaga*, and †*Hypsiprisca* n. gen.). Some fishes present in the mid-lake 18-inch layer are much rarer or absent in the nearshore sandwich beds (large †*Asineops* and †*Notogoneus*).

It is one of the marvels of the Fossil Butte assemblage that we can compare large samples from both mid-lake and nearshore environments for interpreting the paleoecology of this early Eocene lake in detail.

What Is Missing, and What Species Could We Still Expect to Find in the FBM Fish Assemblage?

It is odd that neither the catfish order Siluriformes nor the carp and sucker order Cypriniformes is present in the FBM (or anywhere in Eocene Fossil Lake as far as I know). In the late early Eocene Laney Member deposits of Lake Gosiute, catfishes are abundant and represented by two families: †Hypsidoridae and Ictaluridae (Grande 1987; Grande and Lundberg 1988). †Hypsidoridae is an extinct family known only from the middle Eocene, and Ictaluridae is survived by 46 species living in North America. Suckers (family Catostomidae) are also abundant in the Laney Member Lake Gosiute deposits represented by the genus †*Amyzon* (Grande, Eastman, and Cavender 1982). Catostomids are survived today by over 70 species in North America and one in Asia. There are two possible explanations for the absence of catfishes and suckers from the FBM: time differences or ecological differences. With regard to time, the FBM deposits (52 million years old) are about 3 million years younger than the Laney Member Lake Gosiute deposits (49 million years old) (for dating of these members see Smith, Carroll, and Singer 2008; Smith et al. 2010). It may be that suckers and catfishes did not invade the Green River Lake Complex until well after the FBM phase of Fossil Lake had ended. With regard to ecology, the lake-bottom ecology of Fossil Lake may not have been as suited to suckers and catfishes as the lake-bottom ecology of Lake Gosiute. I would favor the time argument over the ecological one, because if it were only an ecological difference I would have expected to see at least one accidental immigrant into the lake from one of its tributaries, particularly in the Thompson Ranch sandwich beds, where we see other occasional immigrants such as *Esox*. Of course, time will tell, and it is possible that we may one day find such an accidental invader. After all, it was not until 1999 that the first esociform was reported (see page 149).

Other major elements in today's freshwater fish fauna that are lacking in the FBM or other Green River Formation deposits include sturgeons (family Acipenseridae), salmon and trout (family Salmonidae), minnows (family Cyprinidae, subfamily Leuciscinae), darters (family Percidae, subfamily Etheostoma-tinae), and sunfishes (family Centrachidae). Sturgeons are known since the Cretaceous in freshwater fossil deposits in western North America, although usually in river deposits. I predict they will turn up someday in the Thompson Ranch deposits to the north, near the northern tributaries to Fossil Lake (FBM Localities H and M). Salmonids are abundant in the Eocene freshwater deposits of Canada. Like the sturgeons of early days, early salmonids may have preferred more of a stream environment, so could one day be discovered as an accidental immigrant into Fossil Lake. The subtropical climate of Fossil Lake may also have been too warm for them.

The absence of fossil cyprinids, particularly the minnows, from the FBM (and from the entire Green River Formation) is puzzling because they are today so abundant in the freshwaters of North America. Minnows are one of the most diverse and common elements in North American and Asian freshwaters, with over 230 living species north of the Mexican border; yet they are absent from all of the Green River Lakes. The explanation could be one of timing and evolutionary history. Minnows are thought by some fish biologists to have evolved in the later Eocene and quickly diversified to occupy their present distributions across much of the North American continent. The oldest known cyprinid fossils are from the middle Eocene of Asia.

The absence of sunfishes and darters in the FBM may be another timing issue. These two groups are very common and diverse in North American freshwaters today, but they appear to have evolved after the time of Fossil Lake. Although there are nearly 200 living species of darter in North America today, the oldest known fossil darters are from the late Oligocene (about 25 million years before present). There are about 30 living species of sunfishes in North America today, but the oldest known fossil sunfishes are Early Oligocene in age (about 30 million years before present). For a review of the fossil record of living North American freshwater fish families, see Cavender (1998).

The absences and unpredicted distributions are part of what makes the continued work in the FBM still exciting. New discoveries are still made there every year. Some of the "missing" families discussed above may still be discovered in the FBM one day, thus extending their record back in time. The job of sampling biodiversity, whether fossil or living, is never finished, and the need to explore and discover nature is part of human culture.

Fossil Lake and the Evolution of the Modern North American Fish Fauna

The Fossil Butte Member of the Green River Formation comprises an incomparable sample of the early evolution of today's North American fish community. This has not always been recognized, because until recently few ichthyologists were familiar with the true potential of the FBM species. Earlier ichthyologists thought the FBM fossils (and the rest of the Green River Formation fish fauna) were too "primitive" to be of much use to the study of modern fish faunas, and FBM species were all but omitted from such studies. The famous twentieth-century ichthyologist Robert Rush Miller once wrote that the Green River Formation fishes (including the FBM species) were "a fauna too early in fish evolution to give much help in interpreting the modern forms" (Miller 1959, 192). Other ichthyologists of the time concurred (e.g., Uyeno and Miller 1963, 3). But over the last several decades, there has been the discovery of several additional modern families present in the FBM (e.g., the mooneye family Hiodontidae and the pike and pickerel family Esocidae), the recognition of other FBM species belonging to modern families (e.g., the percoids belonging in Moronidae, Latidae), and a better integration of FBM fish species with living species in comprehensive evolutionary studies (e.g., Grande 1999, 2010; Grande and Bemis 1991, 1998; Grande and Grande 2008; Hilton and Grande 2008; Carvalho, Maisey, and Grande 2004; Rosen and Patterson 1969; Whitlock 2010). Although some families present in the FBM are now extinct in North America and living elsewhere (e.g., Osteoglossidae and Gonorynchidae) or extinct worldwide (†Asterotrygonidae, †Heliobatidae, †Paraclupeidae, †Asineopidae), at least nine families present in the FBM still survive in North America today (Polyodontidae, Lepisosteidae, Amiidae, Hiodontidae, Clupeidae, Esocidae, Percopsidae, Latidae, and Moronidae). In addition to being taxonomically relevant and diverse, all FBM species are represented by well-preserved complete skeletons that can easily be incorporated into comparative evolutionary studies of living species. It is clear that the FBM species are critical to any comprehensive study of the origin and affinities of today's modern North American freshwater fish fauna.

Tetrapods

(Superclass Sarcopterygii)

TETRAPOD animals (animals with two sets of paired limbs) include amphibians, mammals, and reptiles (including birds). Birds are usually classified with reptiles today because they have been found to share unique features with other species in Reptilia that they share with no other animal species. The explanation for this is that birds evolved from a species of "reptile." Because I am linking all of the FBM species into a single web of life, I am considering the bird group Aves to be a superorder of the class Reptilia. Most bird specialists working only within the context of Aves consider the group to be of "class" rank. As I discussed above on page 49, the ranks above family level can vary from study to study and should remain flexible for two reasons. First, rank assignments are only a tool used to group species hierarchically according to uniquely shared characteristics. These characters are thought to indicate close evolutionary relationship between the species that possess them; strong, non-random patterns of such characters are the primary line of evidence in evolutionary studies. The second reason that ranks above family level are flexible is that the official rules of nomenclature do not apply to ranks above family-group names (see page 49). There are strict official rules guiding the construction of names from the level of species to superfamily, but no such rules exist for the levels of suborder through kingdom.

Tetrapods as a group are relatively scarce components of the vertebrate fossils in the FBM. The most common tetrapods in the FBM are, of course, aquatic forms (e.g., turtles) or species that flew over the lake and fell into the lake upon death (birds and bats). Mammals in the FBM other than bats are extremely rare, with most species known by only a single specimen. The known FBM tetrapods (also called sarcopterygians) as follows:

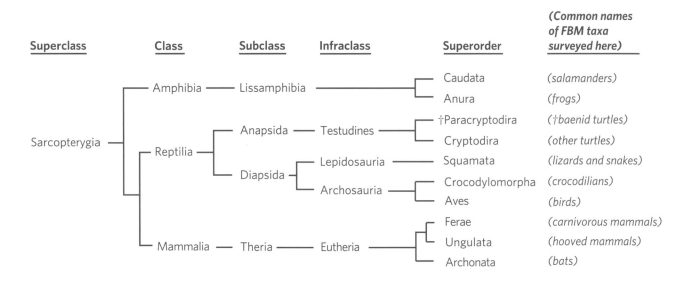

Superclass	Class	Subclass	Infraclass	Superorder	(Common names of FBM taxa surveyed here)
Sarcopterygia	Amphibia	Lissamphibia		Caudata	(salamanders)
				Anura	(frogs)
	Reptilia	Anapsida	Testudines	†Paracryptodira	(†baenid turtles)
				Cryptodira	(other turtles)
		Diapsida	Lepidosauria	Squamata	(lizards and snakes)
			Archosauria	Crocodylomorpha	(crocodilians)
				Aves	(birds)
	Mammalia	Theria	Eutheria	Ferae	(carnivorous mammals)
				Ungulata	(hooved mammals)
				Archonata	(bats)

Amphibians

Amphibians in the FBM are exceedingly rare, known so far by only a single adult frog and a single adult aquatic salamander. This might seem peculiar given that amphibians are such water-loving animals. They are water-bound during at least part of their life cycle because their eggs have no shells and must be laid and develop in freshwater. One possible explanation for the scarcity of amphibians in the FBM is the alkalinity of the lake water. Being supersaturated with calcium carbonate (see pages 9–10), the pH of Fossil Lake during deposition of the FBM would have been relatively high (estimated 8 to 8.5). Most amphibians prefer water of or near neutral pH (e.g., near 7). Both FBM specimens are from the nearshore Thompson Ranch quarries thought to be close to one of Fossil Lake's tributaries. They may have washed in from a connected river or stream. Scarcity of amphibians in non-**FLUVIAL** Paleogene North American localities is not unusual. In Green River Formation deposits outside of the FBM, there are two amphibian specimens (both frogs) known so far from the middle Eocene Laney Member deposits of Lake Gosiute (e.g., Grande 1984, 187) and two specimens (again, both frogs) known from the middle Eocene Parachute Creek Member deposits of Lake Uinta (e.g., Grande 1984, 188). There are no amphibians so far known from the late Eocene Florissant Formation of Colorado (Meyer 2003, 179).

Salamanders (Class Amphibia, Superorder Caudata)

There is only a single known specimen of salamander known from the FBM, †*Paleoamphiuma tetradactylum*. This species is known only from the FBM, where it was first described in 1998. It is now thought to belong in the aquatic salamander family, Sirenidae. †*Paleoamphiuma* is one of the earliest known occurrences of

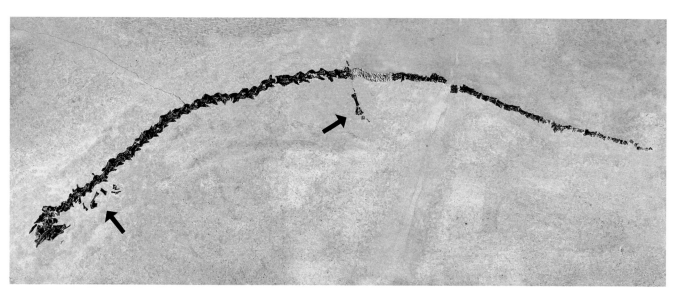

the family in the fossil record. Today living species of the Sirenidae are found only in the warm climates of the southeastern United States and northern Mexico. These are four living species, and they are all eel-shaped aquatic species with tiny arms and no legs. They have the ability to come on land on rainy nights, but they live primarily in slow-moving freshwater. They have no teeth in their jaws but they possess a horny sheath resembling a beak, and feed through SUCTION FEED-ING. They feed on small snails, shrimps, worms, and algae. The only known specimen of *†Paleoamphiuma* is from the sandwich beds of the Thompson Ranch (fig. 96), which is also the quarry where small gastropods and shrimps occur. Thus, the diet of *†Paleoamphiuma* was likely similar to that of living sirenids.

†Paleoamphiuma tetradactylum was originally described as a species of the family Amphiumidae. Later work (yet in progress) has shown that *†Paleoamphiuma* should instead be classified as a member of the family Sirenidae. There is only one species in this genus (below). This species resembles living sirenid species in having toothless jaws and very tiny front limbs (fig. 96, *arrows*). Unlike living sirenids, which have only forelimbs, *†P. tetradactylum* has both forelimbs and hind limbs. The possession of hind limbs is considered a primitive trait within the family. *†Paleoamphiuma tetradactylum* is described in detail in Rieppel and Grande (1998) and is being further studied as a sirenid by Gardner, Rieppel, and Grande.

Frogs (Class Amphibia, Superorder Anura)

The name Anura is Greek for "without tail," which describes one of the characteristics of frogs. There is a single frog specimen known from the FBM (fig. 97). It is a new genus and species currently being described. The specimen is a "split fossil" from Thompson Ranch FBM Locality H, and the best side of the two opposing slabs is in the collection of the Field Museum, while the counterpart slab remains in a private collection. No tadpoles have been reported from the FBM so far. There are over 5,000 species of frogs living today, ranging from the southern tip of South America to the subarctic region, but the majority of species live in tropical environments. Like salamanders, frogs need to lay their eggs in freshwater, and the FBM species was probably a freshwater aquatic form as an adult. Adult frogs have a carnivorous diet, feeding mostly on insects, worms, and small snails. The FBM frog is clearly different from the Lake Gosiute Laney Member frog *†"Eopelo-bates"* reported in Grande (1984, 187). The FBM specimen represents a small species, measuring less than 50 millimeters (2 inches) in length (fig. 97). It appears to be a new species and genus closely related to Pelodytidae (parsley frogs) or Pelobatidae (European spadefoot toads) (Amy Henrici, pers. comm., 2010). Today the Pelodytidae include about three living species, which occur in Europe and the eastern edge of Asia. They are FOSSORIAL, burrowing in soft or sandy soils. They prefer to lay their eggs in streams and weedy ponds. The Pelobatidae are represented today by four living species that are native to Asia, Africa, and Europe.

They lay their eggs in temporary ponds that dry up regularly, so the tadpole stage is brief, lasting as little as two weeks. Frogs are carnivores, feeding on insects, worms, and snails. The description of the FBM specimen is forthcoming.

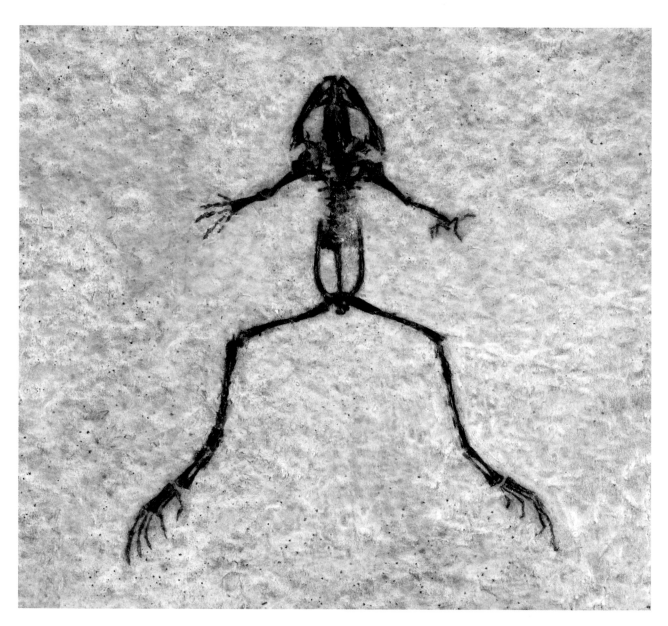

FIGURE 97
An undescribed frog (family either Pelodytidae or Pelobatidae) from the sandwich beds of FBM Locality H. This is the only known frog from the FBM. Specimen is 40 millimeters (1.6 inches) long (FMNH PR2384).

Non-Avian Reptiles

(Class Reptilia; Superorders †Paracryptodira, Cryptodira, Squamata, and Crocodylomorpha)

"Non-avian reptiles" comprise an unnatural group used only for chapter convenience here. It is unnatural because excluding birds from Reptilia makes it **NON-MONOPHYLETIC**. I will discuss non-avian reptiles first (turtles, lizards, snakes, and crocodilians) and leave description of the avian reptiles (birds) until the next major section.

Modern reptiles are characterized as animals that lay tough-shelled **AMNIOTIC EGGS** and have skin covered with scales, or bony scutes, or feathers (in the case of birds and certain dinosaurs). Unlike amphibians, which lay their eggs in water, reptiles lay their eggs on land or buried in the ground, usually near water. The eggs are protected from drying out by their tough shell. Turtles, crocodilians, and aquatic lizards formed a significant part of the Fossil Lake community. The most basal group of reptiles in the FBM are the turtles.

Turtles (Infraclass Testudines, Superorders †Paracryptodira and Cryptodira)

Turtles are characterized by a special bony shell that acts as a shield that develops from a unique combination of the ribs, parts of the shoulder girdle, and mineralized skin. The dorsal or "back" part of the shell is called the carapace, and the belly part of the shell is called the plastron. The turtles of the FBM are all freshwater aquatic forms and are classified in two superorders: the extinct †Paracryptodira (represented in the FBM by the extinct family †Baenidae) and the extant Cryptodira (represented by the extant families Trionychidae, Dermatemydidae, and Emydidae). The Trionychidae (the soft-shelled turtles) are by far the most common cryptodire family in the FBM, and the Dermatemydidae (the Central American river turtles) are the rarest.

One of the difficulties with classifying fossil turtles, including those from the Cretaceous-Eocene of western North America, is the fragmentary nature of turtle fossils from many localities. Some localities produce mostly shell fragments, other localities produce isolated skulls, but very few produce complete skeletons. Many species names are based only on shell or skull fragments, and it is not always clear between localities which skull material belongs with which body material. Adding to the confusion, some fossil turtle specialists have focused only on skull characteristics, while others have focused mostly on shell characters to build their classifications, and the skull-based classifications sometimes do not match the shell-based classifications. Localities like the FBM that usually produce complete skeletons can help decrease this type of confusion. There is nothing like a complete articulated skeleton to show what the complete skeleton looked like! Turtles from the FBM include the following families:

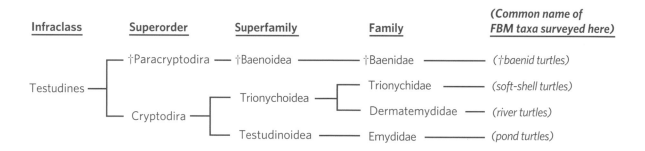

Infraclass	Superorder	Superfamily	Family	*(Common name of FBM taxa surveyed here)*
Testudines	†Paracryptodira	†Baenoidea	†Baenidae	*(†baenid turtles)*
	Cryptodira	Trionychoidea	Trionychidae	*(soft-shell turtles)*
			Dermatemydidae	*(river turtles)*
		Testudinoidea	Emydidae	*(pond turtles)*

The †Baenidae are known only from western North America and date back at least to the Early Cretaceous (at least 120 million years before present). They were most diverse and abundant in the Late Cretaceous, and the FBM species represent some of the last surviving members of the family, which appears to have become extinct in the late Eocene. †Baenids were water-loving, sometimes described as "bottom-walking" forms in lakes and channels (Hutchison 1998). They probably fed mainly on fishes, crayfishes, shrimps, and clams. The †baenid fossil record is relatively good, and most species are known from complete shells or skulls. However, the FBM is one of only a few localities to produce complete skeletons of these animals, and therefore the FBM specimens are extremely important in deciphering the complete anatomy, evolution, and ecology of this group. One of the features revealed by the complete skeletons is that †baenids had extremely long tails (see figs. 98, 99, 100). Unlike other families of turtles represented in the FBM, †baenids were incapable of retracting their heads into their shells, so all the complete skeletons of †baenids from the FBM have their heads and necks extended.

There are about 20 species in the family †Baenidae, of which two are known from the FBM: †*Chisternon undatum* and †*Baena arenosa*. The most common species is †*C. undatum*, a species that was originally described by Leidy in 1871 based on a partial carapace from the lower Bridger Formation of Wyoming. The FBM has produced more than a dozen nearly complete to complete skeletons assigned to this species (identification by Hutchison 2010). †*Chisternon undatum* is known in the FBM by both large individuals (fig. 98) and very young juveniles (fig. 99). No one has yet studied the early development of the skeleton using this complete skeletal material. The several known large individuals are all from the mid-lake deposits (e.g., FBM Localities B and C) and the young juveniles, of which there are about a dozen known specimens, are all from the nearshore deposits of Thompson Ranch (FBM Localities H and M). The other known †baenid species in the FBM is †*Baena arenosa*. This species, like †*C. undatum*, was also described by Leidy based on a partial shell from the lower Bridger Formation of Wyoming. It is represented in the FBM by several complete skeletons from the Thompson Ranch nearshore quarries (e.g., fig. 100). †*Baena arenosa* and †*C. undatum* are very closely related to each other and therefore show much resemblance to each other. †*Chisternon undatum* can be recognized by its rounded shell and its more lightly built skull with large eyes and upper temporal emarginations. †*Baena arenosa*, by contrast, has a more elongate shell, relatively smaller eyes, and no upper temporal emarginations (e.g., compare fig. 100 with figs. 98 and 99). Although the †baenid material from the FBM has yet to be studied in detail, the shells and skulls of these two species were described in detail by Gaffney (1972) based on fossils from another locality of the same age as the FBM.

Snapping turtles (family Chelydridae) were reported from the FBM by Grande (1980, 1984, fig. III.3b), but the specimen this was based on was later determined to be a †baenid. There are no true chelydrids known so far from the FBM, although they are generally known to occur in the Eocene of North America.

The Trionychidae (the soft-shelled turtles) comprise the most common and most species-rich turtle family in the FBM. They are represented there by three

FIGURE 98

†*Chisternon undatum* (Leidy, 1871), a turtle from the extinct family †Baenidae. Large adults from the FBM 18-inch layer. *Top:* An individual in dorsal view measuring 1,270 millimeters (4.2 feet) in total length, on a slab with a large †*Diplomystus dentatus*. Original specimen from FBM Locality C in a private collection; photo courtesy of Ron Mjos. A cast of this specimen is also at the Field Museum (FMNH PR2451). *Bottom:* An individual measuring 1,321 millimeters (4.3 feet), with shell in dorsal view but head in ventral view; from FBM Locality B (CGR-114).

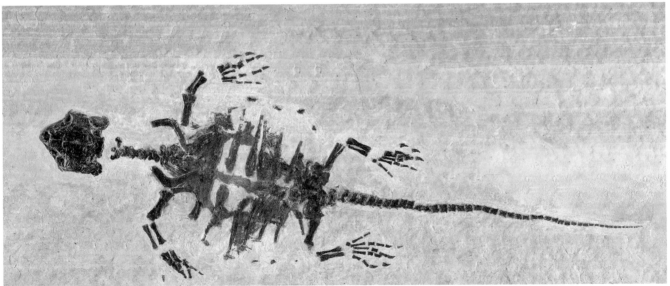

FIGURE 99

†*Chisternon undatum* (Leidy, 1871). Two baby individuals from the sandwich beds of FBM Locality M. *Top:* Specimen is only 110 millimeters (4.3 inches) long (WDC CGR-32). *Bottom:* Specimen is only 115 millimeters (4.5 inches) long (FMNH PR2829). Note that the top shell (carapace) is mostly undeveloped at these young stages.

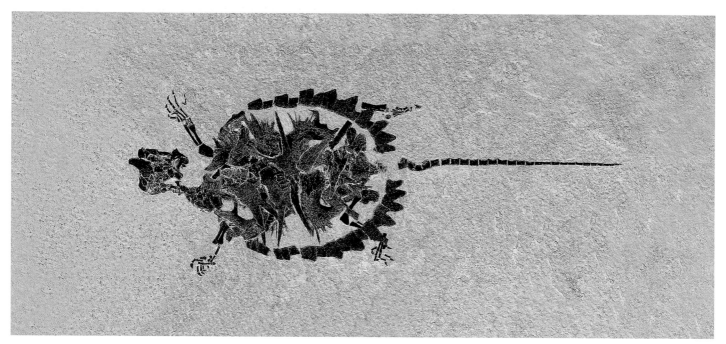

FIGURE 100

†*Baena arenosa* Leidy, 1870a, another species of †baenid turtle. Two specimens from the FBM sandwich beds. *Top:* Young individual in ventral view, measuring 220 millimeters (8.7 inches) long, from FBM Locality L. Specimen in a private collection. *Bottom:* Older individual in dorsal view, measuring 370 millimeters (1.2 feet) long, from FBM Locality H (FMNH PR2391).

species: †*Platypeltis heteroglypta* (fig. 101), †*Axestemys byssinus* (figs. 102, 103), and †*Hummelichelys guttata* (fig. 104). Trionychids occur both in the mid-lake and the nearshore localities, although the very young juveniles (e.g., fig. 103) are known only from the nearshore Thompson Ranch quarries (FBM Localities H and M). Even as the most common of the turtle groups preserved in the FBM, they are still highly coveted by the commercial quarries because they are much rarer than the fishes. I estimate that there is only about one turtle of any type found in the FBM for every 10,000 fishes.

†*Apalone heteroglypta* is known by several specimens, mostly from the mid-lake deposits. One of my favorite FBM pieces is a specimen of †*A. heteroglypta* collected in the late 1980s that was preserved on a large slab together with three species of fishes and several insects, illustrating the community richness of the FBM 18-inch layer (fig. 101). †*Apalone heteroglypta* was originally described as †*Trionyx heteroglpytus* by Cope (1873) on the basis of a partial carapace from the lower Bridger Formation of Wyoming. This species can be distinguished relative to other Eocene trionychids by the rounded carapace, the near-complete reduction of the eighth costal bones, and the unique shell sculpturing that evenly covers the entire carapace. The FBM specimens of this species currently remain largely unstudied.

†*Axestemys byssinus* is one of the more common species of trionychid in the FBM, and the one that reaches the greatest size there. The specimen in figure 102 is over 1.5 meters (5 feet) in length, and I have examined others from FBM Locality B measuring 1.8 meters (6 feet) in length with the heads only partly extended. The largest specimens are all from the mid-lake quarries (e.g., FBM Localities A, B, C, and F). Young juvenile specimens with incompletely formed shells are also known from the nearshore Thompson Ranch quarries (e.g., fig. 103). †*Axestemys byssinus* was originally described on the basis of a partial, disarticulated skeleton from the lower Bridger Formation of Wyoming. The specimens illustrated here are currently unstudied. In addition to its large size, †*A. byssinus* can be recognized by the lack of ornamentation on the surface of the plastral bones, a lack of sculpturing along a broad band that surrounds the carapace, and the broad nuchal that almost lacks a callosity and is only poorly connected with the remaining carapacial disk.

†*Hummelichelys guttata* is known by several specimens from the mid-lake quarries, in the 18-inch layer and in the beds near the K-spar tuff bed. One particularly interesting specimen is a beautifully complete articulated skeleton with tooth marks in its carapace, indicating it narrowly escaped being dinner for a crocodilian at some time before its eventual demise (fig. 104). This species was originally described as †*Trionyx guttatus* by Leidy in 1869 on the basis of a partial shell from the Bridger Formation of Wyoming. This species is best distinguished from other FBM trionychids by its more elongate shell, the narrow nuchal that is well connected to the remaining carapacial disk, and the raindrop-like carapace ornamentation.

⚱ FIGURE 101
†*Apalone heteroglypta* (Cope, 1873), a member
of the soft-shelled turtle family, Trionychidae.
Adult individual 680 millimeters (2.2 feet)
long on a slab with fishes (†*Knightia eocaena*,
†*Diplomystus dentatus*, and †*Phareodus testis*)
and three flies (†*Plecia pealei*). Slab is from
the 18-inch layer of FBM Locality A (FMNH
PF13057). This is a natural assemblage slab
and not an inlay.

⟶ FIGURE 102
†*Axestemys byssinus* (Cope, 1872), another species of the
soft-shelled turtle family, Trionychidae. This magnificent
giant individual measures 1,600 millimeters (5.3 feet) in
length and is from the 18-inch layer of FBM Locality A.
Original is in a private collection in Japan but a cast is at
the Field Museum (FMNH PR3039). Although collected in
FBM Locality A by Jim Tynsky, the specimen was prepared
by Tom Lindgren (in photograph) and his preparators.
Photo courtesy of Tom Lindgren.

⇑ FIGURE 103
†*Axestemys byssinus* (Cope, 1872), a
species of the soft-shelled turtle family,
Trionychidae. A young juvenile only 101
millimeters (4 inches) long, showing an
early growth stage with an incompletely
developed top shell; from the sandwich
beds of FBM Locality M (WDC-CGR 111).

⇢ FIGURE 104
†*Hummelichelys guttata* (Leidy, 1869), another species of the soft-
shelled turtle family, Trionychidae. A beautifully articulated speci-
men 250 millimeters (9.8 inches) long, from the 18-inch layer of
FBM Locality B. Cast of the specimen in the collection of Fossil
Butte National Monument. Note that this specimen has three tooth
punctures in the shell, indicating that it was bitten by a crocodilian.
Top: Entire specimen. *Bottom:* Close-up of bite punctures (arrows).

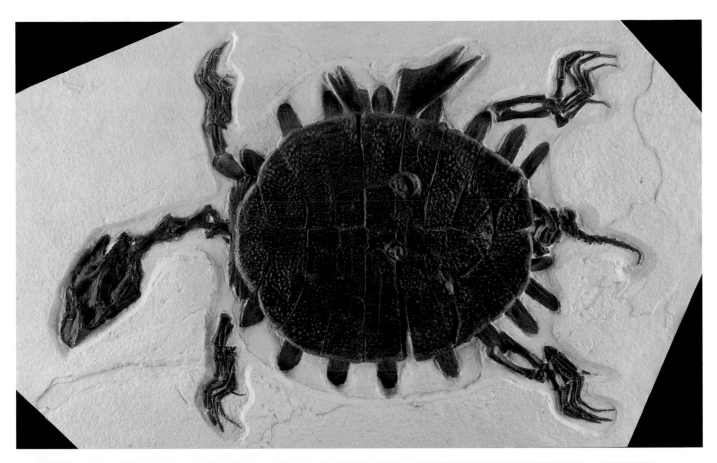

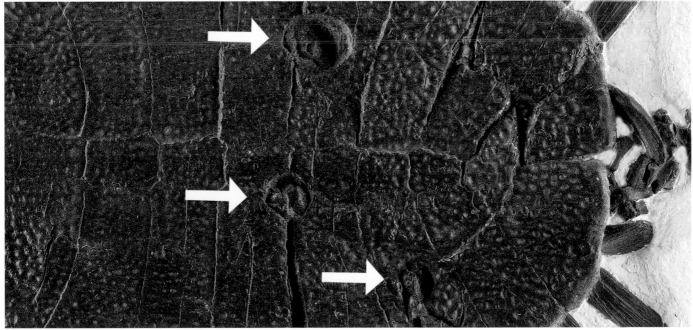

Today there are about 25 living species of trionychid turtles, and they can be found in the freshwaters of North America, Asia, and Africa. They are called "soft-shelled" because the carapace of most species lacks the horny scutes present in other turtles. The central part of the carapace has a solid shield of bone, but the ribs extend beyond the edges of that bony region, supporting an outer carapace margin of thick skin. The light flexible shell allows trionychids to move easily in open water and through muddy lake bottoms. They sometimes lie motionless, buried in mud with only their eyes and mouth protruding, lying in ambush for swimming prey.

Trionychidae contains the longest living species of freshwater turtles in the world today. This giant is *Pelochelys cantorii*, an Asian species that lives in slow-moving freshwater rivers and streams, and reaches a total length of about 2 meters (just over 6 feet). Similarly, the largest turtles in the FBM are trionychids, with †*Axestemys byssinus* reaching 2 meters or more in length (discussed on page 195).

Trionychid turtles today are strict carnivores, feeding mainly on fishes, amphibians, crayfishes, shrimps, snails, and sometimes even birds and small mammals. Their special anatomy requires them to be submerged in order to swallow food. They have long, snake-like necks capable of extending the head quite far to breathe air from the surface or to defend themselves, but normally they keep most of their long necks retracted under their carapace. The FBM specimens illustrated here show examples of individuals with the neck well extended (fig. 104) and with the neck well retracted (fig. 101). The hands and feet of trionychid turtles are webbed. Only three claws are developed on the hands, hence the name *Trionyx* ("three-claw").

There are over 200 named fossil species of Trionychidae, but most are based on fragmentary material. Many of these fragment-based names are invalid and UNDIAGNOSABLE, because they cannot be clearly distinguished from other named species. Trionychids are an ancient group, with fossils as old as the early Late Cretaceous of China (†*"Aspideretes" maortuensis* and †*"Aspideretes" alashanensis*). The FBM trionychid material mostly remains unstudied (at least unpublished on), but living trionychids of the world were reviewed in Meylan (1987) and many of the fossil species are described in Hay (1908).

The Dermatemydidae (Central American river turtles) are among the rarest turtles in the FBM. This family is so far represented in the FBM by only two specimens of †*Baptemys wyomingensis*, one adult and one juvenile (fig. 105). The two specimens show a nice transition of shell development between two different growth stages (note the shell of the small individual was not yet fully developed before it died). The juvenile FBM specimen is the most complete specimen of †*Baptemys* known to date. Although partial shell and skull fragments of †*Baptemys* are very common in Cenozoic localities of western North America, complete shells associated with skulls are virtually nonexistent except for in the FBM. Today the Dermatemydidae are represented by only a single living species,

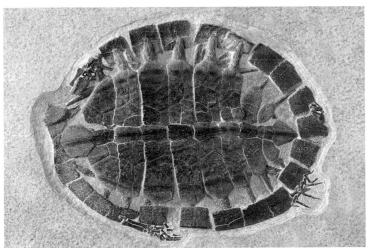

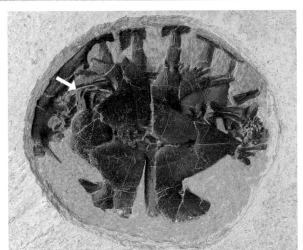

FIGURE 105

†*Baptemys wyomingensis* Leidy, 1870b, a member of the river turtle family Dermatemydidae, which is very rare in the FBM. *Top:* A complete specimen with three limbs and the head retracted inside the body. Shell measures 610 millimeters (2 feet) in length and is from the sandwich beds of FBM Locality M. Specimen currently in a private collection. *Bottom:* Dorsal and ventral view of a juvenile specimen that died before its shell was completely developed. Head retracted inside of shell, visible in ventral view (arrow). Shell measures 145 millimeters (5.7 inches) in length and is from the 18-inch layer of FBM Locality B (FMNH PR3033).

Dermatemys mawii. This species is a nocturnal, freshwater aquatic species that inhabits tropical rivers and lakes of Central America and southern Mexico. It feeds primarily on aquatic vegetation. This family is thought to be closely related to the mud turtle and musk turtle family Kinosternidae, which today inhabits North and South America with about 25 living species. Kinosternids inhabit slow-moving bodies of water with muddy bottoms. The FBM specimens of †*B. wyomingensis* each have their heads and one or more of their limbs retracted inside their shell. The FBM species is nearly identical to its extant relative *D. mawii*, and therefore likely had similar habitat preferences. This taxon can best be distinguished from all Eocene turtles by its elongate shell, its distinct mid-dorsal keel, and the presence of four inframarginal and three gular scutes along its plastron. It is sometimes confused with pond turtles of the genus †*Echmatemys* (see below). Large individuals can reach lengths of more than 600 millimeters (2 feet).

The Emydidae (the pond turtle family) are the largest and most diverse family of turtles living today, with over 40 species. They are primarily freshwater species, although a few species are either brackish water-tolerant or terrestrial. A good example is the pond slider *Trachemys scripta*, which is a common turtle in the pet trade. The emydids date back to the Late Cretaceous and have a good fossil record in western North America. Today the family is found throughout North America, northern South America, Europe, northwestern Africa, and Asia. Some species are herbivorous, and others feed on worms, crayfishes, shrimps, and small fishes.

In the FBM, there is one possible emydid (although this assignment is not conclusive, and some turtle specialists will only assign it to a larger group for now, the Testudinoidea (pond turtles and land tortoises). These turtles are much rarer than either †baenids or trionychids, suggesting that the preferred habitat of †*Echmatemys* was not lacustrine. They are so far represented in the FBM by a yet undetermined single species of the genus †*Echmatemys* (fig. 106). †*Echmatemys* is a genus that first appears in the late Paleocene (about 55 million years before present) and became one of the dominant turtles during much of the Eocene. The last of the †*Echmatemys* species disappeared near the end of the Eocene. †*Echmatemys* is more common in terrestrial and channel deposits than in lacustrine deposits, so its rarity in the FBM is probably due to ecological preferences. The only three FBM specimens of †*Echmatemys* I have seen so far all came from the nearshore Thompson Ranch quarries (FBM Localities H and M), which was probably near one of the tributary streams to the lake. Approximately 20 species of †*Echmatemys* were described during the late nineteenth and early twentieth century based on poorly provenienced and fragmentary material. Until a thorough review of the group has been undertaken, it is not possible to determine the exact species to which the FBM †*Echmatemys* belongs. †*Echmatemys* can be recognized by the presence of a single pair of gulars and the concentric rings that sculpt the surface of the shell.

FIGURE 106

†*Echmatemys* sp., a member of the pond turtle family Emydidae. Pond turtles are much rarer than †baenids or triionychids in the FBM, probably due to the more terrestrial habitat preference of the family. From the FBM sandwich beds. *Top:* A well-preserved complete individual in dorsal view, measuring 200 millimeters (7.9 inches) in length; from FBM Locality H (TMP 1983.017.0001). *Bottom:* Another specimen approximately the same size as the one above but prepared in ventral view to show the "plastron" (the belly shell); from FBM Locality M (WDC-CGR-113).

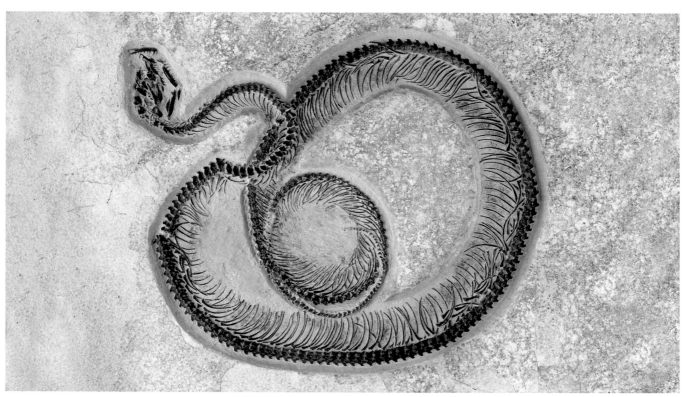
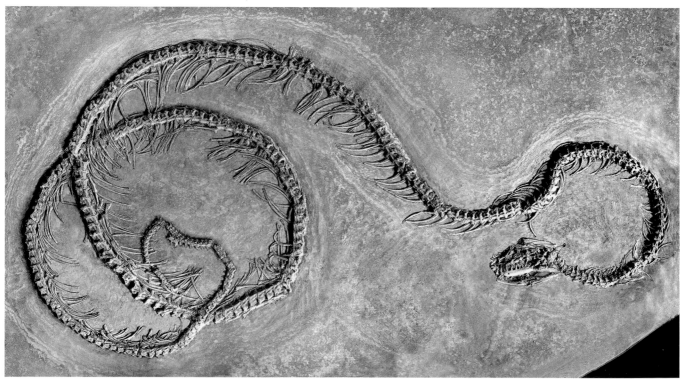

The species of †*Echmatemys* have so far mostly been described based on fragmentary material prior to discovery of complete skeletons in the FBM. None of the FBM specimens have been studied in detail yet, but fossil emydids and other testudinoids from various other localities are described in Hay (1908).

Snakes (Infraclass Lepidosauria, Superorder Squamata, Suborder Serpentes)

Snakes are exceedingly rare in the FBM, known so far by only two prepared specimens of the same species, †*Boavus idelmani* (fig. 107) and one yet unidentified and unprepared snake found in the 18-inch layer deposits in 2012. †*Boavus idelmani* appears to be a member of the family Boidae, a group that contains about 44 living species found mostly in tropical and subtropical regions of Central and South America, Africa, Europe, Asia, the western United States, and several of the Pacific Islands. One of the best-known living members of the family is the boa constrictor. Boids are carnivorous animals, and some, like the anaconda, are aquatic. The FBM species was a small boid and is occasionally classified in the neotropical subfamily Tropidophiinae (the dwarf boas). The two known specimens of *Boavus idelmani* are 960 and 1,250 millimeters (38 and 49 inches) in total length. The holotype specimen (fig. 107, *bottom*) was discovered in the early part of the twentieth century. It was in a private collection and is now lost, but Gilmore (1938) described this specimen in great detail, and casts of it are distributed in several museums including the Field Museum (FMNH PR 1457). In 2007 a second specimen was discovered in the nearshore quarry FBM Locality L. This specimen is now at the Houston Museum of Natural Science. A cast of this second specimen is also at the Field Museum (FMNH PR 2833). The third snake found in 2012 was not prepared or identified in time to be included in this book.

Lizards (Infraclass Lepidosauria, Superorder Squamata, Suborder Lacertilia)

Lizards are represented by three different families in the FBM: the Varanidae (monitor lizards), the Shinisauridae (crocodile lizards), and the Polychrotidae (anoles, which comprise a specialized group of iguanas). Each of these families is known by two or three specimens from the FBM that I am aware of. All of these specimens have been discovered only within the last several decades.

Varanid lizards (also called monitor lizards) are represented in the FBM by †*Saniwa ensidens* (figs. 108, 109). This is the largest lizard in the FBM, with the two specimens measuring 1,310 and 1,320 millimeters (about 52 inches) in length. Both specimens are from the nearshore sandwich beds, but one is from the southwestern part of the lake (FBM Locality K), while the other is from the northwestern part of the lake (FBM Locality H). This species was originally described by

←·· FIGURE 107

†*Boavus idelmani* Gilmore, 1938, a species of neotropical wood snake in the boid subfamily Tropidophiinae. *Top:* A specimen with an axial length of 1,250 millimeters (4.1 feet) discovered in 2007 in the sandwich beds of FBM Locality L (uncataloged specimen in the collection of the Houston Natural History museum; cast of this specimen is in the Field Museum collection [FMNH PR3040]). *Bottom:* The HOLOTYPE specimen from the 18-inch layer of FBM Locality B, formerly in a private collection and now lost. Photograph of original taken from Gilmore (1938). Axial length is 960 millimeters (3.2 feet). Casts of this specimen are widespread, including FMNH PR1457 at the Field Museum.

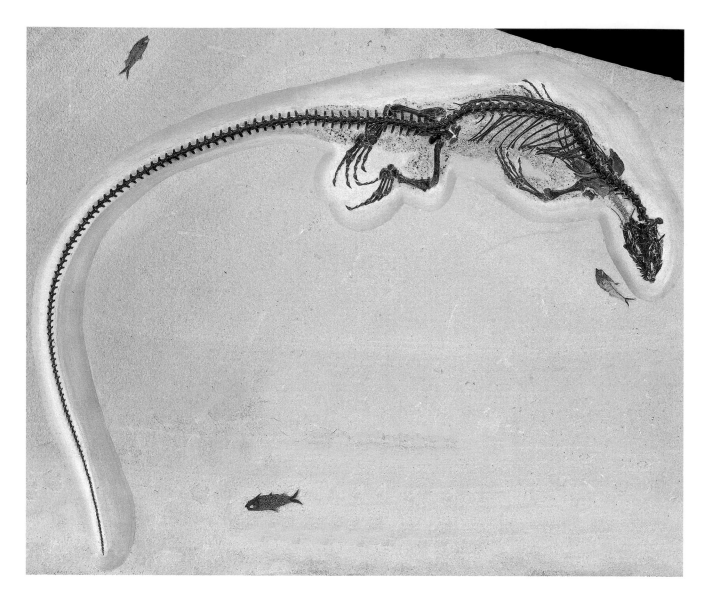

↑ FIGURE 108

†*Saniwa ensidens* Leidy, 1870a, a primitive monitor lizard (family Varanidae). Beautifully preserved and finely prepared complete skeleton with skin impressions preserved. Specimen is 1,310 millimeters (4.3 feet) long, from the sandwich beds of FBM Locality H (FMNH PR2378). Fishes on the same slab are †*Knightia eocaena*.

⤏ FIGURE 109

†*Saniwa ensidens* Leidy, 1870a. *Top:* Close-up of head region from the specimen in fig. 108. *Bottom:* A second specimen with a partly disarticulated skull, from the sandwich beds of FBM Locality K. Specimen measures 1,320 millimeters (4.3 feet) in length (BHI 1285).

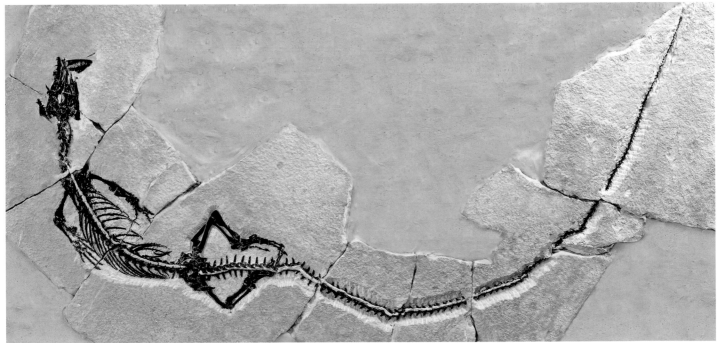

Leidy (1870) from the middle Eocene Bridger Formation of Sweetwater County, Wyoming, based on a partial skeleton.

Today there are about 50 living species of varanids, including the largest living lizard, the Komodo dragon. Varanids live primarily in tropical rainforests and freshwater swamps. They are mostly strict carnivores that feed on other small vertebrates including fishes, but at least one living species is known to consume fruit. Varanids are thought to be among the most intelligent lizards. One of the FBM †*Saniwa* specimens has the skin and neck cartilages preserved (figs. 108, 109, *top*) and appears to have swallowed one of its own teeth. †*Saniwa ensidens* is described in detail in Rieppel and Grande (2007) based on complete skeletons from the FBM. The evolutionary relationships are discussed at length in Conrad, Rieppel, and Grande (2008). The Fossil Butte material indicates that †*Saniwa ensidens* is a primitive species of the family that is key to understanding the post-Eocene diversification of monitor lizards.

Lizards in the family Shinisauridae are known as the "crocodile lizards" (although they are lizards and not crocodiles). In 2006 the first fossil shinisaurid lizard from anywhere, †*Bahndwivici ammoskius*, was described from the FBM, based on a complete skeleton (fig. 110, *top*). Since its description, two additional specimens have been found (one illustrated in fig. 110, *bottom*). All three specimens are from the nearshore Thompson Ranch sandwich beds (FBM Locality H). The second specimen is a beautifully articulated skeleton for which the head was unfortunately not collected.

Shinisaurid lizards are today represented by a single living species in the Guizhou provinces of China, *Shinisaurus crocodilurus*. The living species carries the common name "Chinese crocodile lizard." Its discovery in the FBM is another remnant of the land connection of Asia to western North America during the Cretaceous. There are no known shinisaurids in North America after the early Eocene. Crocodile lizards are freshwater aquatic lizards that spend much of their time in shallow water or in overhanging branches. The Chinese crocodile lizard is a carnivore, feeding on snails, insects, and fishes, and is an endangered species. A 2008 study estimated that there are fewer than 1,000 crocodile lizards left in the wild. The FBM species, †*Bahndwivici ammoskius*, is described in detail in Conrad (2006).

There is also an iguanid lizard described from the FBM, †*Afairiguana avius* (fig. 111). This iguanid is an anole (family Polychrotidae) and is so far the only known specimen of this species. It comes from the nearshore sandwich beds of Warfield Springs (FBM Locality K). This is currently the earliest known nearly complete skeleton of an iguanid fossil from the Americas.

Anoles today have about 380 living species that can be found throughout the Caribbean, in the southeastern United States, Central America, and the northern region of the South American continent. Anoles are mostly small lizards that feed on insects, spiders, and other invertebrates. They are subtropical to tropical, and terrestrial, living in vegetation. The FBM anole must have fallen into the water

FIGURE 110

†*Bahndwivici ammoskius* Conrad, 2006, a semi-aquatic lizard in the family Shinisauridae. Both specimens are from the sandwich beds of FBM Locality H. *Top:* HOLOTYPE, a complete skeleton (FMNH PR2260) with a total length of 330 millimeters (1.1 feet). *Bottom:* A second skeleton that is missing only the skull, currently in a private collection. Length of body is 230 millimeters (9.1 inches).

FIGURE 111

†Afairiguana avius Conrad, Rieppel, and Grande, 2007, an iguanian lizard from the family Polychrotidae. Specimen is the HOLOTYPE for the species and measures 98 millimeters (3.9 inches) long; from the sandwich beds of FBM Locality K (FMNH PR2379).

accidentally and drowned. *†Afairiguana avius* is described in detail in Conrad, Rieppel, and Grande (2007).

Crocodilians (Infraclass Archosauria, Superorder Crocodylomorpha)

Crocodilians are represented in the FBM by at least two forms. One is a primitive crocodile called *†Borealosuchus wilsoni* (fig. 112), and the other is an alligator or caiman called *†Tsoabichi greenriverensis* (fig. 113). The most common crocodilian fossils in the FBM are isolated teeth. Unlike mammals, crocodilians shed and replace their teeth throughout life. Juvenile crocodilians can replace their teeth as often as one tooth per socket each month. They also frequently break off teeth in the process of feeding. Complete skeletons are much rarer, with about a dozen complete skeletons or skulls known from the FBM.

FIGURE 112

†*Borealosuchus wilsoni* (Mook, 1959), a primitive crocodilian of the family †Borealsuchidae, new family. *Top:* A large nearly complete skeleton from the 18-inch layer of FBM Locality A on a slab with two small fishes (†*Knightia eocaena* and †*Diplomystus dentatus*). Crocodilian skeleton is about 4,000 millimeters (13 feet) long (FMNH PR1674). *Center left:* Field photograph of a skull in the 18-inch layer of FBM Locality G, unprepared and as found in the quarry floor. Skull length is 430 millimeters (1.4 feet) long (UW 20531). *Center right:* Specimen from left prepared completely out of the limestone matrix, in dorsal view. *Bottom left:* Same prepared skull in lateral view illustrating the degree of dorsal compression present. *Bottom right:* Same prepared skull in ventral view.

FIGURE 113

†*Tsoabichi greenriverensis* Brochu, 2010, a species of caiman in the family Alligatoridae. *Top:* Specimen is 620 millimeters (2 feet) long on a slab with three †*Priscacara liops,* from the 18-inch layer of FBM Locality C. Private collection; photo courtesy of Ron Mjos. Cast is in the Field Museum (FMNH PR1793). *Bottom:* Specimen is 750 millimeters (2.5 feet) long, from the sandwich beds of FBM Locality C (GAAM-1). Cast is in the Field Museum (FMNH PR3050).

Crocodilians first appear in the fossil record about 84 million years ago in the Late Cretaceous. Today there are about 23 living species. Crocodilians are more closely related to birds and dinosaurs than they are to lizards, which is counterintuitive to some people, because birds have become so specialized that they do not superficially resemble crocodilians. Nevertheless, birds are a form of reptile because they are the only dinosaurian lineage to have survived the mass extinction at the end of the Cretaceous. Because birds are so diversely represented in the FBM, they will have a section of their own immediately following this section on crocodilians.

†*Borealosuchus wilsoni* (formerly in the genus †*Leidyosuchus*) is a primitive basal member of Crocodylomorpha—neither an alligator nor a true crocodile (i.e., not in the family Crocodylidae). It instead forms a sister group to a group containing both alligators and crocodiles. †*Borealosuchus wilsoni* is known by several nearly complete skeletons and at least one skull in the FBM, mostly from the 18-inch layer (e.g., fig. 112). This species was first described by Mook (1959) from a crushed skull from Lake Gosiute (Laney Member?) beds near Wamsutter, Wyoming. The FBM material remains largely undescribed, although it is referred to by Brochu (1997). †*Borealosuchus wilsoni* was the largest of the FBM crocodilians by far, with known specimens up to 4 meters (13 feet) in length. This species was the top predator in Eocene Fossil Lake, feeding on turtles, fishes, and probably any other sizable animals it could catch in its strong toothy jaws. I have seen several turtles with partly healed crocodilian tooth marks in their carapace, indicating a successful getaway (e.g., fig. 104). The largest crocodilians are found in the mid-lake, 18-inch layer deposits (e.g., FBM Localities A, B, and G). Large crocodilian coprolites and isolated teeth are also relatively abundant in the mid-lake deposits (e.g., fig. 208, *top*). Crocodilians are an excellent indicator of the subtropical or tropical conditions that existed in Fossil Lake. For additional information on †*Borealosuchus wilsoni,* see Brochu (1997).

The other known crocodilian from the FBM is a small caiman alligator (family Alligatoridae, subfamily Caimaninae). It was described and named by Brochu (2010) as †*Tsoabichi greenriverensis*. The referred specimens in the type description included a cast of the specimen in figure 113 (*top*). Since that description, another FBM specimen has come to light (fig. 113, *bottom*). This species is slightly less common than †*Borealosuchus wilsoni*. One of the two known specimens is from a mid-lake 18-inch layer deposit, and one is from a nearshore sandwich bed.

Caiman are represented today by six living species. They live in slow-moving streams, lakes, ponds, and flooded wetlands. Small caiman feed on insects, amphibians, crayfishes, shrimps, and birds. They are today native to freshwaters of Central and South America, mostly in tropical or subtropical regions. The earliest known fossil caiman are from the Upper Paleocene of South America. The family Alligatoridae (alligators plus caimans) are today restricted to the Americas and eastern China, although fossil alligatorid material is also known from Europe. For a more detailed description of †*Tsoabichi greenriverensis*, see Brochu (2010).

Birds

*(Class Reptilia,
Superorder Aves)*

Birds are not the first thing one thinks of when the word "reptile" comes to mind. But as the kangaroo and the giraffe are part of the mammalian lineage, so the bird and the crocodile are part of the reptilian lineage. In fact, crocodiles are currently thought to be more closely related to birds than they are to lizards. Birds comprise a highly successful branch of dinosaurs that survived the Late Cretaceous mass extinction event, flourished, and diversified to more than 10,000 species living today. In the Mesozoic, there were still **PROTO-FEATHERED** tyrannosaurid dinosaurs and toothed birds. Today all surviving birds are toothless, and birds have become the most species-rich group of land vertebrates. Many are strong fliers, while others are primarily ground birds or even flightless. Some live in trees or shrubs, while others are aquatic. Birds play a critical role as pollinators, seed dispersers, prey, and predators.

Recently there have been significant efforts to incorporate more of the FBM birds into the evolutionary tree of modern birds, and it has been both enlightening and very challenging. Many of the identifications here are provisional until more in-depth studies can be undertaken. Much of the challenge is due to a lack of knowledge about the evolution of modern bird skeletons and also to the current lack of adequate descriptions for so many of the FBM bird species. Nevertheless, it appears that although the FBM is most famous for its fish fossils, the diversity of bird fossils there exceeds even that of the fishes by a considerable margin (see appendices B and C). There are 34 species (several yet undescribed) belonging to at least 27 different families reported and illustrated here. In addition, there are several other undescribed

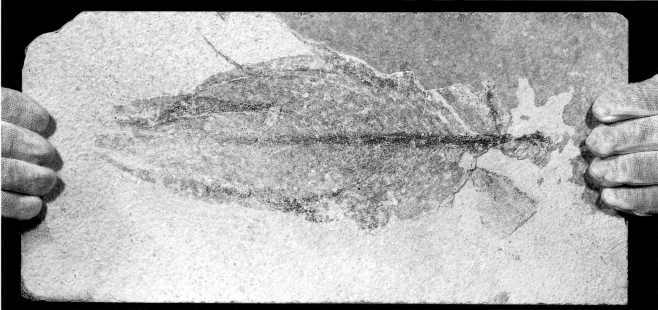

Bird feathers from several FBM localities. *Top left:* Specimen measuring 52 millimeters (2 inches) in length, from the 18-inch layer of FBM Locality A (FMNH PA727). *Top middle:* Two feathers separated by a substantial period of time (11 laminar couplets probably representing several years) on the same small slab—a rare coincidence. Specimen is from the sandwich beds of FBM Locality L (FMNH PA785). Smaller feather is 40 millimeters (1.6 inches) long. *Top right:* Specimen from the 18-inch layer of FBM Locality G; feather is 70 millimeters (2.8 inches) long (FMNH PA784). *Bottom:* Large feather measuring 240 millimeters (9.4 inches) long from an unknown giant bird, possibly †*Gastornis* (Galloanserae). Specimen is from just below the K-spar tuff beds at FBM Locality A (FOBU 13445A).

species currently represented in private collections that I did not have access to for this book. The FBM contains some of the best fossil bird localities in the world, and we are only beginning to understand its scope. It clearly documents early diversification of the major lineages of living birds after the Late Cretaceous mass extinction. The presence of birds such as jacamars, mousebirds, and several species of parrot in the FBM are suggestive of a subtropical forested environment.

Isolated feathers are the most common of all bird fossils in the FBM. Some are intriguing because of their large size, indicating a giant bird lived near Fossil Lake that we have yet to find as a skeleton (fig. 114, *bottom*). This could possibly be a feather from the giant ground bird †*Gastornis* (also known as †*Diatryma*) found in other early and middle Eocene fossil localities in Wyoming. This giant bird stood over 1.8 meters (6 feet) in height and possessed powerful legs, large taloned feet, a massive head, and a fearsome beak several times larger than that of an ostrich. These huge birds are thought by some paleontologists to have been carnivores that fed on mammals and other vertebrates. Other paleontologists contend that they may have been herbivores that fed on large seeds, twigs, and other plant materials, although the large fearsome beak seems excessive for a strict herbivore. If †*Gastornis* was indeed a predator, it would have been at the top of the terrestrial food chain around Fossil Lake, because large predatory mammals had not yet evolved. These avian giants are currently thought to be closely related to Galloanserae (fowl) and evolved from birds that could fly. No bones of †*Gastornis* have yet been reported from the FBM.

The FBM bird skeletons are occasionally preserved with feathers, particularly in the mid-lake deposits (e.g., figs. 130, 143B, and frontispiece), but preparation of these feathers requires extreme meticulousness if they do not split out naturally when excavated. Occasionally, commercial preparators will enhance the damaged feathers with paint, which is easy to detect under magnification (e.g., fig. 131, *left*). The birds in figures 130 and 143B each took several months of full-time preparation under a microscope to properly reveal the feathers without requiring any restoration.

In the FBM, bird skulls and relatively complete articulated bird skeletons make up about .001 percent of the vertebrate fossils being excavated (about 1 in 10,000 vertebrates). Because there have been millions of vertebrate specimens mined from the FBM to date, there are well over 200 bird skulls and/or relatively complete articulated skeletons that have been discovered, some in museums and some in private collections. In addition, there are many smaller fragments of birds (e.g., isolated feet, wings, or other body parts) that are the result of postmortem disarticulation in water or possibly the remnants of a predator's meal.

The most common bird species in the FBM appear to be fish-eating, wading, and otherwise aquatic birds (e.g., frigatebirds and †presbyornithids), rail-like forms (e.g., †messelornithids), PALAEOGNATHOUS BIRDS (†lithornithids), and roller-like birds (†primobucconids); but the greatest amount of diversity appears to be among the small aerial species. However, because the sample size of birds studied from the FBM is still relatively small compared to that of the fishes, there is a great deal of diversity remaining to be discovered and described in the FBM bird fauna. The classification tree for living orders of birds that I follow here is after Hackett et al. (2008). The FBM birds presented below, in context with living relatives, are as follows:

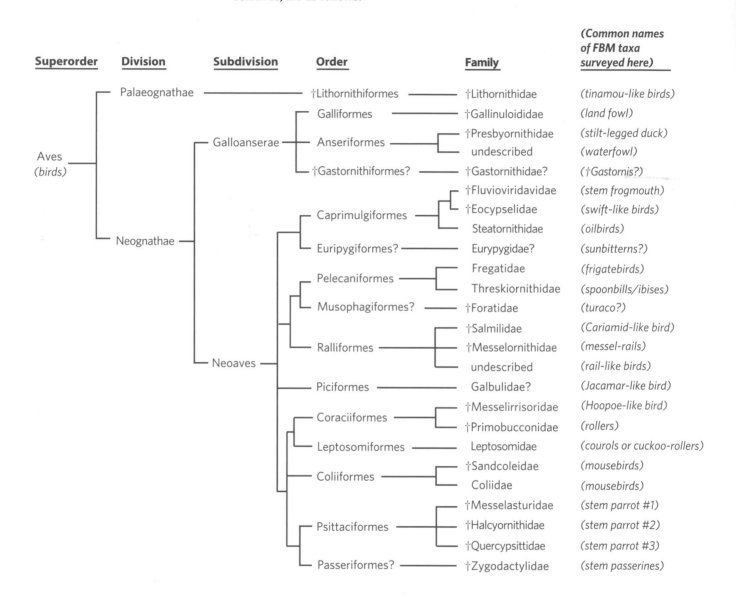

Superorder	Division	Subdivision	Order	Family	*(Common names of FBM taxa surveyed here)*
	Palaeognathae		†Lithornithiformes	†Lithornithidae	*(tinamou-like birds)*
		Galloanserae	Galliformes	†Gallinuloididae	*(land fowl)*
			Anseriformes	†Presbyornithidae	*(stilt-legged duck)*
				undescribed	*(waterfowl)*
			†Gastornithiformes?	†Gastornithidae?	*(†Gastornis?)*
Aves *(birds)*			Caprimulgiformes	†Fluvioviridavidae	*(stem frogmouth)*
				†Eocypselidae	*(swift-like birds)*
				Steatornithidae	*(oilbirds)*
			Euripygiformes?	Eurypygidae?	*(sunbitterns?)*
	Neognathae		Pelecaniformes	Fregatidae	*(frigatebirds)*
				Threskiornithidae	*(spoonbills/ibises)*
			Musophagiformes?	†Foratidae	*(turaco?)*
			Ralliformes	†Salmilidae	*(Cariamid-like bird)*
				†Messelornithidae	*(messel-rails)*
				undescribed	*(rail-like birds)*
		Neoaves	Piciformes	Galbulidae?	*(Jacamar-like bird)*
			Coraciiformes	†Messelirrisoridae	*(Hoopoe-like bird)*
				†Primobucconidae	*(rollers)*
			Leptosomiformes	Leptosomidae	*(courols or cuckoo-rollers)*
			Coliiformes	†Sandcoleidae	*(mousebirds)*
				Coliidae	*(mousebirds)*
			Psittaciformes	†Messelasturidae	*(stem parrot #1)*
				†Halcyornithidae	*(stem parrot #2)*
				†Quercypsittidae	*(stem parrot #3)*
			Passeriformes?	†Zygodactylidae	*(stem passerines)*

Tinamou-like Birds (Order †Lithornithiformes, Family †Lithornithidae)

This extinct order and family of birds contains the only members of the Palaeognathae in the FBM. Today the Palaeognathae, named for their distinctive palate morphology, include ostrich, kiwis, rheas, cassowaries, emus, and tinamous. None of these groups occurs in North America today. The †lithorniforms superficially

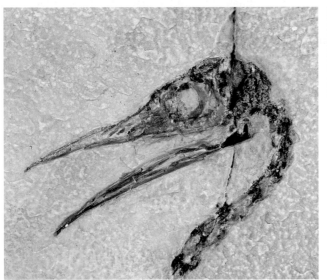
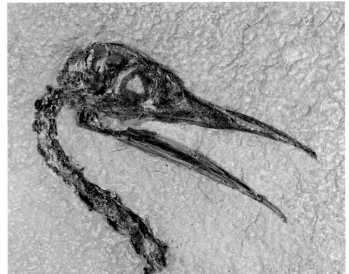

FIGURE 115

†*Pseudocrypturus cercanaxius* Houde, 1988, from the extinct tinamou-like family †Lithornithidae; from the FBM sandwich beds. *Top:* Casts of the part and counterpart of "split-skull" from FBM Locality H before the two sides were glued together and prepared as a single piece (casts of original split slab are both numbered FOBU 11712). Head length is 97 millimeters (3.8 inches). These two splits were later glued together and prepared from both sides (see fig. 116). *Bottom:* A complete skeleton from FBM Locality M. Original specimen WDC CGR-108 with a head length of 99 millimeters (3.9 inches). A cast of this specimen is FMNH PA732.

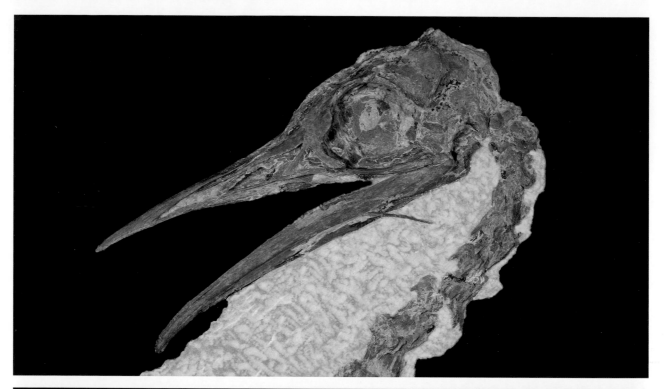

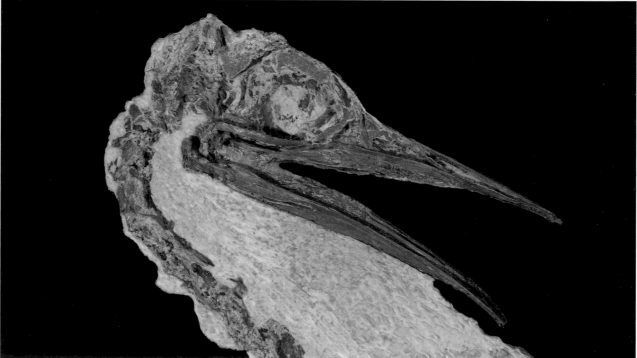

FIGURE 116

†*Pseudocrypturus cercanaxius* Houde, 1988, HOLOTYPE, prepared by gluing the two split skull pieces together (see fig. 115, *top*). Once the two sides were reconnected to each other, the skull was acid-prepared from both sides (UNSM 336103). Skull length is 98 millimeters (3.9 inches). *Top:* Left side. *Bottom:* Right side.

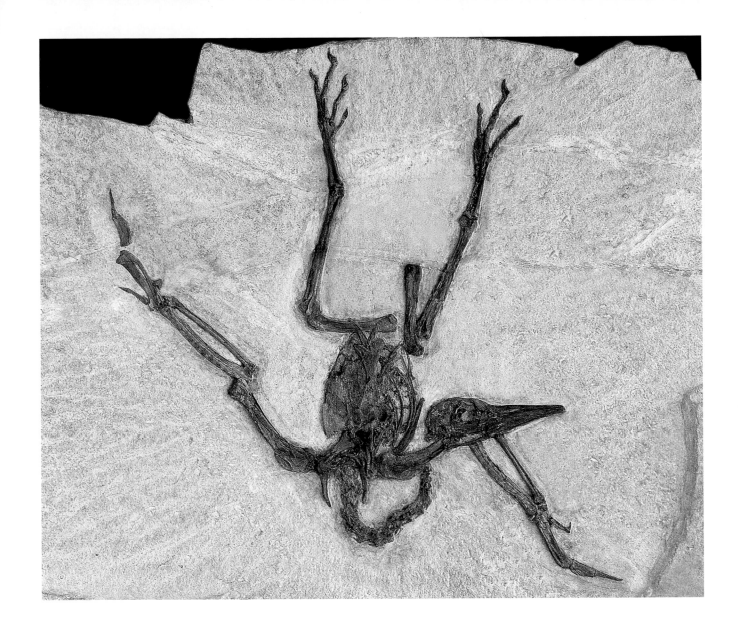

FIGURE 117

†*Pseudocrypturus* sp., an undescribed tinamou-like species of †Lithornithidae. Beautifully preserved complete skeleton, from the sandwich beds of FBM Locality H, with a skull length of 98 millimeters (3.9 inches) (SMA 0186).

resemble tinamou, a neotropical group of the Americas. They are so far known by seven species from the late Paleocene to mid-Eocene of North America, and one species from the early Eocene of Europe. †Lithornithid species are known as early as late Paleocene, but appear to have become extinct by the late middle Eocene. In the FBM, there is one described species of †lithornithid, †*Pseudocrypturus cercanaxius* (figs. 115, 116). The holotype for this species is a skull and neck prepared from both sides (fig. 116), which was made after gluing two sides of a Thompson Ranch "split-skull" together (fig. 115, *top*). Later a complete skeleton of this species was discovered (e.g., fig. 115, *bottom*). †*Pseudocrypturus cercanaxius* had long legs and a relatively long, narrow beak. It may have used this beak for probing along shorelines and shallow waters for insects and other invertebrates. Its wing structure indicates that in contrast to living tinamous, it was probably more capable of sustained flight, and the foot structure indicates that this bird was also capable of perching (Houde 1988; Mayr 2009). So far this species is known only from the nearshore sandwich beds of Thompson Ranch.

†*Pseudocrypturus cercanaxius* is described in detail in Houde (1988) and Houde and Olson (1981). There is also an undescribed FBM †lithornithid species that may belong to a new genus (fig. 117).

Land Fowl (Order Galliformes, Family †Gallinuloididae)

The first bird species described from the FBM was †*Gallinuloides wyomingensis* in 1900 (fig. 118, *right*). This species is a member of the extinct family †Gallinuloididae and represents the earliest known occurrence of the order Galliformes

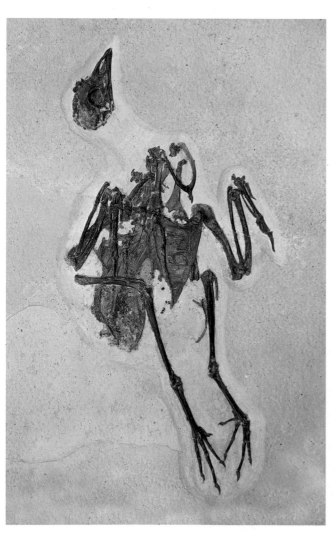

("Pangalliformes," *sensu* Ksepka 2009). Today there are over 250 living species of galliform birds divided into eight families, including pheasants, turkeys, grouse, chickens, quail, partridges, and related forms. Most species are herbivorous, heavy-bodied, ground-feeding birds with short, thick bills adapted for foraging on the ground for fruits, seeds, and other plant materials. Occasionally galliforms will also eat insects. Given that several skeletons of the FBM species are now known (mostly from the nearshore Thompson Ranch quarries), it probably was a mixed feeder foraging along the shore of Fossil Lake. †*Gallinuloides wyomingensis* had a smaller **CROP** than living galliforms, suggesting that the vegetable part of its diet probably consisted more of fruits and other easily digestible plant material rather than seeds (Mayr 2009). Currently, most paleo-ornithologists consider †Gallinuloididae to be basally located within the order Galliformes (Mayr and Weidig 2004; Mayr 2009).

This species occurs in both the nearshore and mid-lake deposits. Because it is a non-aquatic bird that is known by complete articulated skeletons from the mid-lake deposits (e.g., fig. 118, *right*), it must have flown out over the lake on occasion. Individuals may have been killed by volcanic gases from the active volcanoes near the lake at the time.

†*Gallinuloides wyomingensis* is described in detail by Eastman (1900), Lucas (1900), and Mayr and Weidig (2004). The phylogenetic relationships of †*Gallinuloides* are also discussed in Ksepka (2009).

Waterfowl (Order Anseriformes, Family †Presbyornithidae and Family Unknown)

A major group of early Anseriformes present in the FBM, and throughout the Green River Formation as well, is the extinct †Presbyornithidae. This family is represented in the FBM by a yet unidentified species of †*Presbyornis* (fig. 119). This genus has been described as having the body of a flamingo and the head of a duck (Feduccia 1999), which is a good visual description of this animal. In some of the shoreline deposits of the FBM, †*Presbyornis* has been found in mass-mortality beds containing thousands of isolated bones along with eggshell fragments, probably representing large nesting sites of †*Presbyornis* (Leggitt, Buchheim, and Biaggi 1998). In spite of the abundance of isolated bones (a rare mode of preservation within the FBM), no articulated skeletons of †*Presbyornis* have yet been found in the FBM. A composite skeleton of the FBM †*Presbyornis* was constructed from isolated bones at the Utah Museum of Natural History (fig. 119, *left*). Bones from this bird are very abundant in mass-mortality beds of both the upper and lower units of the FBM (Leggitt, Buchheim, and Biaggi 1998).

†Presbyornithids were web-footed filter-feeding birds that fed on planktonic microorganisms along the shoreline with their broad, flat, duck-like bill. Large **ALGAL BLOOMS** within the Green River lake system were probably an important

←— FIGURE 118

†*Gallinuloides wyomingensis* Eastman, 1900, in the extinct land fowl family †Gallinuloididae. *Left:* Specimen from the sandwich beds of FBM Locality H, with a head length of 44 millimeters (1.7 inches) (WDC-CGR 012). *Right:* Specimen from the 18-inch layer of FBM Locality B, with a head length of 48 millimeters (1.9 inches) (MCZ 2221). Specimen is the **HOLOTYPE** for the species. The MCZ specimen (*right*) was originally a nearly perfect, beautifully preserved specimen when it was excavated in the late nineteenth century, but it was then very crudely prepared, probably with sandpaper and knives. That was how most of the FBM fossils were prepared up to the 1950s. Most scientific studies today require much more careful preparation, resulting in maximum morphological detail.

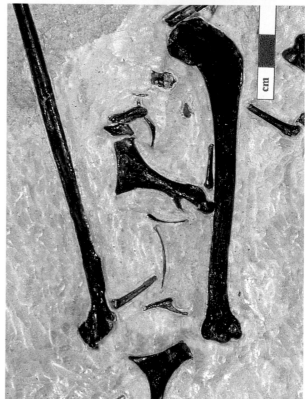

FIGURE 119

†*Presbyornus* sp., an as yet-undescribed long-legged waterfowl species of the extinct family †Presbyornithidae, from shoreline deposits of FBM Locality N. This species is so far known only on the basis of bone beds from Eocene nesting sites. Although no articulated skeletons have yet been found in the FBM, a composite skeleton has been made based on hundreds of isolated bones prepared out of the bone beds. *Left:* A composite skeleton mount made from the bone bed material (UMNH uncataloged). Skull length is 90 millimeters (3.5 inches). *Right top:* A skull 89 millimeters (3.5 inches) long still in matrix from the bone bed (UMNH uncataloged). *Right bottom:* A slab from the bone bed (bones were later prepared out and are in the collection of UMNH). Photos courtesy of Mark Loewen.

FIGURE 120
An undescribed possible waterfowl, family unknown. Specimen is from the sandwich beds of FBM Locality H and has a head length of 75 millimeters (3 inches) (FMNH PA725).

part of its diet (see page 283). It is thought that †*Presbyornis* lived in large colonies along the southern shore of Fossil Lake. Large shoreline nesting sites with bones and eggshells of this bird occur at several levels and localities throughout the Green River Formation. In the Lake Gosiute deposits of the Green River Formation, there is a trackway locality with footprints indicating that †*Presbyornis* had webbed feet (e.g., Grande 1984, 206). Fossils assigned to †Presbyornithidae have also been reported from South America, Asia, and Europe, but some of this material still needs further study to verify its true affinities (Mayr 2009). The earliest proposed †presbyornithid fossils are Late Cretaceous in age, and the family is not known to have survived past the middle Eocene. †Presbyornithids are reviewed and discussed further in Olson and Feduccia (1980) and Mayr (2009).

There is also an anseriform-like bird from FBM Locality H that is currently undescribed (fig. 120). Work on this specimen is currently in progress (Clarke et al.). As discussed above on page 217, an anseriform/galliform-like bird that was likely present around Fossil Lake was †*Gastornis* (known from several other early Eocene fossil localities in Wyoming), but so far in the FBM this giant bird is indicated only by very large feathers (fig. 114, *bottom*).

Anseriforms today are represented by 150 living species, including ducks, geese, swans, and screamers. All species are web-footed forms highly adapted to an aquatic existence. The Anseriformes are thought to be most closely related to the Galliformes (see above). Their origin goes back at least as far as the Late Cretaceous, with the extinct †*Vegavis* (Clarke et al. 2005) from Antarctica (which was much warmer in the Late Cretaceous; Tarduno et al. 2011).

Frogmouths, Swifts, and Oilbirds (Order Caprimulgiformes, Families †Fluvioviridavidae, †Eocypselidae, and Steatornithidae)

The Caprimulgiformes (commonly called "goat-suckers") are an order of birds with a global distribution today, touching every continent except Antarctica. There are several caprimulgiform families present in the FBM, including a probable frogmouth (family †Fluvioviridavidae), swift-like birds (family †Eocypselidae), and oilbirds (family Steatornithidae).

In the FBM, †*Fluvioviridavis platyrhamphus* appears to be allied with the frogmouths (living family Podargidae) but has been placed in its own extinct family †Fluvioviridavidae (fig. 121) (Nesbitt, Ksepka, and Clarke 2011). This is the earliest known species of frogmouth, and the family today has 13 living species found in tropical and subtropical regions of Asia, India, and Australia. The name

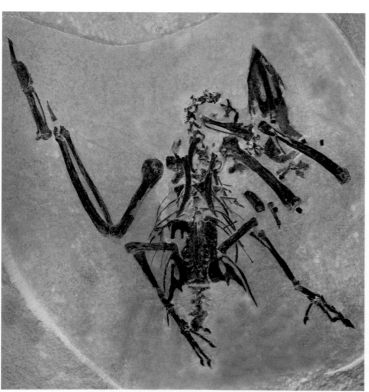

"frogmouth" comes from their large flattened bills with a huge frog-like gape, which they use to catch insects and even small vertebrates. The living species are nocturnal birds and weak fliers that rest on tree branches during the day. †*Fluvioviridavis platyrhamphus* was first described and named on the basis of a nearly complete skeleton from FBM Locality H (fig. 121, *left*), but its affinities were unknown. Later a second specimen was assigned to this species by Nesbitt, Ksepka, and Clarke (2011) (fig. 121, *right*), who also more definitively identified this species as a frogmouth. †*Fluvioviridavis platyrhamphus* clearly has the wide beak and very short legs of living frogmouths and was probably a major predator of insects around Fossil Lake. So far this species is known only from the near-shore Thompson Ranch quarries (FBM Locality H). Another species of the genus †*Fluvioviridavis* has also been reported from the Eocene of Europe. The anatomy of †*Fluvioviridavis platyrhamphus* is described in Mayr and Daniels (2001) and Nesbitt, Ksepka, and Clarke (2011).

Other caprimulgiforms in the FBM include at least one species (probably two or more) of swift-like birds, all yet undescribed and probably belonging in the extinct family †Eocypselidae and possibly the genus †*Eocypselus* (fig. 122, identification by D. Ksepka). This family appears to be closely related to the extant swift family Apodidae although it also shows some hummingbird characteristics (family Trochilidae). The wing anatomy of †*Eocypselus* is intermediate between the short wings of hummingbirds and the hyper-elongated wings of extant swifts, and it shows neither modifications for the continuous gliding of swifts or modifications for hovering flight used by hummingbirds (Ksepka, Clark, Nesbitt, and Grande, submitted). This genus appears to be an early part of a lineage that later diversified into swifts and hummingbirds (Mayr 2009). Long feet also suggest stronger perching capabilities than present in living swifts. These species are some of the smallest birds known from the FBM. Their beaks are short and not modified for drinking nectar like hummingbirds. Modern hummingbirds, like swifts, eat insects for protein, and it seems likely that †*Eocypselus* also fed on insects. Modern swifts are highly adapted to catching insects in flight, and †eocypselids were probably frequent fliers over Fossil Lake in search of food. Most of the known specimens are from the mid-lake deposits, suggesting that they died miles from shore. Several specimens of FBM †eocypselids have been found with the feathers preserved (e.g., fig. 122). The FBM †eocypselids are among the earliest known swift-like birds. The Eocene was evidently the beginning of diversification for this group as for many others, and today there are about 100 living species of swifts and 340 living species of hummingbirds around the world.

The oilbirds are another caprimulgiform family present in the FBM. One nearly complete but disarticulated skeleton from FBM Locality H was described in 1987 by Storrs Olson as †*Prefica nivea* (fig. 123). The oilbird family, Steatornithidae, was monotypic (i.e., it contained only a single living species) until the FBM species was described. The living oilbird, *Steatornis caripensis*, is found in

← FIGURE 121
†*Fluvioviridavis platyrhamphus* Mayr and Daniels, 2001, a stem-frogmouth bird species in the extinct family †Fluvioviridavidae, from the FBM sandwich beds. *Left:* HOLOTYPE specimen from FBM Locality H, with a skull length of 62 millimeters (2.4 inches) (SMNK-PAL.2368). *Right:* Specimen from FBM Locality H, with a skull length of 47 millimeters (1.9 inches) (FMNH PA607).

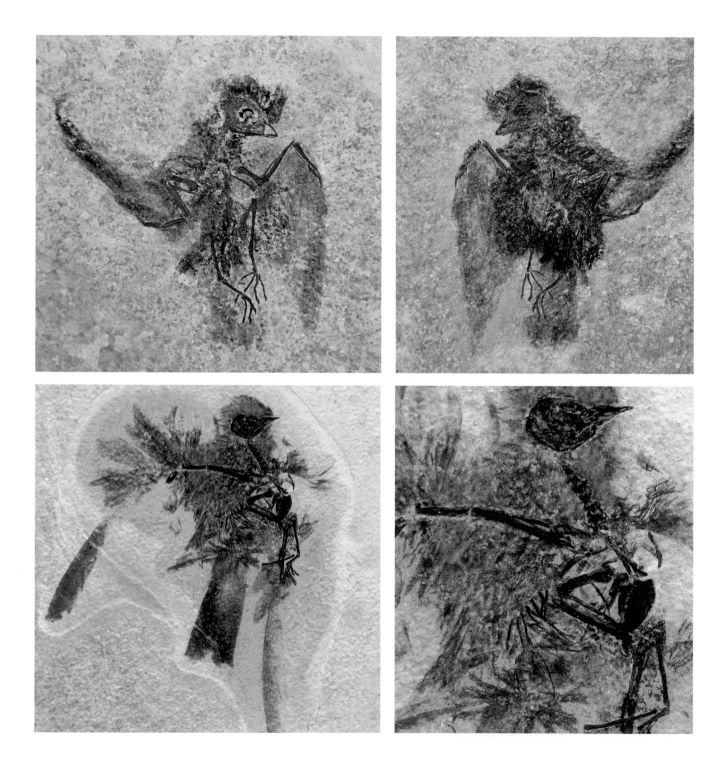

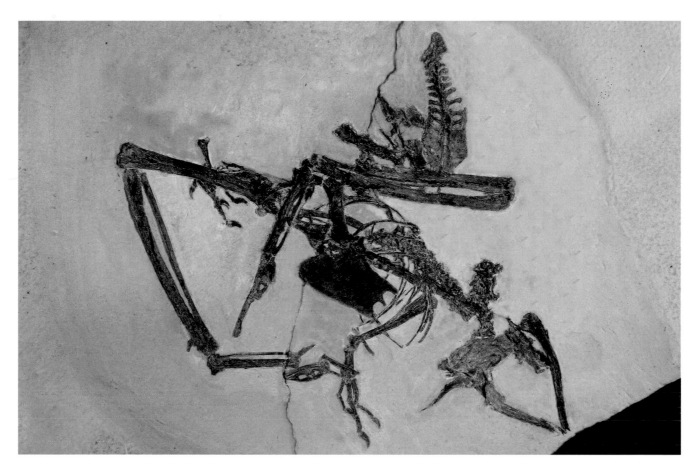

↑ FIGURE 123
†*Prefica nivea* Olson, 1987, an oilbird (family Steatornithidae) from the sandwich beds of FBM Locality H. Specimen is a **HOLOTYPE** (USNM 336278), and the right mandible measures 50 millimeters (2 inches) in length.

← FIGURE 122
†*Eocypselus* sp. nov., a bird somewhat intermediate between swifts and hummingbirds, in the extinct family †Eocypselidae, with feathers preserved. *Top:* Specimen from FBM Locality B, with a head length of 23 millimeters (0.9 inch) (WDC CGR-109). Part and counterpart of a split. *Bottom:* Specimen of another species of swift-like bird from the 18-inch layer at FBM Locality B, with well-preserved long feathers (feathers all natural with no artificial restoration). Head length is 24 millimeters (0.9 inch). Full image with feathers and close-up of skeletal region. Private collection; photo courtesy of Dan Ksepka.

tropical and subtropical regions of northern South America, Central America, and the island of Trinidad. It is a large nocturnal bird that feeds primarily on palm fruits. There were plenty of palms surrounding Fossil Lake for †*Prefica nivea* to have fed on, like its modern counterparts. The feet of †*Prefica* are very small and may have been used to cling to vertical surfaces of trees and cliffs. Living oilbirds roost and nest almost exclusively in caves. Like bats, oilbirds have evolved navigation by echolocation through the use of sharp audible clicks. †*Prefica nivea* is described in detail in Olson (1987).

Sunbittern (Order Eurypygiformes?, Family Eurypygidae?)

There is one species of possible sunbittern from the FBM, which remains undescribed and unnamed, but was first reported as an "undescribed gruiform" in

Grande (1980, fig. III 23). The specimen was later identified as a close relative of the sunbittern by Olson (1989) and Weidig (2003). This identification remains tentative (Clarke, pers. comm., 2011). The comprehensive DNA study of birds by Hackett et al. (2008) determined that sunbitterns are not really gruiforms but are more closely related to Caprimulgiformes. The position of sunbitterns within Aves is still under study by ornithologists.

The FBM specimen is a complete skeleton (fig. 124). It was damaged somewhat during excavation and preparation in the late 1970s when it was found in the southern shoreline quarry of Warfield Springs (FBM Locality K). No additional specimens of this species are known from the FBM.

There is one living species of sunbittern today, *Eurypyga helias*, which lives in tropical and semitropical regions of Brazil to southern Mexico. It inhabits weeded areas near water where it hunts fish and other small vertebrates, much like herons. The FBM species had a relatively long, narrow bill, similar to the living species, and may have been a major predator of †*Knightia* and other small fishes along the shores of Eocene Fossil Lake.

Eurypyga helias is the only living species in the family Eurypygidae, and the FBM species (yet undescribed) is the only known fossil record of the family. The family's closest relative is the kagu, *Rhynochetos jubatus*, which today lives in the Pacific Island chain of New Caledonia (Hackett et al. 2008).

The FBM sunbittern is currently under study.

Frigatebirds (Order Pelecaniformes, Family Fregatidae)

There are two described species of frigatebirds known from the FBM. One is †*Limnofregata azygosternon*, described in 1977 based on a specimen from FBM Locality E, and the other is †*Limnofregata hasegawai*, a larger species based on a specimen from FBM Locality H (see figs. 125, 126). Both species are known from both nearshore and mid-lake FBM localities, and this genus is the most common large bird from the FBM, currently known by over 20 specimens.

The FBM species are the earliest known occurrences of the frigatebird family, which today includes five species that live over tropical and subtropical oceans worldwide. Living frigatebirds are all seabirds, so the FBM species are unusual in that they lived over a freshwater lake. Given that Fossil Lake had a large surface area and was in a subtropical environment, the environmental conditions were similar. Frigatebirds are piscivores (fish eaters). They swoop down over the water and take food while flying by, dipping their long beaks in the water and scooping up surface fishes (Calixto-Albarrán and Osorno 2000). They also harass other piscivores like boobies, terns, and gulls into giving up prey they have caught. No doubt the large schools of †*Knightia* in Eocene Fossil Lake were a primary food source for †*Limnofregata*. Frigatebirds also feed on hatchling turtles, amphibians, and occasionally the chicks of other birds. They are

←··· FIGURE 124
An undescribed species previously proposed as a possible sunbittern (family Eurypygidae) from the sandwich beds of FBM Locality K, with a head length of 71 millimeters (2.8 inches) (USNM 336377). Photo courtesy of Steve Jabo, Hans-Sues, and the USNM.

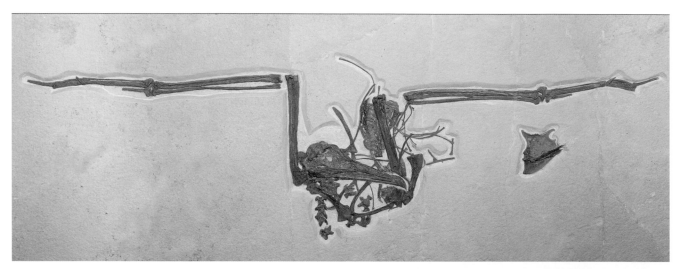

FIGURE 125

†*Limnofregata azygosternon* Olson, 1977, a probable fish-eating species in the family Fregatidae. Frigatebirds are among the most abundant birds in the FBM. *Top:* Complete skeleton with a skeletal wing span of 1,090 millimeters (3.6 feet) and a head length of 140 millimeters (5.5 inches), from the sandwich beds of FBM Locality M (WDC CGR-106). *Bottom left:* HOLOTYPE specimen (USNM 22753) from the 18-inch layer of FBM Locality E, with some feathers preserved (feathers all natural with no artificial restoration) and a head length of 132 millimeters (5.2 inches). *Bottom right:* Specimen from the sandwich beds of FBM Locality H, with a head length of 138 millimeters (5.4 inches) (FMNH PA755).

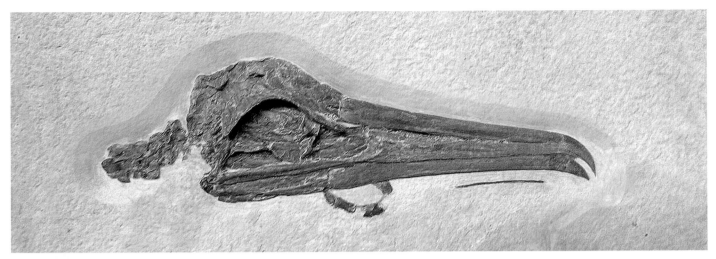

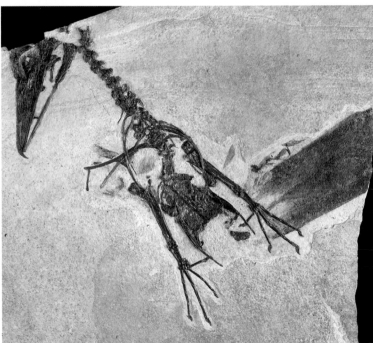

FIGURE 126

†*Limnofregata hasegawai* Olson and Matsuoka, 2005, a second species of frigatebird (family Fregatidae) from the FBM. This species is larger in size than †*L. azygosternon. Top:* Skull with several neck vertebrae. Skull is 153 millimeters and is from the sandwich beds of FBM Locality H. Private collection. *Lower left:* Specimen from the 18-inch layer of FBM Locality E, with feathers preserved and wings missing (feathers all natural with no artificial restoration) (WSGS U1-2001). Lower jaw length is 152 millimeters (6 inches). Extrapolation of lower jaw length indicates that the skull length would have been at least 160 millimeters (6.3 inches). *Lower right:* Head region with lower jaws detached, with a head length of 156 millimeters (6.1 inches) and a lower jaw length of 49 millimeters (1.9 inches). Most of the posterior third of the skull roof was preserved as a natural endocast of missing overlying bone. The surface of the endocast region was colored to more clearly show the entire area of the skull roof. Specimen is **PARATYPE** (FMNH PA719) from the sandwich beds of FBM Locality H.

sometimes called "frigate pelicans" because of a previously presumed relationship, but recent studies show them not to be closely related to pelicans (Hackett et al. 2008). Frigatebirds do not swim, nor do they walk on land very well. They have a huge wingspan-to-body-weight ratio that allows them to stay aloft for days at a time, landing only to roost or breed on trees or cliffs. †*Limnofregata azygosternon* is described in detail in Olson (1977), and †*L. hasegawai*, which is a larger species with a proportionately longer bill, is described in detail in Olson and Matsuoka (2005).

Spoonbills and Ibises
(Order Pelecaniformes, Family Threskiornithidae)

There is one undescribed species of threskiornithid-like bird present in the FBM. It is known so far by only a single headless specimen currently under study by Nate Smith (fig. 127). It was found in the nearshore Thompson Ranch Quarry FBM

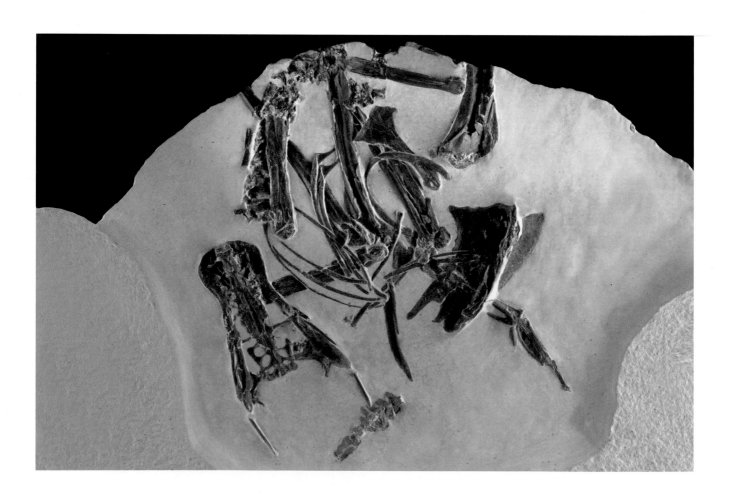

Locality H. This is the oldest known occurrence of the family Threskiornithidae (or STEM GROUP threskiornithid), which includes both spoonbills and ibises. Fossil ibises are also known from the middle Eocene deposits of Messel, Germany, and from other middle Eocene through early Oligocene deposits in Europe, and from the late Eocene of China and the middle Eocene of Myanmar (Mayr 2009). Fossil spoonbills are unknown other than some Pleistocene/Holocene material from Australia.

Today there are six living species of spoonbill and about 30 species of ibises distributed primarily in tropical and subtropical environments of the world. They are long-legged wading birds with either flat spatulate bills (the spoonbills) or long, narrow, curved bills (ibises). They are DIURNAL, spending most of their days in shallow waters searching for food. Ibises probe the mud for insects, worms, and other aquatic organisms to eat, while spoonbills sweep their partly opened bills from side to side in the water until they sense a small fish or other aquatic creature and snap their bills shut to capture the prey. The family is found in both freshwater and saltwater, but they prefer freshwater. Hopefully a skull of this species will soon be discovered to indicate if the FBM species had an ecological niche similar to that of living spoonbills or ibises, or if it instead had a very different and distinct ecology. Current molecular analyses indicate that threskiornithids are closely related to pelicans (Hackett et al. 2008).

Turaco? (Order Musophagiformes?, Family †Foratidae)

There is yet another mystery genus and family of birds from the FBM called †*Foro panarium*. It was described by Olson (1992) based on a beautifully prepared, near-perfect skeleton from FBM Locality H (fig. 128). So far this is the only known specimen of this species. It is currently thought to be a stem turaco (order Musophagiformes) or possibly stem hoatzin (order Opisthocomiformes), but these two orders are not currently thought to be closely related, so further study is needed to fully resolve the relationships of this fossil species.

The body of †*Foro panarium* is somewhat stocky with weakly developed wings, and it is probable that this species was not a good flier. It is a pheasant-size bird and has longer legs than either turacos or hoatzins. Both turacos and hoatzins are highly ARBOREAL birds, feeding on fruits, flowers, and leaves. Both groups are relatively poor fliers and live among the branches of trees and other vegetation in tropical to semitropical regions. †*Foro panarium* would be the earliest representative of either the Musophagiformes or the Opisthocomiformes. There are about 24 living species of turacos, and there is only one living species of hoatzin. †*Foro panarium* is the only known species in the family †Foratidae, and the type specimen of †*Foro panarium*, as well as the PHYLOGENETIC RELATIONSHIPS of †Foratidae, are in much need of further study.

FIGURE 128

†*Foro panarium* Olson, 1992, a turaco-like bird in the extinct family †Foratidae. HOLOTYPE, a complete skeleton from the sandwich beds of FBM Locality H with a skull length of 60 millimeters (2.4 inches) (USNM 336261).

Rails (Order Ralliformes, Families †Messelornithidae, †Salmilidae, and *incertae sedis*)

The rails today comprise a large group of small to medium-size birds commonly associated with wetlands and forested environments. There is only one species of rail-like bird described so far from the FBM, †*Messelornis nearctica* (fig. 129), although more species await description. †*Messelornis* is thought to belong to an extinct family of birds called the "Messel rails" (†Messelornithidae) because the first species was described from the middle Eocene deposits of Messel, Germany. †*Messelornis* is one of the more common birds in the FBM, but well-articulated specimens are rare. The FBM species has also been placed by other paleo-ornithologists as close relatives to living rails plus finfoots (Heliornithidae plus Rallidae; Mayr 2004) or the sunbittern (Eurpygidae; Hesse 1988). The original description for the FBM species was based on a nearly complete skeleton from FBM Locality H (Hesse 1992). Hesse mistakenly dated this specimen as "middle Eocene." There are several partial skeletons of this species in the collection of the Field Museum. They are much more common in the nearshore localities of Thompson Ranch than in the mid-lake quarries, and the one mid-lake specimen I am aware of is disarticulated, suggesting possible postmortem transport. The name †*Messelornis* comes from the fact that the first †messelornithid species came from the middle Eocene deposits of Messel in Germany (†*Messelornis cristata* Hesse, 1988).

Another FBM rail-like bird that appears to link the FBM to the Messel locality of Germany is a new undescribed species tentatively allied with the enigmatic family †Salmilidae (reported from the FBM here for the first time). The specimen is represented by a nearly complete skeleton with feathers preserved (fig. 130). Previously, †Salmilidae was known only by a single species from Messel, Germany. This family of birds was flightless, as indicated by its very small wings. There is also another larger undescribed species of rail-like bird from the FBM awaiting description (fig. 131). The family to which this species belongs has yet to be determined.

Rails were formerly included in an order named Gruiformes, which contained not only rails and cranes but also other families not closely related to rails (Hackett et al. 2008). Subsequently, rails were placed in their own order, Ralliformes, along with cranes, finfoots, and certain fossil taxa (e.g., †messelornithids). The FBM †*Messelornis nearctica* is the earliest known, confidently classified rail in the fossil record. Today there are about 150 living species of rails, and they inhabit every continent except Antarctica. Rails typically occupy damp environments near lakes, swamps or rivers, or dense forests; and reed beds are a favored habitat. Most species are considered weak fliers and prefer to walk or run using their strong legs, and some are flightless during molting season. Some have lost the ability to fly completely. Most species have moderately elongated legs with long toes that are well adapted to soft, uneven surfaces, and some even walk on

⇡ ↗ FIGURE 129

†*Messelornis nearctica* Hesse, 1992, a rail-like bird in the extinct family †Messelornithidae. Two skeletons from FBM Locality H. *Left:* Specimen from the sandwich beds of FBM Locality H; length of each leg and foot is about 195 millimeters (7.7 inches) (BHI 6398). Note the splint-like ossified tendon bones near the top of the long bone (tibiotarsus) of the leg. These are characteristic of this FBM species. *Right:* HOLOTYPE specimen (SMF AV406), a poorly prepared but complete skeleton, with a head length of 26 millimeters (1 inch).

⇢ FIGURE 131

A nearly complete skeleton of another rail-like bird with feathers preserved (although many of the feathers in this commercially prepared specimen have been enhanced or restored with paint). Specimen is from the 18-inch layer of FBM Locality B (WDC CGR-107). The specimen split apart when it was excavated, so pieces that split off were later inlaid back into the main slab. *Left:* Image of entire piece. *Right:* X-ray of specimen on right clearly showing the three pieces that were inlaid back into the main slab.

◄⋯ FIGURE 130
A new species of the flightless bird family †Salmilidae (note the very small wings) with feathers preserved (feathers all natural with no artificial restoration). Specimen is from FBM Locality A and has a head length of 74 millimeters (2.9 inches) (FMNH PA778).

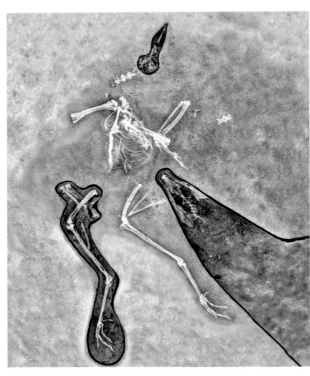

the surfaces of lily pads (e.g., *Nelumbo*, a water plant that occurred in the FBM). The FBM species has both long legs and a fairly long neck but short feet more typical of wood rails. One of the characteristics of Messel rails in the FBM is an abundance of splint-like ossified tendon bones in the upper long bones of the legs (see fig. 129, *left*).

Rails are mostly OMNIVORES, and some eat seeds. The FBM species, †*Messelornis nearctica*, is known by one specimen (currently in a private collection) preserved with GASTROLITHS in the ABDOMINAL REGION. Although †*Messelornis nearctica* is a relatively common bird in the FBM, known by 20 or more specimens, all known specimens are adults. This size distribution of FBM specimens was proposed by Weidig (2010) to indicate that the nesting sites were located inland rather than near the shore of Fossil Lake.

Further description of †*Messelornis nearctica* can be found in Hesse (1992) and Weidig (2010); and the †Messelornithidae are reviewed in Mayr (2004).

Jacamars (Order Piciformes, Family Galbulidae?)

There is one species of jacamar-like bird known from the FBM, †*Neanis kistneri* (fig. 132), although the type specimen in the original description is from the Lake Gosiute Tipton Shale Member of the Green River Formation. Since the original description of the Tipton Shale Member specimen, several specimens of this species have also been discovered in the nearshore Thompson Ranch deposits of the FBM. This species was originally described by Feduccia (1973) and referred to the genus †*"Primobucco,"* but that genus was later found to be a wastebasket group containing species belonging to several different families and orders (e.g., see page 251). Feduccia (1977) later transferred this species to a new genus, †*Neanis,* which was placed in the jacamar family Galbulidae by Houde and Olson (1989). Further studies by Weidig (2010) supported its relationship with that family, although Mayr (2009, 200) questioned this placement.

†*Neanis kistneri* is the earliest known species proposed to be a jacamar. Today this family has 18 species in tropical regions of South America, Central America, and Mexico. Modern jacamars are mainly birds of low-altitude tropical forest edges and canopies. They are insectivores that specialize on moths and butterflies that they catch while on the wing. The largest species also feeds on small lizards and amphibians. The feet of †*N. kistneri* are ZYGODACTYL, which is a condition seen in species with enhanced perching abilities and other FBM birds of uncertain affinity. †*Neanis kistneri* is described and discussed by Feduccia (1973) and by Weidig (2010).

Hoopoe-like Bird (Order Coraciiformes, Family †Messelirrisoridae)

There is a hoopoe-like bird from the FBM that has long been known (originally reported as a kingfisher in Grande 1980, 212) but still remains unnamed and undescribed (fig. 133). There is only a single known specimen of this species so far, and it is from the nearshore FBM Locality H of Thompson Ranch. This species is currently considered to be a member of the extinct hoopoe-like family †Messelirrisoridae, which ranges from early Eocene to early Oligocene, and is known from Europe and North America. The family appears to consist of perching birds, as opposed to the hoopoe (family Upupidae), which inhabits bare ground or vertical surfaces with cavities. Extant hoopoes feed primarily on insects and are widespread in Asia, Africa, and Europe.

A description of the FBM hoopoe-like bird is currently in progress. The family †Messelirrisoridae includes one of the predominant small forest birds of central Europe during the middle Eocene.

← FIGURE 132

†*Neanis kistneri* (Feduccia, 1973), a species previously proposed to be related to jacamars (family GALBULIDAE). Specimen is from the sandwich beds of FBM Locality H and has a head length of 32 millimeters (1.3 inches) (USNM 336268).

FIGURE 133

An undescribed hoopoe-like bird possibly in the order Upupiformes. Specimen from the sandwich beds of FBM Locality H, with a head length of 45 millimeters (1.8 inches) (BHI-GR 207). *Top:* Entire specimen. *Bottom:* Close-up of skull.

Rollers (Order Coraciiformes, Family †Primobucconidae)

One of the more common small birds in the FBM is a species of roller, †*Primobucco mcgrewi* (fig. 134 and frontispiece). Currently, the FBM rollers are in the extinct family †Primobucconidae. Rollers include this family plus two extant families. At one time, †Primobucconidae was a wastebasket group containing most "small birds" from the Green River Formation, but more recently the systematics of the group was improved by removing the stem parrots and jacamars, and leaving only the rollers. †*Primobucco mcgrewi* is represented by over 20 specimens, and there are 10 specimens in the Field Museum's collection alone. This species occurs in both the nearshore and mid-lake deposits, but it is more common near shore. Other species in the family †Primobucconidae occur in the middle Eocene of Germany and the early Eocene of France. The FBM roller is among the oldest known species of the group. Today there are 11 living species of "true rollers" in the family Coraciidae that live in tropical to warm temperate regions of Africa, Europe, Asia, and Australia. They are strong fliers with broad bills, large heads, weak feet, and short legs. They take their prey on the wing, feeding primarily on insects, but also on small amphibians and small snakes. The other extant family of rollers, the "ground rollers" or Brachypteraciidae, are short-winged, long-

FIGURE 134

†*Primobucco mcgrewi* Broadcorb, 1970, a species of roller-like bird in the extinct family †Primobucconidae. This is among the most common bird species in the FBM. Both specimens are from the FBM sandwich beds. *Left:* Specimen from FBM Locality H, with a head length of 39 millimeters (1.5 inches) (FMNH PA724). *Right:* Specimen from FBM Locality H, with a head length of 40 millimeters (1.6 inches) (USNM 336284).

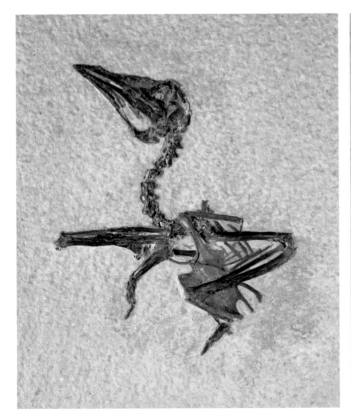

legged birds that nest in holes in the ground, feed on insects and small reptiles, and are endemic to Madagascar. Some †primobucconid species have been found with seeds preserved in the stomach region, indicating a more generalized diet than living rollers (Mayr, Mourer-Chauviré, and Weidig 2004). The beak and head shape is also not modified as in extant rollers and other fossil rollers.

†*Primobucco mcgrewi* from the FBM represents the earliest known occurrence of rollers in the New World, and the last known occurrence of rollers in North America is the broad-headed †*Paracoracias occidentalis* from the middle Eocene Laney Shale Member of the Green River Formation (Clarke et al. 2009). For a detailed description of †*Primobucco mcgrewi*, see Ksepka and Clarke (2010b).

Courols, or "Cuckoo-rollers" (Order Leptosomiformes, Family Leptosomidae)

The family of the courols (also known as the cuckoo-roller) is the only family in the order Leptosomiformes. It is known by two described species from the

FBM: †*Plesiocathartes wyomingensis* (fig. 135) and †*Plesiocathartes major*, both described by Weidig (2006).

†*Plesiocathartes wyomingensis* is known by a single specimen from the sandwich beds in the nearshore FBM Locality H. This specimen is a nearly complete skeleton with the right humerus and skull heavily restored. The wings are relatively shorter and the legs relatively longer than the living species, *Leptosomus discolor*. The second species, †*Plesiocathartes major*, is known by two specimens: a headless skeleton from FBM Locality H and a foot from FBM Locality B, both illustrated in Weidig (2006). The Leptosomidae (sometimes incorrectly spelled Leptosomatidae) is known by several other fossil species from the Eocene of Europe, but the FBM species are the earliest known occurrences of the family and order. By the Oligocene, leptosomids appear to have disappeared from the Northern Hemisphere. Today there remains a single living species of leptosomid, *Leptosomus discolor*, which is a carnivorous, tropical forest-dwelling bird in Madagascar and the Comoro Islands. It has a large head and feeds on insects and small lizards.

The two species of †*Plesiocathartes* from the FBM are described in detail by Weidig (2006), but more skull material is still needed, particularly for †*P. major*, for a more complete description.

Mousebirds (Order Coliiformes, Families †Sandcoleidae and Coliidae)

There are two mousebird species described from the FBM: †*Anneavis anneae* in the extinct family †Sandcoleidae (fig. 136) and †*Celericolius acriala*, which is closely related to the extant family Coliidae (fig. 137). So far, there is only a single known specimen of each species.

The specimen of †*Anneavis anneae* is a complete articulated body skeleton with feathers preserved, missing only the head. It comes from the southern nearshore deposits of FBM Locality K and was originally a complete skeleton, but the head was lost during excavation of the specimen in the field. This species is in the extinct family †Sandcoleidae, which is known from the early and middle Eocene of North America and Europe. The family is thought to be a STEM GROUP of basal Coliiformes, and sister group to the extant family Coliidae. One specimen of †Sandcoleidae from the middle Eocene deposits of Messel, Germany, was discovered with seeds in its stomach region, indicating a diet similar to that of mousebirds living today.

The species of FBM mousebird, †*Celericolius acriala*, is based on a specimen from just above the K-spar tuff bed in FBM Locality A. This specimen was damaged as a "split bird" for which the counterpart was evidently lost, but there is enough of the specimen preserved to indicate it is a mousebird. There are even faint traces of feathers (although not well preserved). This species, like other

←··· FIGURE 135

†*Plesiocanthus wyomingensis* Weidig, 2006, a species of possible courol or cuckoo-roller in the family Leptosomidae. Parts of head and right humerus restored. Specimen is the HOLOTYPE and is from the sandwich beds of FBM Locality H; it has a head length of 75 millimeters (3 inches) (WDC-2001-CGR-021). Photo courtesy of Ilka Weidig.

†*Anneavis anneae* Houde and Olson, 1992, a species of mousebird from the extinct family †Sandcoleidae. Complete body with feathers preserved, missing head and on slab with two †*Knightia eocaena*. Note preservation of feathers in the two wings and the long tail (feathers all natural with no artificial restoration). Slab (BMS E 25337) is from the sandwich beds of FBM Locality K. Width of slab as shown in figure measures 230 millimeters (9.1 inches). Specimen is the HOLOTYPE for the species. Photo courtesy of Richard Laub.

 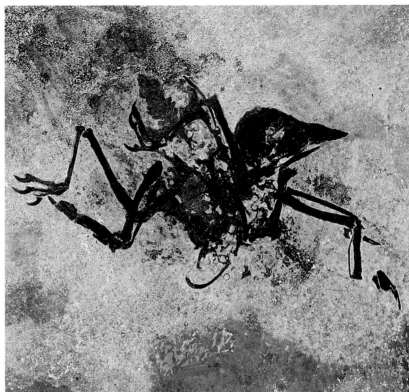

FIGURE 137

†*Celericolius acriala* Ksepka and Clarke, 2010a, a mousebird in the family Coliidae. HOLOTYPE specimen is from just below the K-spar tuff bed in FBM Locality A, with a head length of 36 millimeters (1.4 inches) (FMNH PA730). *Left:* Entire specimen showing carbonized traces of tail and left wing feathers. *Right:* Close-up of skeleton. This specimen was prepared by a commercial fossil company, and some of the bone impressions in the matrix have been colored.

Coliidae and like †Sandcoleidae, had a very long tail. It had more elongate wings than extant mousebirds and may have been a better flier. In its original description by Ksepka and Clarke (2010a), †*Celericolius acriala* was described as being more closely related to Coliidae than to †Sandcoleidae. Therefore, I include this species in Coliidae here. †*Celericolius acriala* is among the oldest known fossils of the family Coliidae. Today there are six living species of mousebirds, all confined to sub-Saharan Africa. These are small birds with soft, loose body feathers, and they are not strong fliers. They are **ARBOREAL**, clambering through leaves and branches like mice in search of berries, fruit, and buds (thus the name "mousebird"). The latest known occurrence of mousebirds in North America was in the late Eocene (Ksepka and Clarke 2009). Mousebirds are considered to be a relic, "living fossil" group because the order, which today contains only six species, was more diverse in the past. There are 19 named fossil species from the Cenozoic of North America, Europe, and Africa.

For a detailed description of †*Anneavis anneae*, see Houde and Olson (1992), and for †*Celericolius acriala*, see Ksepka and Clarke (2010a).

Parrots (Order Psittaciformes, Families †Messelasturidae, †Halcyornithidae, and †Quercypsittidae)

There are at least four primitive parrot species described from the FBM: two in the family †Halcyornithidae (†*Cyrilavis colburnorum* and †*Cyrilavis olsoni*), one in the family †Quercypsittidae (†*Avolatavis tenens*), and a fourth, referred to as the family †Messelasturidae (†*Tynskya eocaena*), which might also be a primitive parrot. These represent some of the earliest known occurrences of parrots in the fossil record and show that the parrot fauna of the FBM was diverse. None of the FBM species have the distinctive beak seen in living parrots.

†*Cyrilavis colburnorum* (family †Halcyornithidae) is a primitive, or "stem," parrot from the FBM (fig. 139). This species is one of the more common varieties, known by at least two complete skeletons, one complete skull with neck vertebrae, and several partial specimens. It is known primarily from FBM Localities C and H. In Locality C, a mid-lake quarry, it occurs primarily in the mass-mortality layer of small †*Knightia*, near the K-spar tuff layer. In the nearshore Thompson Ranch Locality H quarry, it occurs in the sandwich beds. †*Cyrilavis* is in the ex-

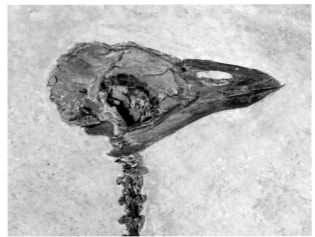

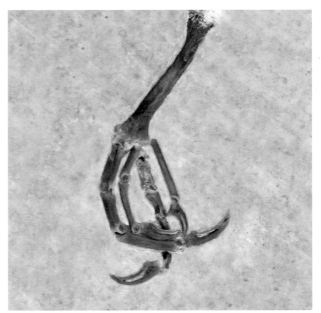

⤒ FIGURE 139

†*Cyrilavis colburnorum* Ksepka, Clarke, and Grande, 2011, a remarkably well-preserved parrot in the extinct family †Halcyornithidae, from the FBM sandwich beds. *Left:* HOLOTYPE (FMNH PA754), a complete skeleton from FBM Locality C, with a head length of 46 millimeters (1.8 inches). *Top right:* Skull and neck vertebrae, from FBM Locality H, with a head length of 43 millimeters (1.7 inches) (FMNH PA722). *Bottom right:* Right foot of FMNH PA766 from FBM Locality H.

⤎ FIGURE 138

†*Tynskya eocaena* Mayr, 2000, a stem parrot from the extinct family †Messelasturidae. HOLOTYPE specimen (BSP 1997 I 6), a split fossil from the sandwich beds of FBM Locality H, with a head length of 42 millimeters (1.7 inches). Photos courtesy of Gerald Mayr and Oliver Rauhut.

⸸ FIGURE 140

†*Cyrilavis olsoni* (Feduccia and Martin, 1976), another species
of primitive parrot from the extinct family †Halcyornithidae.
HOLOTYPE specimen (UAM PV 2005.6.196) with feathers preserved (feathers all natural with no artificial restoration), from
the 18-inch layer of an unknown FBM quarry; head length is
47 millimeters (1.9 inches). Photo courtesy of Dan Ksepka and
Jim Parham.

⤑ FIGURE 141

†*Avolatavis tenens* Ksepka and Clarke (2012), a species of parrot
from the family †Quercypsittidae. HOLOTYPE from FBM
Locality E (UW 39876). Left leg and foot complex measures
110 millimeters (4.3 inches). *Left:* Positive side of specimen.
Right: Negative counterpart with image flipped horizontally to
face same direction as the positive side.

tinct family †Halcyornithidae. In the original description of †*Cyrilavis colburno-rum* by Ksepka, Clarke, and Grande (2011), the species †*Cyrilavis olsoni* (fig. 140) was also shown to belong in †Halcyornithidae. This species is also from the FBM and was originally mistakenly described in the genus †*Primobucco*. (†Primobuc-conids, including the true species of the genus †*Primobucco*, were later found to be rollers; see above). †*Cyrilavis colburnorum* and †*C. olsoni* are described in detail in Ksepka, Clarke, and Grande (2011).

The most recently described species of parrot from the FBM is †*Avolatavis tenens*. This species appears to belong in (or is closely related to) the family †Quercypsittidae and is represented in the FBM by a single articulated skeleton missing the head and wings (fig. 141). The foot structure indicates that this was an **ARBOREAL** bird adapted to climbing and grasping. †*Avolatavis tenens* is described in detail in Ksepka and Clarke (2012), although the head morphology remains unknown.

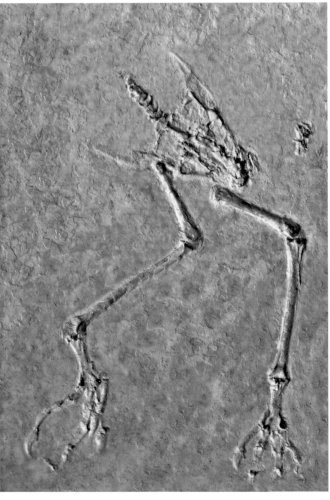

Somewhat more problematic is †*Tynskya eocaena*, which is so far known by only a single specimen from the FBM (fig. 138). It is a "split-bird," from the near-shore FBM Locality H of Thompson Ranch. It was first described as a "raptor-like bird" in 2000 by Mayr, but he later referred this species to the extinct family †*Messelasturidae*. †Messelasturids are currently thought to possibly be primitive parrots, although this is still controversial (Clarke, pers. comm., 2012). This family is also known from the middle Eocene of Germany and the early Eocene of England. This species is unusual among parrots in having a hooked beak and raptorial claws, suggesting that it was a bird of prey. †*Tynskya eocaena* is described and discussed in detail in Mayr (2000, 2005, 2009).

The FBM provides the earliest look at well-preserved parrots in the fossil record. It appears that there was a diversity of parrot species present around Eocene Fossil Lake.

Today there are nearly 400 living species of parrots, which live primarily in tropical to subtropical regions of the Caribbean and Pacific Islands, South America, Central America, southern North America, southern Asia, Australia, and Africa.

Living parrots are characterized by their strong, broad, curved bill, which allows them to exert tremendous biting pressure in order to feed on nuts and large seeds. The diet of parrots consists primarily of seeds, nuts, fruit, and sometimes insects; but seeds and nuts are the most important food source for most parrots. None of the FBM parrots, or other described Eocene parrots, show the modified beak of living parrots, perhaps suggesting a less specialized diet.

Passerine-like Birds
(Order Passeriformes?, Family †Zygodactylidae)

There are at least two species of the extinct family †Zygodactylidae present in the FBM. One is †*Eozygodactylus americanus* (fig. 142A), and the other is a yet undescribed †zygodactylid species (fig. 142B). These are small delicate perching birds that use a different foot structure for perching than living passerines. †Zygodactylids have a ZYGODACTYL foot (hence the family name), with two toes in front and two in back. In contrast, living passerines have three toes in front and one in back. The FBM †zygodactylids are among the earliest known members of the family, and the last known members are from the middle Miocene of France. Other described species of †Zygodactylidae are from Eocene, Oligocene, and Miocene deposits of Europe (Mayr 2009).

†Zygodactylid birds may be stem passerines (Mayr 2008), although this is controversial and in need of further study. The †zygodactylids may have been the perching songbirds inhabiting the trees and hills surrounding Eocene Fossil Lake. Passerines today include half of all living bird species, with over 5,000 described species inhabiting nearly all regions of the world. The zygodactylous

†*Eozygodactylus americanus* Weidig, 2010, a species of stem passerine bird ("songbird") from the extinct family †Zygodactylidae. Beautiful complete skeleton from the sandwich beds of FBM Locality H, with a head length of 34 millimeters (1.3 inches) (FMNH PA726).

An undescribed species of stem passerine bird from the extinct family †Zygodactylidae. Two examples from the sandwich beds of FBM Locality H. *Left:* Small specimen with a head length of 32 millimeters (1.3 inches) (NAMAL 2000-0217-004). Photograph taken through glass of an exhibit case and is courtesy of Gary Hyatt. *Right:* Small specimen with a height of 60 millimeters (2.4 inches), still unprepared under a thin layer of matrix (BHI 1285). Photo courtesy of BHI and Tim Larson.

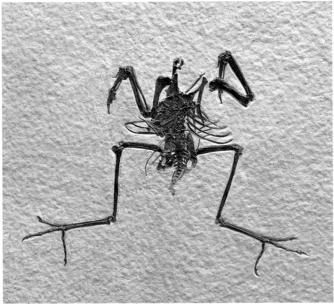

method of perching may have made †Zygodactylidae more adapted for clinging to tree trunks or other near-vertical surfaces than modern passerines. The FBM †zygodactylids have a delicate-looking bill structure and probably fed on insects, although †zygodactylids from middle Eocene deposits of Messel, Germany, have been found with seeds in their intestinal area.

For more detailed morphological description of †*Eozygodactylus americanus*, see Weidig (2010). For additional discussion of †Zygodactylidae, see Mayr (2009). The second FBM species (first reported by Weidig 2010) is being described (Clarke, pers. comm., 2012), but a few details are given in Weidig (2010, 39–40).

Other Bird Skulls and Skeletons Currently of Uncertain Identification

There are a number other FBM bird skeletons in museum collections representing additional taxa not mentioned above. Some of these are illustrated here (fig. 143A).

There is a skull from the FBM that clearly shows a sharply hooked beak resembling that of a raptor (fig. 143A, *top left*). Sometimes the order "Falconiformes" has been used for these birds, but this group was later found to be an artificial group (i.e., NON-MONOPHYLETIC) with two independent evolutionary origins. Identification of this skull as a raptor is very provisional at this time. Much work is yet needed to better resolve the systematics of raptors, and this skull along with additional material is currently being studied (Julia Clarke, pers. comm., 2011).

There is a new species of "long beak" bird known from the FBM whose family relationships are still unknown (fig. 143A, *top right* and *bottom right*). Although there is a superficial resemblance to †*Pumiliornis* from Messel, that relationship is tenuous at best. This species is currently under study by Clarke et al. (in progress).

There is a beautifully preserved bird specimen recently discovered in the FBM that the Field Museum received totally unprepared with only a thin layer of limestone covering the fossil. Museum preparators spent months doing a careful job of preparation on the specimen and found nearly all the feathers still in articulation with the skeleton (fig. 143B). This specimen appears to belong to a species that is closely related to another fossil species from the early Eocene Fur Formation of Denmark, †*Morsoravis sedilis*. The FBM specimen may even belong in a new family together with the Denmark species (Daniel Ksepka, pers. comm., 2011). In the original description of †*M. sedilis* by Bertelli et al. (2010), the genus was left largely unclassified as a "charadriiform-like species." Other ornithologists (Mayr 2009) later expressed doubt of that ordinal assignment. The FBM specimen may shed some new light on the relationships and classification of this group once the fossil has been described in detail (work in progress).

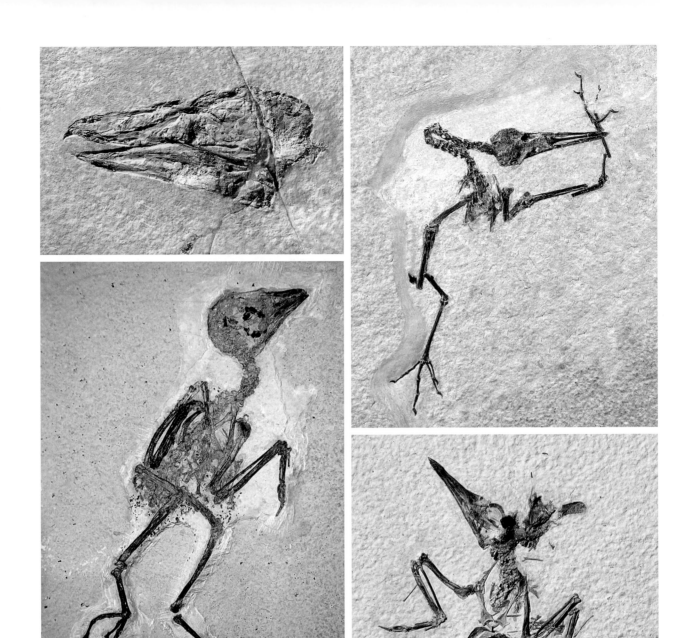

FIGURE 143A

Various bird skulls and skeletons from the sandwich beds of FBM Locality H in need of further study and/or material in order to be adequately identified. *Top left:* Raptor (?) skull of undetermined order or family, with a head length of 55 millimeters (2.2 inches) (FMNH PA777). The identification of this poorly preserved skull as a raptor is only provisional, and additional, better-preserved material is needed to describe and verify taxonomic placement of this species. *Bottom left:* A small ZYGODACTYL bird of unknown affinity measuring 144 millimeters (5.7 inches) in height (TMP 1983.17.2). Possibly another new species of parrot. *Top right:* An undescribed new species of "long-beak" bird whose family relationships are yet unknown. Specimen has a head length of 44 millimeters (1.7 inches) (FMNH PA728). *Bottom right:* Another specimen of the "long-beak" bird shown above, with a head length of 42 millimeters (1.7 inches) (FMNH PA756).

FIGURE 143B
A beautifully preserved bird skeleton of uncertain relationship. There is no restoration on this piece, with remarkable preservation of the feathers in articulation with the bones. Although this species remains undescribed, it is similar to the species †*Morsoravis sedilis* (unknown order and family) from the early Eocene of Denmark. Specimen is from the sandwich beds of FBM Locality L, with a head length of 27 millimeters (1.1 inches) (FMNH PA789). Description in progress.

What Is Most Obviously Missing from the FBM Bird Fauna?

The order Passeriniformes, or the "passerines" as they are commonly called, make up well over one-half of all the living bird species, including 110 families and over 5,000 described species. The group is distributed all over the world today, from the tropics to the arctic and well into the Southern Hemisphere. But among the diverse assemblage of birds in the FBM, not a single undisputed passerine has yet been discovered. Some authors have suggested that the FBM †Zygodactylidae might be a primitive or stem passerine (Mayr 2008), but this is very controversial because the foot structure appears to be wrong. Even if these birds are eventually found more conclusively to be passerines, they make up only a tiny percentage of the bird diversity in the FBM. Passerines are currently thought to have evolved in the late Paleocene or early Eocene. The earliest known fossil passerines have been identified from isolated bones discovered in the early Eocene of Australia. The earliest proposed fossil passerine from North America was late Eocene in age, but this fossil was later identified as a mousebird rather than a passerine (Ksepka and Clarke 2009). Undisputed fossil passerines occur in the early Oligocene of Europe (Mayr and Manegold 2004). Early fossil passerines are very small delicate birds, so exceptional preservation would be necessary to produce fossils readily identifiable as passerines. The FBM is just the right type of locality where some of the earliest representatives of this order may be discovered. Whether or not a strong relationship between passerines and †zygodactylids exists, the FBM bird fauna is clearly very different from the modern North American bird fauna.

Other major bird groups that seem oddly missing from the FBM (considering the composition and distribution of the modern bird fauna) include gulls, grebes, swimming ducks, loons, and herons. Perhaps one or more of these groups will yet be discovered in the FBM.

Mammals

There are literally hundreds of Eocene mammal localities in North America, but the FBM localities have one very significant advantage over most of the others: the FBM fossils are almost always complete, articulated skeletons. Even pieces that are now incomplete appear to be that way because the quarrier did not collect the rest of the specimen in the field (e.g., fig. 151). Most other Eocene mammal localities were formed in high-energy environments such as rivers and streams, and produce mainly isolated teeth, jawbones, and other fragments. In fact, many Eocene mammal species have been described and named based solely on teeth, and the form of the skeleton remains completely unknown. And many isolated bones have been found that cannot be assigned to a tooth-based species because no complete skeletons exist to show which body and limb bones go with which teeth. Thus there is much to be learned about extinct mammalian groups from the beautifully preserved complete skeletons from the FBM. Some of the FBM skeletons provide a Rosetta-stone-like key to correlating teeth and isolated body bones for certain extinct families of mammals. That said, mammal fossils are extremely rare from the FBM, except for one of the bat species †*Icaronycteris index*, now known by about 50 skeletons. Most other mammal species from the FBM are known by only one or two specimens each (most illustrated in this book), and mammal skeletons other than bats from the FBM have come to light only within the last several decades (e.g., after publication of Grande 1984). However, with the intense level of quarrying activity currently under way, I am sure that the FBM will yield many more mammal species in the years to come.

The most common mammals in the FBM by far are bats (many of these in private collections). The second most common form is the otter-like †*Paleosinopa*, currently known by three skeletons. †Brontotheres are also known from the FBM by two partial skeletons and one articulated foot. Other mammal taxa known from the FBM are each represented by a single specimen as far as I was aware of at the time of writing this. The broader mammal classification used here roughly follows that of Rose (2006). The FBM mammals presented below are as follows:

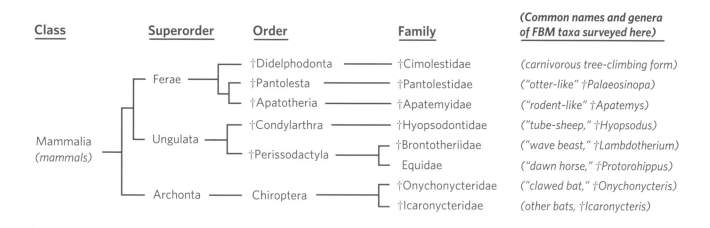

Class	Superorder	Order	Family	*(Common names and genera of FBM taxa surveyed here)*
Mammalia (mammals)	Ferae	†Didelphodonta	†Cimolestidae	*(carnivorous tree-climbing form)*
		†Pantolesta	†Pantolestidae	*("otter-like" †Palaeosinopa)*
		†Apatotheria	†Apatemyidae	*("rodent-like" †Apatemys)*
	Ungulata	†Condylarthra	†Hyopsodontidae	*("tube-sheep," †Hyopsodus)*
		†Perissodactyla	†Brontotheriidae	*("wave beast," †Lambdotherium)*
			Equidae	*("dawn horse," †Protorohippus)*
	Archonta	Chiroptera	†Onychonycteridae	*("clawed bat," †Onychonycteris)*
			†Icaronycteridae	*(other bats, †Icaronycteris)*

Carnivorous Tree-Climbing Form (Order †Didelphodonta, Family †Cimolestidae)

In the 1980s there was a very peculiar but beautifully preserved complete mammal skeleton discovered in the nearshore Thompson Ranch FBM Locality H (figs. 144, 145). It is evidently a tree-climbing carnivorous form not seen before. It has an extremely long and PREHENSILE tail that adapted it to be able to grasp tree branches. It is the earliest prehensile tail known for any placental mammal, and the tail has more vertebrae than in any other known mammal (Michelle Spaulding, pers. comm., 2011). Together with its long fingers and toes, the tail would

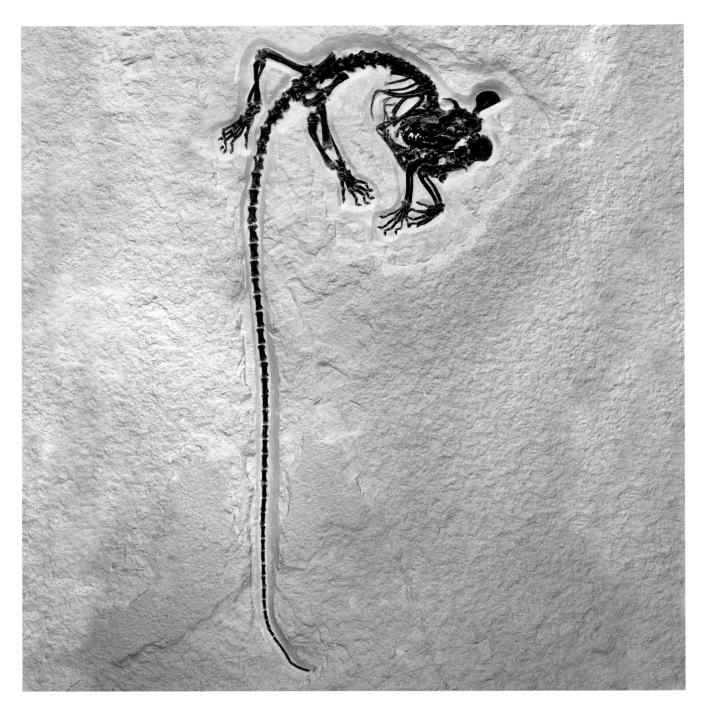

FIGURE 144

An undescribed species of tree-climbing carnivore in the extinct family
†Cimolestidae. Beautifully preserved complete skeleton (partly enlarged in
the next figure). Note the extremely long tail, which has the largest number of vertebrae of any known mammal (Michelle Spaulding, pers. comm.,
2011) and is the earliest known example of a prehensile tail in mammals.
Specimen is from the sandwich beds of FBM Locality H and has a head
length of 56 millimeters (2.2 inches) (FMNH PM1095).

FIGURE 145
Close-up of body and head region of
specimen from fig. 144.

have allowed it to move with great speed and agility through trees. Although
monkeys and a few other mammal groups have prehensile tails, this feature is
unusual for carnivorous mammals.

This Field Museum specimen is unique as of this writing, although it is possible that the teeth of the same species have already been described somewhere.
It will involve significant research to determine which isolated fragments discovered elsewhere might belong to this species or genus. As mentioned above, many
Eocene mammals are known only by their teeth, and some have been named and
classified on this basis only, leaving paleomammalogists to speculate on what the
entire animal may have looked like. That is one reason why specimens like this
are so important.

This animal is here provisionally placed in the extinct order †Didelphodonta,
an order of mammals thought to be closely related to the order Carnivora. Much
further study is needed to confidently determine the precise relationships and
accurate classification of this peculiar species. Like so many other FBM fossils,
this species provides a unique look at mammal species usually known only by
tooth and jaw fragments disassociated from the rest of the skeleton. It is unlike

any other known mammal species. The specimen is currently under study by Spaulding, Flynn, and Grande.

Otter-like †*Palaeosinopa* (Order †Pantolesta, Family †Pantolestidae)

One of the better-described mammalian species from the FBM is the †pantolestid, †*Palaeosinopa didelphoides*, which is known in the FBM by three skeletons: one articulated complete skeleton (fig. 146), one disarticulated complete skeleton (fig. 147), and one partial articulated skeleton (fig. 148). All three skeletons are from the nearshore Thompson Ranch localities in the sandwich beds. The FBM specimens are currently the oldest known articulated †pantolestid skeletons in the world (although there are some fragmentary tooth-based species from the Paleocene). The FBM †*Palaeosinopa* specimens are the most complete †pantolestid remains from North America. †Pantolestid skeletons and even skulls are extremely rare, and of the more than 25 described, named species in the family †Pantolestidae, the vast majority are based only on teeth or jaw fragments, and without any known significant skull or POSTCRANIAL material (Rose and Koenigswald 2005). This is one reason why a species determination has not yet been made for the FBM specimen. All the named "tooth-species" must first be reviewed and compared to the teeth of the FBM species to determine if a valid species name already exists for it. This will also require that the OCCLUSAL surfaces of the teeth on the FBM specimen be better exposed.

†Pantolestids are characterized by robust skeletons and teeth adapted for an omnivorous diet. The FBM †*Palaeosinopa* included fishes in its diet, determined by the fish bones and scales that are present in the stomach region of the better-articulated skeleton. The fish remains in the gut region of the fossil appear to be from †*Knightia* and †*Diplomystus*. †Pantolestid dentition also indicates that their diet could have included hard objects, such as snails and clams. The FBM species is a semi-aquatic form, which probably inhabited an otter-like ecological niche. Its skeleton indicates it had strong pectoral musculature and broad hands adapted for both swimming and digging. The large, heavy claws would also have greatly enhanced the animal's ability to dig. The robust tail and broad feet indicate that when swimming †*Palaeosinopa* probably propelled itself by hind-limb paddling and undulation of the tail, in a manner similar to that of otters today (Rose and Koenigswald 2005).

Given that all three specimens of the FBM †*Palaeosinopa* are from the nearshore Thompson Ranch localities in the sandwich beds, it is possible that †*Palaeosinopa* was an inhabitant of a nearby tributary river or stream, or perhaps it moved along the shoreline of Fossil Lake, although there have been no traces of this animal from the more southern nearshore quarries such as FBM Localities K and L. The extinct family †Pantolestidae is known from the Paleocene through the early Oligocene of North America, Asia, and Europe, and from the late Paleocene through early Eocene of Africa. The family †Pantolestidae is currently

FIGURE 146

†*Palaeosinopa didelphoides* (Cope, 1881), an otter-like mammal in the extinct family †Pantolestidae. Specimen is from the sandwich beds of FBM Locality H and has a head length of 96 millimeters (3.8 inches) (WDC CGR-009). The FBM contains the most complete skeletons known of the family †Pantolestidae. Small bone fragments preserved in the gut region indicate that this species fed on fishes. A cast of this piece also exists at the Field Museum (FMNH PM61122). *Top:* Entire specimen. *Bottom:* Close-up of head and front limbs.

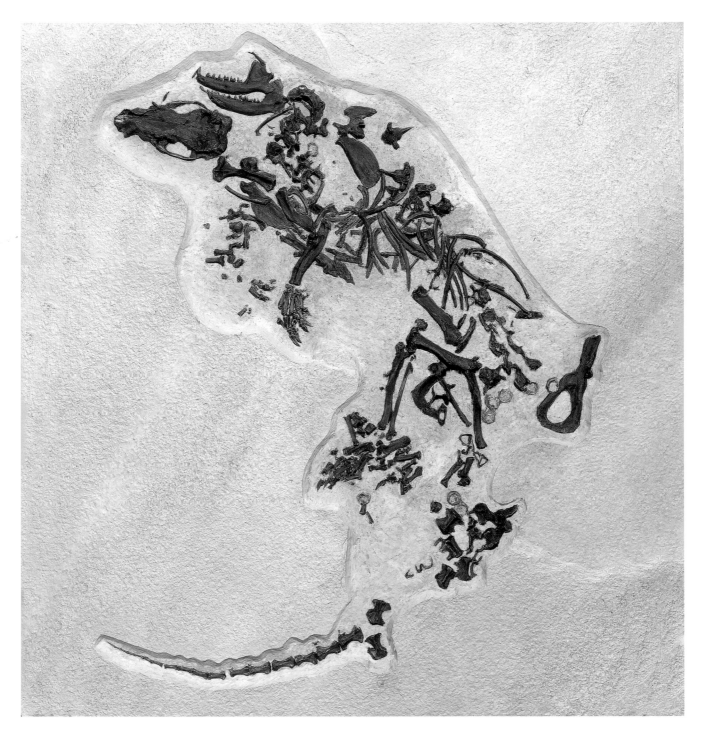

FIGURE 147

A second specimen of †*Palaeosinopa didelphoides* (Cope, 1881), about the same size as the specimen in fig. 146, but this specimen (WDC uncataloged) is disarticulated. Several casts of this specimen also exist (e.g., FOBU 13380). Original specimen is from the sandwich beds of FBM Locality H. Image courtesy of Arvid Aase.

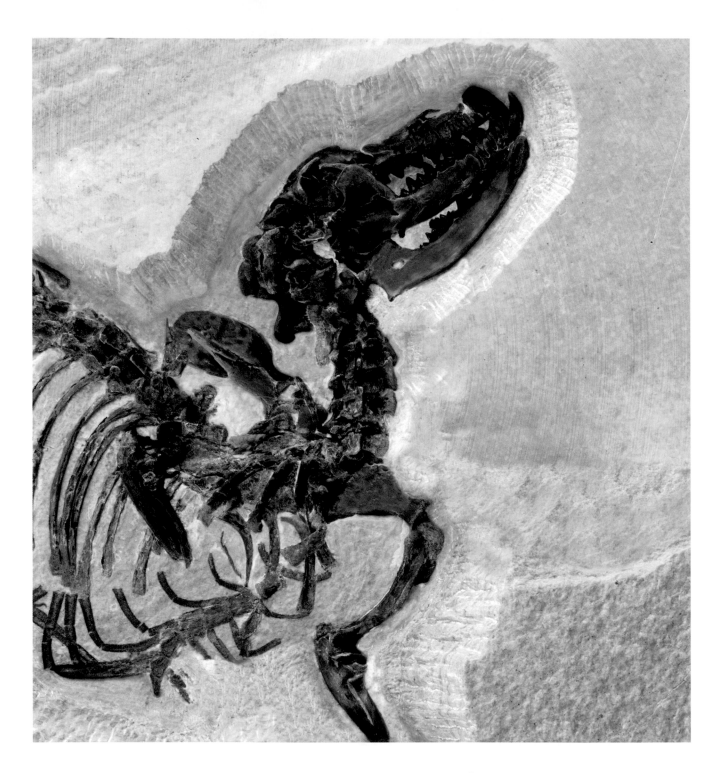

†*Palaeosinopa didelphoides* (Cope, 1881), a juvenile skeleton from the sandwich beds of FBM Locality H. This specimen is much smaller than the ones illustrated in figs. 146 and 147. Head length is 66 millimeters (2.6 inches) long (FMNH PM61144).

included in the extinct order †Pantolesta, whose relationships are very poorly known. The †*Palaeosinopa* from the FBM is described in detail in Rose and Koenigswald (2005), although determination of species was only recently made (Ken Rose and Rachel Dunn, pers. comm., 2012).

Rodent-like †*Apatemys* (Order †Apatotheria, Family †Apatemyidae)

There is one †apatemyid known from the FBM: †*Apatemys chardini*. This species is known by a single specimen from the FBM that was discovered in the late 1990s (fig. 149). The species was originally described in 1930 as a stem primate (†Plesiadapidae) based on a partial right jaw from the early Eocene Gray Bull beds (Willwood Formation) of Bighorn Basin, Wyoming (Jepsen 1930). It was later determined to be an †apatemyid rather than a †plesiadapid. The FBM specimen is the only known nearly complete skeleton of this species.

The †Apatemyidae is the only family in the extinct order †Apatotheria, which includes six genera and is known from the Paleocene and Eocene of Europe, and from the Paleocene, Eocene, and Oligocene of North America. The order does not appear to have survived past the Oligocene. The relationships of †Apatotheria are poorly known, although recent work by Silcox et al. (2010) suggests they might be a primitive STEM GROUP for a larger group containing rodents, rabbits, primates, tree shrews, and a few other closely related species. Clearly, much research is still needed to resolve the relationships of †apatemyids and †Apatotheria. Most species of †Apatotheria are known only from teeth and jaws. The only complete skeletons are the FBM specimen of †*Apatemys chardini* and †*Heterohyus nanus* from the middle Eocene deposits of Messel, Germany (e.g., see Koenigswald et al. 2005a, b).

Detailed study of the FBM specimen of †*Apatemys chardini* is revealing about its probable ecology (Koenigswald et al. 2005). The mobile foot structure, round head, grasping form of feet and claws, and long tail are characteristic of an ARBOREAL life (living in trees). Some researchers have proposed that features of its hands and claws indicate that it fed largely on wood-boring insects extracted from tree bark, similar to certain marsupials in Australia and New Guinea and lemurs in Madagascar (Koenigswald et al. 2005b). They further speculate that the rise of woodpeckers, which competed for this food source, helped lead to the eventual extinction of †Apatotheria.

The unique FBM specimen of †*Apatemys chardini* comes from the nearshore Thompson Ranch locality (FBM Locality M). It may have fallen from its tree into one of Fossil Lake's tributary streams, eventually swept by the current into the lake, or it may have been accidentally dropped into the lake by a predator. In any case, its scarcity indicates it was probably not a normal resident of the Eocene lake shore. A detailed description of the FBM †apatemyid can be found in Koenigswald et al. (2005a).

FIGURE 149

†*Apatemys chardini* (Jepsen, 1930), a small **ARBOREAL** mammal in the extinct family †Apatemyidae, on a slab with a †*Knightia eocaena*. Specimen is currently in a private collection, but a detailed cast of it is FMNH PM61092. Original specimen is from the sandwich beds of FBM Locality M, and the †*Apatemys* has a head length of 38 millimeters (1.5 inches).

Tube-Sheep (Order †Condylarthra, Family †Hyopsodontidae)

There is a single skeleton of †hyopsodontid known from the FBM, †*Hyopsodus wortmani* (fig. 150), which may be the most complete skeleton of a †hyopsodontid ever discovered. This specimen is from the nearshore FBM Locality M on Thompson Ranch. †Hyopsodontids include more than 20 nominal genera, and the most diverse genus by far was †*Hyopsodus*, with about 20 nominal species. †*Hyopsodus wortmani* was originally described in 1902 by Osborn based on isolated teeth and jaw fragments from the early Eocene Wind River Formation of northwestern Wyoming. Nearly all †hyopsodontid species are known only from isolated teeth and jaw fragments, so the discovery of a nearly complete skeleton in the FBM is extremely important to better understand the anatomy, ecological habits, and evolutionary relationships of this poorly known group of mammals. The extreme rarity of articulated †hyopsodontid skeletons is ironic because during the early and middle Eocene, †*Hyopsodus* bones and fragments are the most common mammal fossils of the Rocky Mountain region (comprising over half of all mammal fossils found in the Bridger Formation of Wyoming).

FIGURE 150

†*Hyopsodus wortmani* Osborn, 1902, a small weasel-shaped mammal in the extinct family †Hyopsodontidae (commonly called "tube-sheep" because of their narrow bodies). This is one of the most complete known specimens of the family †Hyopsodontidae. Specimen is currently in a private collection, but a detailed cast of this is FMNH PM61121. Original specimen is from the sandwich beds of FBM Locality M, with a lower jaw length of 51 millimeters (2 inches).

The many species of †*Hyopsodus* were small rat-size to raccoon-size species of mammals ranging from early through late Eocene time in North America, Asia, and Europe. The very few known specimens with postcranial skeletons preserved indicate that †*Hyopsodus* had a strange, weasel-like body shape. This profile has led to the common name of "tube-sheep" occasionally used for these animals, although they were clearly smaller than sheep in size. In addition to its narrow body shape, †*Hyopsodus* had a muscular chest, claws, and short limbs, suggesting a propensity for digging or rooting. It is also thought to have lived part of its life in trees. It may also have been a swimmer because it is commonly found in near-shore deposits of the Bridger Formation. The teeth of †*Hyopsodus* suggest that its diet was that of a generalist. It could have fed on plants and small animals, and because it inhabited lakeshore regions, it may have been another of the many animals that relied on aquatic animals for food. The disappearance of the great Eocene Lake systems may have been part of what led to the decline of the family in the late Eocene. †*Hyopsodus* evidently favored a subtropical climate, and as the climate began to cool in the late Eocene, its numbers declined until the end of the Eocene, when it appears to have became extinct.

Currently the FBM specimen of †*Hyopsodus wortmani* is being studied by Dunn and Rose (Ken Rose, pers. comm., 2010), and a detailed description of it is forthcoming.

Brontotheres and Horses (Order †Perissodactyla, Families †Brontotheriidae and Equidae)

There are two known species and families of perissodactyl that have been found in the FBM: the "wave beast" †*Lambdotherium popoagicum* (family †Brontotheriidae) and the "dawn horse" †*Protorohippus venticolus* (family Equidae). The order Perissodactyla today contains horses, tapirs, and rhinoceroses, but also includes over a dozen extinct families. The order is otherwise known as the "odd-toed ungulates," and ungulates are the "hooved mammals." "Odd-toed" refers to the character of having a single toe or a middle toe that is larger than the toes on either side, rather than having an even number of toes. Perissodactyls are thought to have evolved in Asia, and the earliest known fossils of the order are thought to be early Eocene in age.

There are two partial skeletons of †*Lambdotherium popoagicum* known from the FBM; an 18-inch layer specimen from FBM Locality B (fig. 151, *left*) and another specimen from the sandwich beds of FBM Locality I (fig. 151, *right*). Unfortunately, the skeletons were each only partly collected by the commercial quarrier who found them, because their importance or value was not initially recognized. (Under the covering layer of rock, only a few bumps could be seen, and they were originally thought to be disarticulated fishes, which the commercial quarriers rarely bother to collect). In addition to the two partial skeletons illustrated

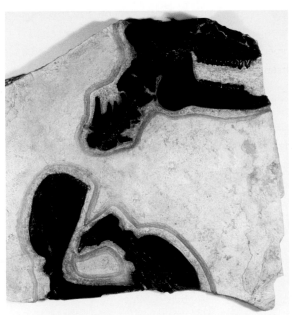

here, there is also an articulated foot from the 18-inch layer in the collection of the Department of Geology at the University of Wyoming (UW 40362).

†*Lambdotherium* is sometimes considered to be in a separate family, †Lambdotheriidae, but Rose (2006) considers it to belong with the †brontotheres in the †Brontotheriidae. The closest living relatives to the †Brontotheriidae are horses, which together with †Brontotheriidae make up the suborder Hippomorpha. The name "wave beast" comes from some of the larger species of the family, the †brontotheres (or †titanotheres), which were immense creatures and some of the largest land animals of Cenozoic time. The family †Brontotheriidae are thought to have become extinct in North America at the end of the Eocene and slightly later in Eurasia. There was a trend for increasing size in the family, with the smaller forms such as †*Lambdotherium* known in the early Eocene and the largest forms in the late Eocene.

†*Lambdotherium* was the earliest genus of the family †Brontotheriidae. The teeth suggest it fed on coarse vegetation such as leaves and twigs. The appearance of the †*Lambdotherium* in a mid-lake deposit is surprising because there is no evidence that †*Lambdotherium* was a swimmer, but the partial disarticulation of the skeleton suggests that the animal's carcass may have been transported some distance to the mid-lake region by currents.

One of the most remarkable finds of recent years in the FBM was the discovery of a complete skeleton of a "dawn horse" (figs. 152, 153). The nomenclature (i.e., taxonomic names) of fossil horses is in flux right now. Previous names that

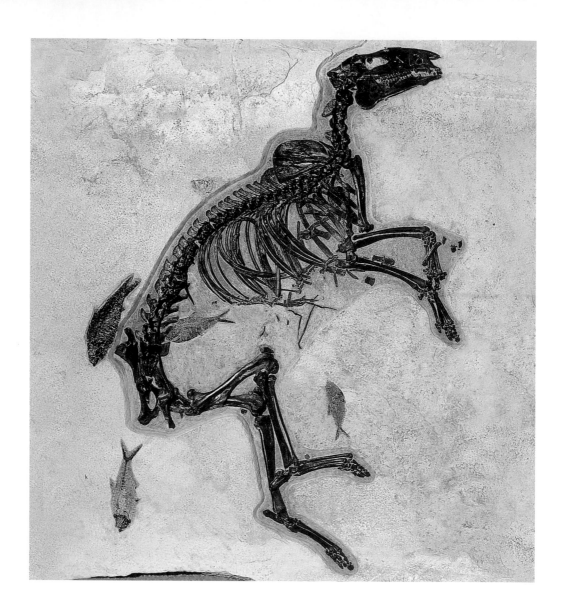

FIGURE 152

†*Protorohippus venticolus* (Cope, 1881), the "dawn horse" in the extant family Equidae. Beautifully preserved specimen on a slab with †*Knightia eocaena* and †*Diplomystus dentatus* is currently in a private collection, but a detailed cast of it is FMNH PM61117. Original specimen is from FBM Locality D; the horse has a head length of 150 millimeters (5.9 inches).

have been used for the dawn horse are †*Eohippus* and †*Hyracotherium*, but both of these names have been found to belong more correctly to other animals, so the most recent name applied to the FBM species is †*Protorohippus* (also occasionally referred to as the "mountain horse") and the species appears to be †*Protorohippus venticolus* (Aaron Wood, pers. comm., 2010). †*Protorohippus* belongs to the horse family, Equidae, which today contains seven living species, including horses, donkeys, and zebras.

The FBM specimen is the most complete articulated skeleton of a dawn horse ever discovered. How it got to the middle region of the Eocene lake (FBM Locality D) is a tantalizing question, particularly given the perfect articulation of all the bones. Horses have the ability to swim, although they are not aquatic

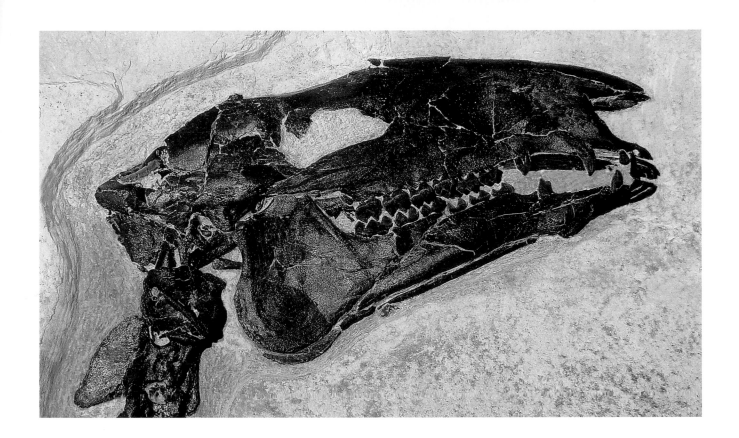

FIGURE 153
†*Protorohippus venticolus,* close-up of skull from fig. 152.

animals. Perhaps this individual went into the lake trying to evade a land predator and got caught in a current, sweeping it out into the lake's center, where it drowned. This is the only trace of †*Protorohippus* known from the FBM, so whatever the circumstances behind its preservation, they were unusual.

†*Protorohippus* was much smaller than today's horses. It stood less than 51 centimeters (20 inches) high at the shoulder and was only about 61 centimeters (2 feet) long as an adult. Unlike modern horses that have a single hooved toe on each limb, †*Protorohippus* had four small hooved toes on each of its front limbs and three on each of its back limbs. Its feet were padded like a dog's, and its toes were tipped with the tiny hooves rather than claws. The back legs were longer than the front legs. The elongate head, slim body, and long hind legs of †*Protorohippus* are all characteristics of a good jumper.

The dawn horse has been widely considered to be the most primitive of horses in the horse family Equidae. †*Protorohippus* lived in areas with a subtropical climate. Its grinding teeth made it adapted to life as a browser, feeding on fruit and soft foliage such as soft leaves and plant shoots. It has been found in Eocene deposits of North America, Europe, and questionably in Asia.

The systematics and nomenclature of dawn horses are still under study, and the description of the FBM specimen is pending.

Bats (Order Chiroptera, Families †Onychonycteridae and †Icaronycteridae)

There are two very important bat species and families known from the FBM: †*Onychonycteris finneyi* (the "clawed bat") described only recently, and †*Icaronycteris index*, a bat named and described in 1966. Both species are known by multiple complete skeletons from the FBM, have been found nowhere else, and are critical in deciphering the early evolution of bats. Although there are thousands of fossil bat specimens known from localities all around the world, almost all of these consist only of isolated teeth or other small bone fragments. Most complete skeletons come either from the FBM or from the middle Eocene deposits of Messel, Germany.

One of the most exciting fossil bats ever discovered is †*Onychonycteris finneyi*. This species—in its own family, †Onychonycteridae—is known only from the FBM and is thought to be the most primitive of all known bats; a "missing link" of sorts, between bats and their non-flying ancestors. Three specimens of this species have come from FBM Locality M, including the complete skeleton in figure 154, which is currently in a private collection. Another complete skeleton in the collection of the Royal Ontario Museum served as the holotype for the original species description in 2008. So far, †*Onychonycteris finneyi* is known only from the northern shoreline localities of Thompson Ranch and not from any of the mid-lake or southern quarries. This may indicate that a colony of †*Onychonycteris* was based near the northeastern shore of Fossil Lake. This species is much rarer than the other FBM bat, †*Icaronycteris*, which is currently known by more than a dozen specimens. †*Onychonycteris finneyi* was named after Bonnie Finney, one of the current leaseholders for the quarry at FBM Locality M.

The claws are part of what makes †*Onychonycteris* so primitive and places it in a family of its own and sister group to a group containing all other fossil and living bats. It has well-developed claws on all ten wing digits (fingers). Unlike birds, in which the arm bones form the wing skeleton, it is the finger bones that support the wings in bats. The extra wing claws of †*Onychonycteris* probably made it an agile climber that could climb with all four limbs. In all other fossil and living bat species, most of the wing digits have lost the claws typical of other closely related mammals, although †*Icaronycteris* also retains a well-developed claw on its index finger. The hind legs of †*Onychonycteris* are also unusually long for a bat.

†*Onychonycteris* had well-developed wings and could fly much like more evolutionarily advanced bats. It had a broad tail membrane that functioned as an airfoil, which probably gave it increased agility in flight. Among mammals, powered flight is unique to bats, and all bats can fly. The ear morphology of †*Onychonycteris* suggests that it lacked the ability for ECHOLOCATION that other bats have for tracking prey. Lacking the ability for echolocation, †*Onychonycteris* would have had to track its prey by vision, suggesting that it may have been DIURNAL rather

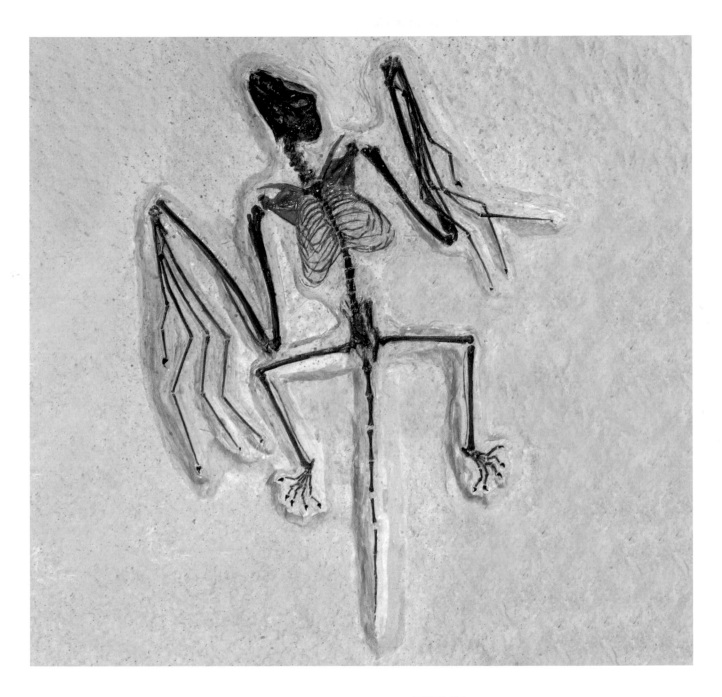

FIGURE 154

†*Onychonycteris finneyi* Simmons et al., 2008, a primitive bat in the extinct family †Onychonycteridae. Specimen is currently in a private collection, but a detailed cast of it is FOBU 13514. Original specimen is from the sandwich beds of FBM Locality M, with a head length of 23 millimeters (0.9 inch).

than NOCTURNAL like living bats. Without echolocation, †*Onychonycteris* would either have had to have been a day flier to find insects on the wing, or it would have had to have very large eyes to see in the dark, and so far no intact eye sockets have been preserved to determine the size of the eye in this species. The teeth of †*Onychonycteris* suggest that it fed on insects like other Eocene bats and many living bats.

†*Onychonycteris* and the other FBM species, †*Icaronycteris*, are among the oldest known bats and are at the base of the group's evolutionary tree. Today bats are one of the most successful groups of mammals and include over 1,100 species (discussed below in relation to †*Icaronycteris*). †*Onychonycteris*, like so many other organisms in the FBM, documents the early beginnings of one of today's most important groups of animals. The fact that the species is represented with complete skeletons makes it all the more valuable to understanding the early development of today's North American biota. Detailed description of †*Onychonycteris* can be found in Simmons et al. (2008).

Another species of bat from the FBM is †*Icaronycteris index* (fig. 155). This species has been known since the discovery of the holotype in the early 1930s and is more closely related to modern bats than to †*Onychonycteris*. It was not named until 1966, when it was described by Glenn Jepsen, who spent many months (perhaps years) preparing this delicate specimen. It does not have claws on all of the wing digits like †*Onychonycteris*, but it does have claws on the index finger. As a stem-group family of modern bats, †Icaronycteridae is another example of how crucial the FBM is to deciphering the early evolutionary history of bats. There are two other species assigned to the †Icaronycteridae: †*I. menui*, from the early Eocene of France, and †*I. sigei*, from the early Eocene of India. But both of these other species are based solely on teeth and bone fragments. The overall known morphology of the family is based primarily on the FBM species †*I. index*. The family †Icaronycteridae is not known to have survived past the early Eocene.

The FBM bats are not only the earliest known complete bat skeletons; they are critical to understanding the evolution of one of today's most successful groups of mammals. The two most basal families of bats are known only from FBM specimens as shown by the following tree:

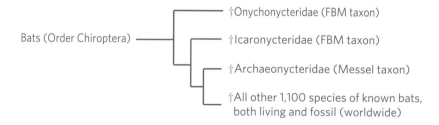

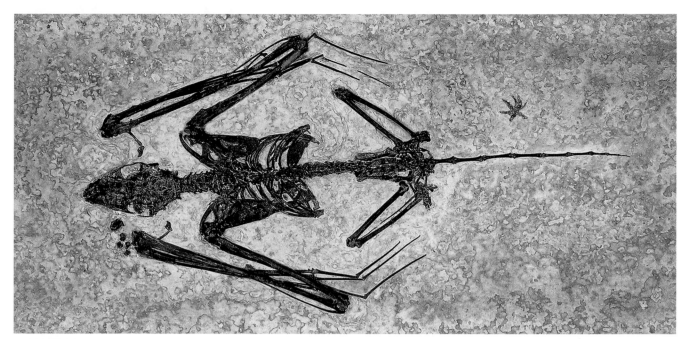

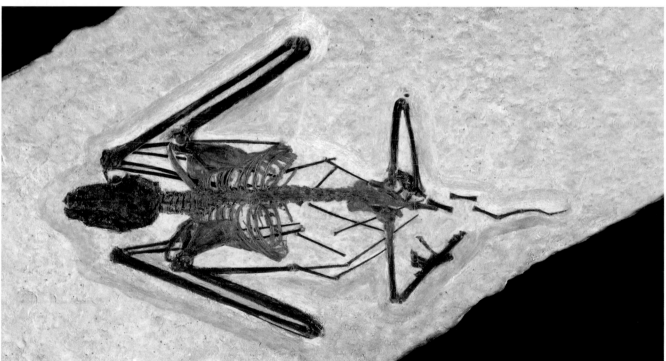

FIGURE 155

†*Icaronycteris index* Jepsen, 1966, a small bat in the extinct family †Icaronycteridae. *Top:* Specimen PU 18150 (**HOLOTYPE**) is from the 18-inch layer of FBM Locality B and is on a slab with small palm flower (just above the tail); head length is 20 millimeters (0.8 inch). Photo courtesy of the Division of Vertebrate Paleontology, Peabody Museum of Natural History, Yale University, New Haven, CT. *Bottom:* Specimen FMNH PM1096 from the sandwich beds of FBM Locality H, with a head length of 18 millimeters (0.7 inch).

Today bats have about 1,100 species and represent about 20 percent of all classified mammal species. Most feed on insects while on the wing, although about 25 percent of bats are fruit eaters and a few feed on the blood of animals.

Most bats (other than †*Onychonycteris*) are capable of echolocation while in flight. This is a perceptual system in which ultrasonic sounds or "clicks" are emitted to produce echoes, like a radar system. This allows them to "see" in the dark while on the wing, even though they may have very tiny eyes. By hunting at night, they can feed on nocturnal insects and avoid competition from birds.

†*Icaronycteris* has relatively unspecialized teeth, much like those of a shrew. Jepsen (1966) also noted a small coprolite in the holotype specimen that may have included insect chitin and other elements. It probably fed on insects like most bats, but the holotype also has a small fish scale in the stomach region (Jepsen 1966), indicating that †*Icaronycteris* included fish in its diet as well. This is not surprising given that today some bats feed on small fishes, gaffing them from the water's surface with the claws of their feet.

The vast majority of the dozen or so known specimens of †*Icaronycteris index* come from the nearshore Thompson Ranch sandwich beds (FBM Localities H and M). The holotype and at least two other specimens are exceptions, coming from the mid-lake 18-inch layer deposits of FBM Localities B and E. Bats are capable of flying up to hundreds of miles in search of food, so the presence of bats in the mid-lake deposits is not surprising. †*Icaronycteris* is the best described of all fossil bat skeletons, and further morphological and other details can be found in Jepsen (1966, 1970) and Simmons and Geisler (1998).

Plants

Although much has been written about the plant fossils from the middle Eocene Parachute Creek Member of Lake Uinta in the Green River Formation (e.g., MacGinitie 1969 and Brown 1929, 1934), and various deposits from Lake Gosiute (Knowlton 1923 and Newberry 1883), very little has been reported about plants from the FBM and Eocene Fossil Lake. A few FBM species were reported in Brown (1937), but these are now mostly known to have been inaccurately assigned in terms of family. Lesquereux (1883) also described a few FBM species that he mistakenly attributed to "Florissant" rather than the FBM (MacGinitie, personal notes, courtesy of Steve Manchester, 2011). The majority of fossil plant species, genera, and families presented in this book are here reported from the FBM for the first time.

The plant fossils excavated from the FBM are almost all "split fossils" (see page 12). Unlike fossil vertebrates, they are normally undetectable if there is even a thin layer of rock covering them. Nevertheless, plants are commonly found in the FBM as the layers

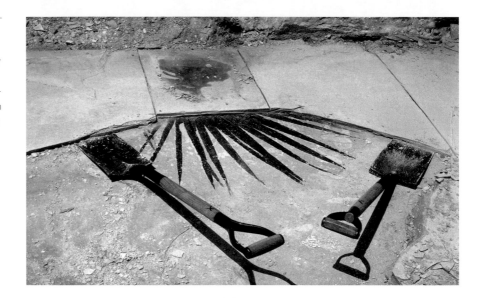

are split apart in the field (e.g., fig. 156, *left*). Rarely, plants are also discovered
back in the laboratory as other fossils are prepared (e.g., the tree of heaven seed
under the fish in fig. 79, *bottom*).

One of the challenges of identifying many of the plant fossils from the FBM is
the fact that they consist mainly of individual leaves or leaflets unassociated with
other parts of the plant. This is because most plants shed their leaves and other
structures separately into the sedimentary environments. A single leaf represents
only a small part of the total organism. A tree or shrub may include leaves of
many very different shapes, fruits, seeds, flowers, pollen, trunks, branches, bark,

FIGURE 157

There is much variation in leaf shape in species of deciduous trees, often making species identification difficult. The example here is three variants of the tree leaf *Gyrocarpus* sp., from the 18-inch layer of the FBM. *Left:* Non-lobed variant; specimen is from FBM Locality A, measuring 260 millimeters (10.2 inches) in length (FMNH PP53395). *Middle:* Two-lobed variant; specimen is from FBM Locality A, measuring 420 millimeters (16.5 inches) in length (FOBU 10668). *Right:* Three-lobed variant; specimen is from FBM Locality A, measuring 300 millimeters (11.8 inches) in length (FMNH PP53567).

and roots. Look at a sassafras tree, maple tree, or an oak tree in the park and see the variation that exists among the leaves of even a single individual. This variation is also apparent in some of the FBM leaves (figs. 157, 183). To further complicate the identification of isolated leaf fossils, most **VASCULAR PLANTS** are classified based on reproductive structures (fruits, flowers, cones, and pollen) that are almost never found attached to leaves in the fossil record. Nevertheless, paleobotanists have made significant strides in identifying many fossil leaves at least to family level, based on scarce specimens with well-preserved venation, branches containing multiple leaves, and, rarely, a few instances of leaves associated with reproductive structures.

In the FBM, plant fossils include well-preserved leaves, pollen grains, flowers, fruits, branches, and root systems. Unfortunately, these parts are rarely, if ever, found attached to each other. Consequently, it is largely unproven which leaves belong with which flowers, seeds, fruits, or pollen. One of the goals of paleobotanists in the FBM is to find branches with attached flowers and leaves together, hopefully even with pollen grains present in the flower stamens.

In general, the fossil plants of the FBM indicate that the shores of Fossil Lake were populated with many palms, huge elephant ear leaves, delicate ferns, ivy, guava, amaryllid flowers, sumac, and other plants indicating a moist, subtropical lowland immediately surrounding the Eocene lake. The marginal lake vegetation included horsetails, cattails, floating ferns, ceratophyllum, and lotus. The surrounding nearby highlands, which would have had a slightly cooler climate, contained many hardwood and other broadleaf trees, conifers, and other more temperate species whose sturdy leaves probably washed into the lake with the tributary streams draining into it. Leaves can be transported into lakes and other drainage basins from many miles away. MacGinitie (1969, 31) reported his observation that in present-day northwestern California, leaves are transported at least 14 miles in streams and rivers. Of the FBM tree species indicated by leaves, conifers were extremely rare while broadleaf trees and palms predominated. Palm fronds are common in most of the FBM localities. The major groups of plants known from the FBM are as follows:

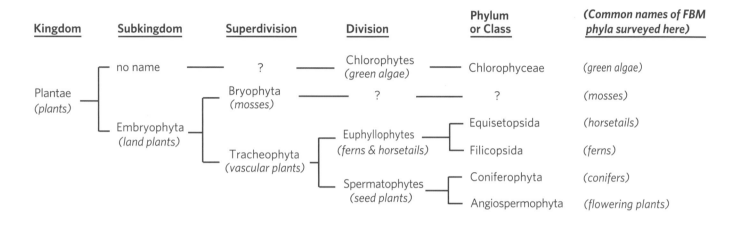

Kingdom	Subkingdom	Superdivision	Division	Phylum or Class	*(Common names of FBM phyla surveyed here)*
Plantae *(plants)*	no name	?	Chlorophytes *(green algae)*	Chlorophyceae	*(green algae)*
	Embryophyta *(land plants)*	Bryophyta *(mosses)*	?	?	*(mosses)*
		Tracheophyta *(vascular plants)*	Euphyllophytes *(ferns & horsetails)*	Equisetopsida	*(horsetails)*
				Filicopsida	*(ferns)*
			Spermatophytes *(seed plants)*	Coniferophyta	*(conifers)*
				Angiospermophyta	*(flowering plants)*

Green Algae

The smallest yet most important plants in Eocene Fossil Lake were the green algae. In the FBM, these were preserved both as the single-celled but often colonial *Pediastrum* sp. (fig. 158). Green algae made up the lifeblood of Fossil Lake, harnessing sunlight through photosynthesis and turning it into food at the base of the food chain. The algae were the main **PRIMARY PRODUCERS** of Fossil Lake that were consumed directly by **PRIMARY CONSUMERS** such as zooplankton, ostracods, snails, aquatic insects, crayfishes, larval stages of frogs and salamanders, and filter-feeding fishes such as †*Knightia*. Primary consumers of algae were then eaten by **SECONDARY CONSUMERS**, transferring energy up the food chain to every animal in the lake and to some animals around the lake. To this day, sunlight processed through green algae comprises the route by which most energy enters our biosphere. Algae also produce a major portion of the oxygen on the planet.

Another probable representation of fossilized green algae in the FBM are the thin dark bands of kerogen in the rock visible in cross

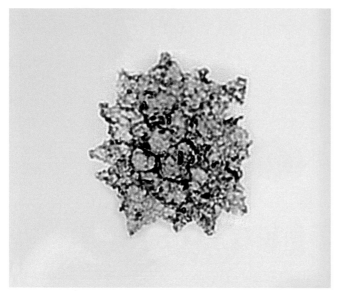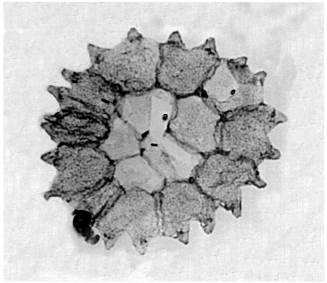

FIGURE 158
Pediastrum sp., a form of microscopic green algae fossilized within the FBM. Left specimen measures approximately 58 micrometers (.002 inch) in diameter, and right specimen measures 75 micrometers (.003 inch) in diameter. Black-and-white photos taken through a light microscope at 1,000× power of specimens immersed in oil. Specimens were dissolved out of lower capping unit beneath the 18-inch layer on Fossil Butte. Mounted slide of specimens is ww FBM-4. Photos courtesy of Dr. Robert Cushman.

section, particularly in the 18-inch layer (see fig. 6A, *bottom left*). These dark organic layers are most likely made of decomposed algae and bacteria, although they are too badly decomposed to allow identification of individual elements.

Fossilized *Pediastrum* sp. (family Hydrodictyaceae) in the FBM requires a microscope to see (fig. 158). The genus *Pediastrum* has about 16 valid living species today, although many more than that have been named. It inhabits mainly freshwater habitats and is fastest growing in warm water that receives much sunlight. Algae such as *Pediastrum* also need nitrogen to grow, and the major source of this element in Fossil Lake would have been dead organisms and animal waste products in the water. By removing animal waste toxins from the lake's water system, green algae also contributed to the health of the lake environment for aquatic vertebrates.

Some of the larger fossils of amorphous plant material in the FBM that resemble little more than brown smears on the rock surface may represent mats of algae. Such mats of algae could have served as a food source for herbivorous "grazers" on the lake bottom.

Algae can sometimes also overproduce in freshwater lakes and become temporarily harmful. A growth explosion of algae (**ALGAL BLOOM**) due to changes in light, temperature, and nutrients in the water can sometimes eventually crash. The resulting abnormally large amounts of dead algae in the water can lead to severe oxygen depletion through the breakdown of the decaying plant matter. This could be one of the explanations for stratigraphic horizons within the FBM that contain mass-mortality layers of fishes (e.g., figs. 67, *bottom*; 71; 72, *bottom*; 83, *bottom*; 93; 101; 214, *top*).

Ferns and Horsetails

(Phylum Filicopsida and Phylum Equisetopsida)

Ferns are less common in the FBM than some species of angiosperm leaves, but this is probably an artifact of the fact that they are **UNDERSTORY** rather than canopy plants. A single angiosperm tree produces many leaves in its lifetime that are shed to become possible fossils. They are shed from heights, can be blown for distances, and are sometimes more rugged than many understory plants. The understory plant leaves are more closely rooted to the ground and often not shed like the large tree leaves are. Horsetails are thought to be closely related to ferns and are grouped together with them in the division Euphyllophytes (also sometimes referred to as Pteridophyta). Ferns and horsetails have neither flowers nor seeds, and they reproduce with spores. Several hundred million years ago, euphyllophyte plants dominated the forests. They were a major component of the land plant community until the Late Cretaceous, when flowering plants became dominant.

There is a diversity of ferns known from Fossil Lake (some families reported from the FBM for the first time here). The major groups of horsetails and ferns known from the FBM are as follows:

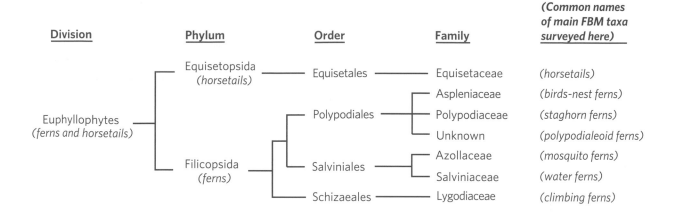

Division	Phylum	Order	Family	(Common names of main FBM taxa surveyed here)
Euphyllophytes (ferns and horsetails)	Equisetopsida (horsetails)	Equisetales	Equisetaceae	(horsetails)
	Filicopsida (ferns)	Polypodiales	Aspleniaceae	(birds-nest ferns)
			Polypodiaceae	(staghorn ferns)
			Unknown	(polypodialeoid ferns)
		Salviniales	Azollaceae	(mosquito ferns)
			Salviniaceae	(water ferns)
		Schizaeales	Lygodiaceae	(climbing ferns)

Horsetails (Phylum Equisetopsida, Order Equisetales)

Equisetum sp. (commonly called "horsetail" or "scouring-rush") is present in the FBM, where it is one of the most primitive large plants of the flora (e.g., fig. 159). Sometimes considered a class, while other times a phylum, and grouped together with ferns based on molecular analysis of living species, Equisetopsida are known as fossils as far back as 375 million years ago, when they included trees that formed a major part of the forest. The order still survives today with one genus (*Equisetum*) and 15 living species, mostly relatively small plants, but some species reaching over 3 meters (10 feet) in height. Because of the great age of the fossil species, the living species are commonly referred to as "living fossils." *Equisetum* is easily identifiable in the FBM by its striated, reed-like rhizome (underground

FIGURE 159

†*Equisetum winchesteri* Brown, 1928, a horsetail in the family Equisetaceae. Specimen is from FBM Locality C (FOBU 13285). Section of underground stem (rhizome) showing whorls of roots at jointed nodes. Rhizome width is 10 millimeters (0.4 inch).

stem), segmented along its length with jointed nodes bearing whorls of small, thin rootlets. This plant today grows in moist to wet environments often on the margins of wetlands and poorly drained areas. There are at least two species present in the FBM: a moderately sized form with a stem and rhizome approximately 10 millimeters (0.4 inch) in diameter (e.g., fig. 159) and a giant species I have seen pieces of in private collections with stalk diameters of nearly 100 millimeters (4 inches).

Polypodioid Ferns (Phylum Filicopsida, Order Polypodiales)

There are a variety of fern species in the order Polypodiales present in the FBM. These include several of the more typical fern morphologies, such as staghorn ferns, and several unidentified kinds. Today the order Polypodiales includes 80 percent of all living fern species.

Staghorn ferns are rare in the FBM (fig. 160, *left*). The FBM form appears to be an undescribed species of *Platycerium*, which today includes about 18 living species. They are commonly called "staghorn" or "elkhorn ferns" because

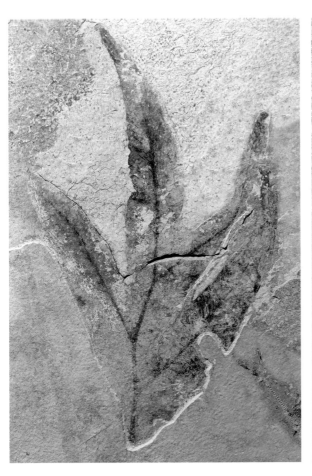

of their uniquely shaped fronds. This genus is native to tropical and temperate areas of Southeast Asia, South America, Australia, New Guinea, and Africa.

There are a variety of other polypodialeiform ferns present in the FBM that need further study to determine their family relationships (e.g., fig. 160, *right*). This is the most diverse order of ferns in the FBM.

Mosquito Ferns and Water Ferns (Phylum Filicopsida, Order Salviniales)

This order of ferns contains a variety of aquatic ferns in the FBM, all of which are floating varieties.

The mosquito fern (family Azollaceae) is represented in the FBM by an undescribed species of the genus *Azolla* (fig. 161). This genus today includes seven living species that flourish mostly in warm (subtropical to tropical) freshwater environments around the world. With its closely overlapping, small, scale-like

FIGURE 161

Azolla sp., a mosquito fern in the family Azollaceae. Scale bar equals 20 millimeters (0.8 inch). Specimen is from the 18-inch layer of FBM Locality B (FOBU 9836).

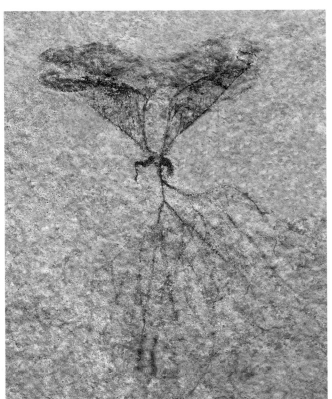

FIGURE 162

†*Salvinia preauriculata* (Hicky, 1977), a floating fern in the family Salviniaceae. Two specimens, each with three leaves: two leaves that floated on the surface of the water and one finely dissected, root-like leaf that hung submerged below the surface. Both specimens are fertile with paired spore-bearing structures at the juncture between the two float leaves and the branched root-like leaf. From FBM Locality A in the upper layers near the K-spar tuff layer. *Left:* Specimen has a total height of 70 millimeters (2.8 inches) (FMNH PP53586). *Right:* Specimen has a total height of 44 millimeters (1.7 inches) (FMNH PP54855).

leaves, it resembles duckweed or moss more than typical ferns. *Azolla* floats on the surface of the water with its roots hanging below. This is a very delicate plant and is a rare component of the known FBM flora.

Another variety of floating fern in the FBM that is closely related to *Azolla* is a member of the family Salviniaceae, †*Salvinia preauriculata* (Berry, 1925). This species differs from *Azolla* in having a larger leaf size, often found with only two floating leaves above the surface and a root-like, finely dissected leaf hanging below the surface (fig. 162). Today there are about 15 living species of *Salvinia* (commonly called "water moss"), mostly in tropical regions of the Americas, Asia, Europe, the West Indies, Africa, and Madagascar. They are restricted to freshwater, intolerant to salinity, and are hardy plants that spread quickly in warm waters. They can sometimes form dense surface mats that can eventually choke off oxygen to the waters below. This is a relatively common water fern in the FBM.

Climbing Fern (Phylum Filicopsida, Order Schizaeales)

A relatively abundant fern of the FBM is the climbing fern †*Lygodium kaulfussi* Heer, 1861. *Lygodium* is the sole genus within the family Lygodiaceae. Today there

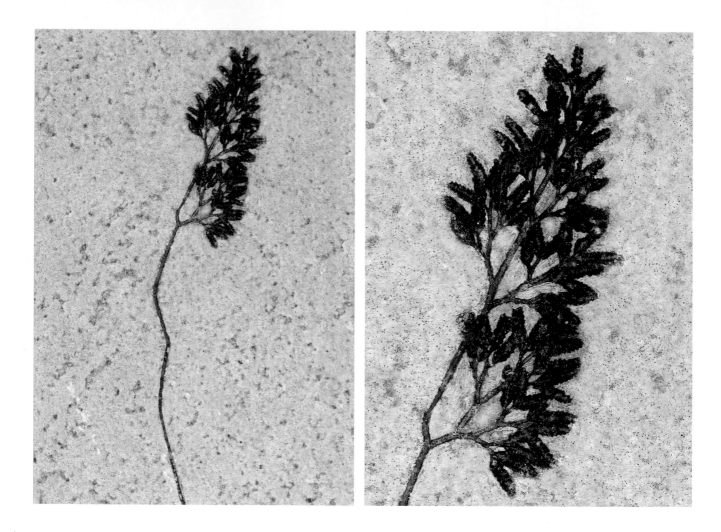

FIGURE 163

†*Lygodium kaulfussi* Heer, 1861, a climbing fern in the family Lygodiaceae. Specimen is from the 18-inch layer of FBM Locality A (FMNH PP54856). *Left:* Entire specimen, with length of 100 millimeters (3.9 inches). *Right:* Enlargement of fertile branches on image to left. This is the branched, spore-bearing fertile leaf of this species.

are about 40 living species of the genus native to tropical regions across the world (mostly Asia). A few species also exist in warm temperate regions of eastern Asia and North America. *Lygodium* is called a climbing fern because it has INDETERMI-NATE GROWTH and a flexible RACHIS that allows the fronds to form climbing or trailing vines. Small parts of *Lygodium* fronds are common in the FBM, but large, more complete fronds such as the one in figure 163 are rare.

Conifers

*(Phylum
Coniferophyta)*

MACROFOSSILS of conifers are very rare in the FBM and are yet only provisionally identified as such. They consist of possible firs and a possible juniper. Living conifers are mostly evergreen trees and shrubs that usually have needle-shaped leaves. Although there are some tropical to subtropical species, most are adapted for cooler climates. Their scarcity in the FBM indicates that these fossils may have been transported into Fossil Lake in tributary streams and rivers flowing from the higher, cooler elevations surrounding the Fossil Lake valley. It may also indicate that these fossils are incorrectly identified.

Pines and Junipers (Order Pinales, Families Pinaceae and Cupressaceae)

Pines (family Pinaceae) are represented in the FBM by winged seeds (fig. 164, *bottom left*) and poorly preserved needles (fig. 164, *top*). These are very rare in the FBM. The winged seeds are very similar in

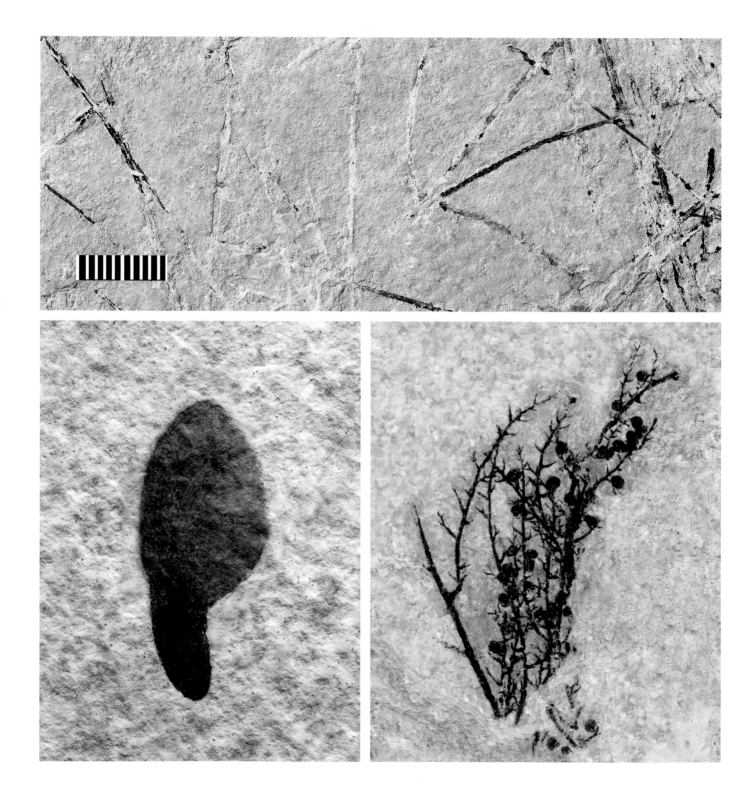

shape to the genus *Abies*, but unlike extant pine seeds, the FBM specimen shown here appears to have veins on the wings. These FBM seeds are in need of further study. The poor preservation of the needles may be indicative of long-distance transport into the lake. Today there are about 250 species of pine in the family Pinaceae, distributed throughout most of the Northern Hemisphere, from tropical to subarctic regions.

There is also a unique specimen of possible juniper (family Cupressaceae) from the FBM (fig. 164, *bottom right*). This specimen is a sprig containing sharp, needle-like leaves and small "berries" (modified cones). Today there are about 65 living species of juniper distributed throughout the Northern Hemisphere. They range from tropical to arctic regions.

&··· FIGURE 164

Conifers (order Pinales) in the families Pinaceae (pines) and Cupressaceae (junipers). *Top:* Pine needles (*Pinus* sp?) from the 18-inch layer of FBM Locality G (BMNH v.62000). Scale equals 10 millimeters (0.4 inch). *Lower left:* Winged seed from a fir. Specimen from the 18-inch layer of FBM Locality A, measuring 20 millimeters (0.8 inch) in length (FMNH PP54665). *Lower right:* A possible sprig of juniper (family Cupressaceae) from FBM Locality A. Specimen measures about 60 millimeters (2.4 inches) in height (FMNH PP54794).

Non-Eudicot Flowering Plants

(Phylum Angiospermophyta; Subclasses Magnoliids, Monocotyledons, and Ceratophyliids)

The flowering plants (angiosperms, phylum Angiospermatophyta) are the most diverse group of land plants living today. They are characterized by a number of uniquely derived features, including the ability to produce flowers, fruit surrounding the seeds, and an **ENDOSPERM** within the seed. By the Late Cretaceous, angiosperms had replaced ferns and cycads as the dominant land plants, and from the Late Cretaceous through early Eocene, large canopy-forming angiosperm trees became more and more dominant in forest environments formerly dominated by conifers. There is such a wide variety of flowering plants represented in the FBM that I divide them into two organizational sections here: non-eudicot flowering plants in this section (subclasses Magnoliids, Monocotyledons, and Ceratophyliids) and eudicot flowering plants in the next section (subclass Tricolpates). The major groups of angiosperms known from the FBM are as follows:

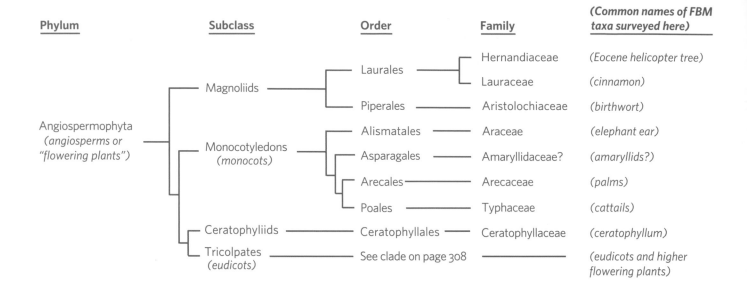

Phylum	Subclass	Order	Family	(Common names of FBM taxa surveyed here)
Angiospermophyta (angiosperms or "flowering plants")	Magnoliids	Laurales	Hernandiaceae	(Eocene helicopter tree)
			Lauraceae	(cinnamon)
		Piperales	Aristolochiaceae	(birthwort)
	Monocotyledons (monocots)	Alismatales	Araceae	(elephant ear)
		Asparagales	Amaryllidaceae?	(amaryllids?)
		Arecales	Arecaceae	(palms)
		Poales	Typhaceae	(cattails)
	Ceratophyliids	Ceratophyllales	Ceratophyllaceae	(ceratophyllum)
	Tricolpates (eudicots)	See clade on page 308		(eudicots and higher flowering plants)

FIGURE 165

Gyrocarpus sp., from the 18-inch layer of FBM Locality A. *Left:* Trilobed variant of a very large leaf (family Hernandiaceae). Total length is 300 millimeters (11.8 inches) (FMNH PP53567). There is much variation in shape for this species (see fig. 157), but note the very long PETIOLE, smooth leaf margins, and PALMATE VENATION. *Right:* Winged seed measuring 60 millimeters (2.4 inches) in length that possibly belongs to this species (FMNH PP 54857).

Laurels (Order Laurales, Families Hernandiaceae and Lauraceae)

One laurel relative that is quite common in the FBM is *Gyrocarpus* sp. in the family Hernandiaceae. These leaves show one-, two-, and three-lobed forms (fig. 157). Some of them are quite large and well preserved (fig. 165), and they were probably growing near the shores of Eocene Fossil Lake. This was one of the species that probably helped form the upper tree canopy of the forests surrounding Fossil Lake. The family Hernandiaceae is today represented by about 60 species living in the tropical regions around the world. The FBM species represents one of the earliest known examples of this family in the fossil record.

Lesquereux (1883, 165) also reported another laurel, the genus *Cinnamomum* (the cinnamon genus), from "Florissant," but the type specimen he used for this (now USNM P1579) was later found to be from the FBM (MacGinitie, personal notes, courtesy of Steve Manchester, 2011). *Cinnamomum* is in the family Lauraceae and is today represented by about 300 extant species living in tropical and subtropical regions of the Americas, Asia, and Australasia. The family Lauraceae was widespread in the Late Cretaceous and Tertiary.

Birthworts (Order Piperales, Family Aristolochiaceae)

An undescribed birthwort species is occasionally found in the FBM (fig. 166). Birthworts, also called "pipevines" and "Dutchman's pipes," are perennial herbaceous plants, shrubs, and vines. Today there are about 400 living species in the

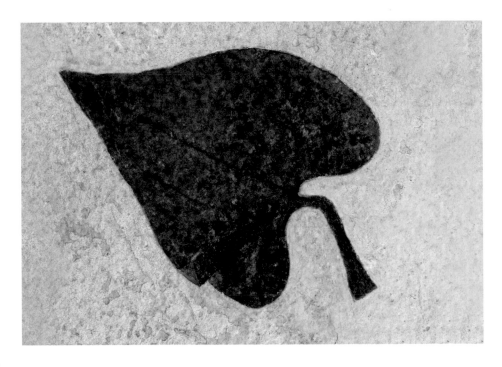

FIGURE 166
An undescribed birthwort leaf (family Aristolochiaceae). Specimen measures 95 millimeters (3.7 inches) in length and is from the 18-inch layer of FBM Locality A (FMNH PP53545).

FIGURE 167

An undescribed species similar to the elephant ear plant, *Colocasia* sp., in the aroid family, Araceae. *Left:* Specimen from the sandwich beds of FBM Locality L, measuring 800 millimeters (2.6 feet) in length (FMNH PP54865). *Right:* Specimen from the 18-inch layer of FBM Locality E; in a private collection. Fish at bottom beneath leaf measures 110 millimeters (4.3 inches) in length.

family Aristolochiaceae. Species from this family live primarily in the tropics, and almost all of them are pollinated by flies.

Elephant Ear (Order Alismatales, Family Araceae)

There are some very large MONOCOT leaves found in the FBM that appear to represent at least one species of aroid, similar to the "elephant ear" plant, *Colocasia* (fig. 167). While large, nearly complete leaves are rare, fragments of this leaf are relatively common elements of the FBM flora. They were probably growing near the Eocene lake's shoreline and possibly in the shallow marshy areas of the lake. The genus *Colocasia* is today native to tropical Asia and Polynesia. There are about 25 living species in the genus today. The FBM species awaits scientific description.

Tube Flowers (Order Asparagales, Family Amaryllidaceae?)

The most common flower in the FBM appears to be an undescribed species of monocot flower, possibly in the family Amaryllidaceae (fig. 168). These fossils are commonly referred to as "tube flowers" by the fossil quarriers. These flowers are among several MONOCOT families found in the FBM. Monocots are angiosperms

FIGURE 168

Undescribed MONOCOT flowers, possibly of the family Amaryllidaceae, from the 18-inch layer of FBM Locality A. This is the most common flower in the FBM 18-inch layer, and it is known by well over 100 specimens. *Top left:* Specimen from FBM Locality A, measuring 57 millimeters (2.2 inches) in height (FMNH PP53390). *Top right:* Specimen from FBM Locality E, measuring 41 millimeters (1.6 inches) in length (FMNH PP 55058). *Bottom left:* Specimen from FBM Locality A, measuring 62 millimeters (2.4 inches) in height along curve (FMNH PP53583). *Bottom right:* Specimen from FBM Locality E, measuring 88 millimeters (3.5 inches) in height along curve (FMNH PP55060).

in which the major veins run parallel to each other (rather than branching) along the length of the leaf. Other monocots reported here from the FBM include palms, cattails, and elephant ear plants.

There have been several hundred specimens of tube flowers found in the 18-inch layer of the mid-lake quarries, and there may even be more than one species there. Amaryllids are a diverse group today, containing over 800 living species distributed worldwide. The common daffodil is a member of this family. The FBM tube flowers are in need of detailed study, and assignment of them to Amaryllidaceae is only provisional.

Palms (Order Arecales, Family Arecaceae)

Palms are another one of the FBM monocots, and one of the most impressive of the FBM plant fossils (fig. 169). One species identified in the FBM is †*Sabalites powelli* (Newberry 1883). There may be other species present too. Palm fronds are very common in the FBM, both in the 18-inch layer and the sandwich beds, although they are often more difficult to see in the nearshore sandwich beds because the plant fossils there often lack contrasting color (i.e., they are the same color as the matrix). They can sometimes be very large. The author observed one perfectly preserved frond from FBM Locality M that was 4 meters (13 feet) in length. Such pieces must have been from trees right at the water's edge. Besides fronds, there are also palm flowers and inflorescences preserved in the FBM (fig. 170). In fact, in one quarry (FBM Locality M), some fronds have been found associated with inflorescences. The abundance of large, complete, delicately preserved palm fronds and flowers suggests that there were palm groves very near the shores of Fossil Lake.

Today there are more than 2,600 species of palms living in tropical to warm temperate regions of the world.

⋯⟶ FIGURE 169

†*Sabalites powellii* (Newberry, 1883), an extinct species of palm frond in the family Arcaceae. Specimen is from the 18-inch layer of FBM Locality E (FMNH PP53399). Frond measures 1,930 millimeters (6.3 feet) in height. Slab also contains eight fishes, including four †*Knightia eocaena*, two †*Diplomystus dentatus*, one †*Knightia alta*, and one †*Cockerellites liops*.

FIGURE 170

Palm flowers and **INFLORESCENCE** (family Araceae). *Left:* Palm inflorescence, mostly stems missing flowers on a slab with †*Knightia
eocaena.* †*Knightia* measures about 110 millimeters (4.3 inches) in
length. Specimen is from FBM Locality G (FMNH PP43943). *Right top:*
Palm flower from the 18-inch layer of FBM Locality A, measuring
8 millimeters (0.3 inch) in diameter (FMNH PP43972). *Right bottom:*
Another palm flower from the 18-inch layer of FBM Locality A, measuring 8 millimeters (0.3 inch) in diameter (FMNH PP55104).

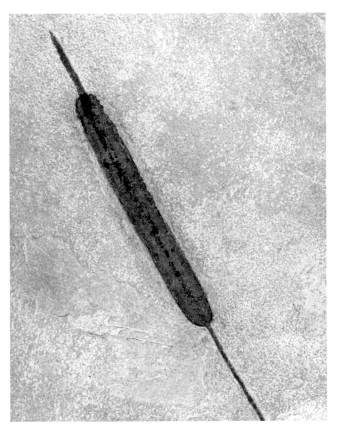
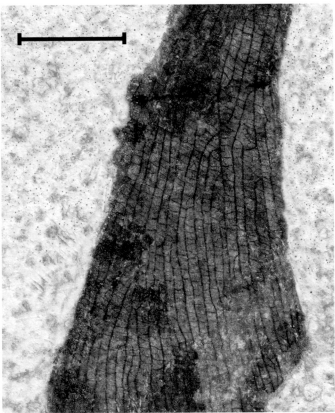

NON-EUDICOT FLOWERING PLANTS

FIGURE 171

Typha sp., an undescribed species of cattail in the family Typhaceae. *Left:* Flowering head. Specimen is from the 18-inch layer of FBM Locality G, and the length of the head is 180 millimeters (7.1 inches) (FMNH PP33654). *Right:* A section of partially rotted leaf from the 18-inch layer of FBM Locality A. Note the characteristic crisscrossing vein pattern (FMNH PP55430).

Cattails (Order Poales, Family Typhaceae)

Cattails are known in the FBM by at least one undescribed species of *Typha*, both as leaves and by its distinctive flowering head (fig. 171, *left*). The leaves, which are occasionally found as fragmented pieces in the FBM, are distinctive in the cross-veined pattern visible under magnification (e.g., fig. 171, *right*). Cattails are wetland plants that grow along freshwater lake margins and in marshes. They often serve as home to insects, birds, and amphibians. The genus *Typha* is today represented by about 11 living species, and it is distributed throughout much of the Northern Hemisphere. Cattails are relatively uncommon in the FBM.

Ceratophyliids (Order Ceratophyllales, Family Ceratophyllaceae)

There is at least one species of *Ceratophyllum* in the FBM (fig. 172). It is a relatively common component of the FBM flora that was described by Herendeen, Les, and Dilcher (1990) as an extinct subspecies of the living *Ceratophyllum muricatum*. This is one of the only FBM species that is supposedly still alive today, although many systematists do not use the classification rank of "subspecies"

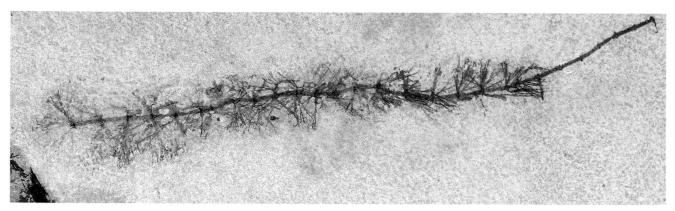

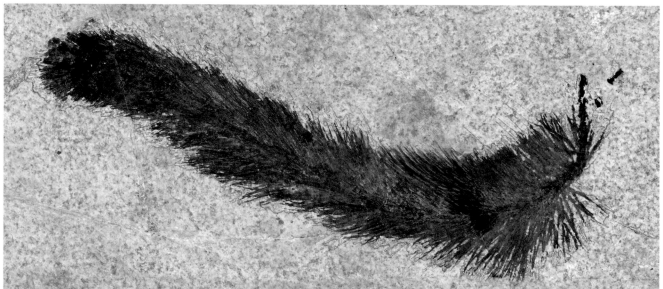

and contend that if a distinct form can be clearly diagnosed, then it should be a new species. Also, limited preservation of the fossil provides insufficient evidence that it is truly the same species (e.g., lack of DNA evidence or certain details of morphology). In any case, the FBM species is clearly in the living genus *Cerato-phyllum*. It is the earliest known occurrence of that genus with well-preserved leaves and stems associated with fruits. The fruits of *Ceratophyllum* are small dark elliptical bodies bearing small spines, although the spines are often not pre-served in the FBM specimens.

Ceratophyllum is a hardy underwater plant that grows in freshwater and in fact is often used in freshwater aquaria today. It is a floating plant without roots, but it can occasionally develop root-like leaves that can anchor it to the bottom. It has short, stiff, hair-like leaves at intervals along nodes of the flexible stem. In the FBM, some specimens show dense bunches of leaves while others have only thin, sparse leaves (fig. 172). In lakes, *Ceratophyllum* offers excellent protection for newly hatched fishes and are an important source of oxygen production for aquatic organisms. Today there are more than 20 living species of *Ceratophyllum*. Fossil species are known as far back as the late Paleocene. The species *C. muricatum* lives today in Asia, North America, Cuba, and Africa, mostly in tropical freshwater environments. For further description of the FBM species of *Ceratophyullum*, see Herendeen, Les, and Dilcher (1990).

←··· FIGURE 172
Ceratophyllum, an aquatic plant in the family Ceratophyllaceae. *Top: Ceratophyllum muricatum* Chamisso, 1829, from the 18-inch layer of FBM Locality G (FMNH PF12455), with a length of 260 millimeters (10.2 inches). This is the only known FBM fossil that is currently classified as a subspecies of a species still living today (Herendeen et al. 1990). On a slab with †*Knightia eocaena* (tip of tail showing in lower left corner). *Bottom: Ceratophyllum* sp., an undescribed species from the 18-inch layer of Locality FBM Locality A, with a length of 200 millimeters (7.9 inches) (FMNH PP 43940).

Eudicot Flowering Plants

(Phylum Angiospermophyta, Subclass Tricolpates)

Eudicots (also called tricolpates, eudicotyledons, and "non-magno-liid DICOTS" by various authors) contain over 70 percent of all living flowering plants, and also make up the majority of the FBM flowering plants. This is a natural (monophyletic) group, unlike "non-eudicot flowering plants" of the last chapter, which is only a group of convenience for the organizational purposes of this book. The major groups of eudicots known from the FBM are as follows:

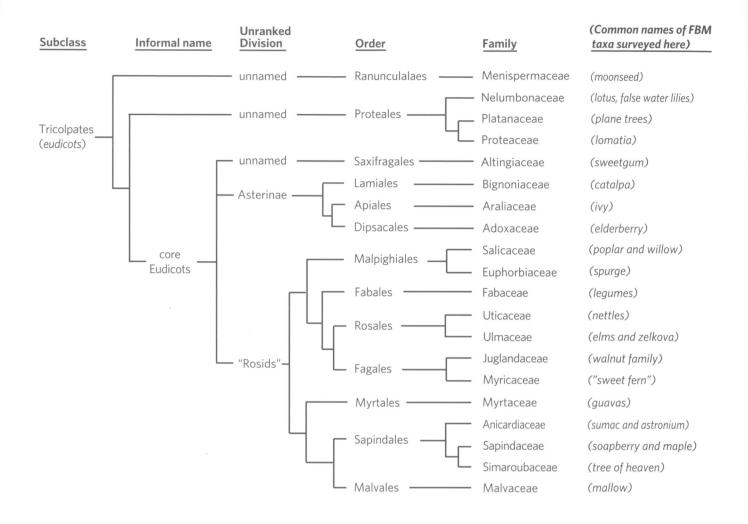

Subclass	Informal name	Unranked Division	Order	Family	(Common names of FBM taxa surveyed here)
Tricolpates (eudicots)		unnamed	Ranunculalaes	Menispermaceae	(moonseed)
		unnamed	Proteales	Nelumbonaceae	(lotus, false water lilies)
				Platanaceae	(plane trees)
				Proteaceae	(lomatia)
	core Eudicots	unnamed	Saxifragales	Altingiaceae	(sweetgum)
		Asterinae	Lamiales	Bignoniaceae	(catalpa)
			Apiales	Araliaceae	(ivy)
			Dipsacales	Adoxaceae	(elderberry)
		"Rosids"	Malpighiales	Salicaceae	(poplar and willow)
				Euphorbiaceae	(spurge)
			Fabales	Fabaceae	(legumes)
			Rosales	Uticaceae	(nettles)
				Ulmaceae	(elms and zelkova)
			Fagales	Juglandaceae	(walnut family)
				Myricaceae	("sweet fern")
			Myrtales	Myrtaceae	(guavas)
			Sapindales	Anicardiaceae	(sumac and astronium)
				Sapindaceae	(soapberry and maple)
				Simaroubaceae	(tree of heaven)
			Malvales	Malvaceae	(mallow)

Moonseed (Order Ranunculales, Family Menispermaceae)

The moonseed family Menispermaceae is a group of tropical flowering plants, some of which produce the famous arrow poison, curare. In the FBM, the family is represented by the genus †*Odontocaryoidea*, which is so far represented only by **ENDOCARPS** (fig. 173), and an unknown genus represented by a leaf (fig. 174). The Menispermaceae are mostly climbing shrubs, and this family is today represented by about 500 living species that live mostly in warm temperate to tropical regions of the Americas, Asia, and Africa. Some fossils from the **CRETACEOUS** may belong in this family, but diversification of the family appears to have happened in the Paleocene and Eocene.

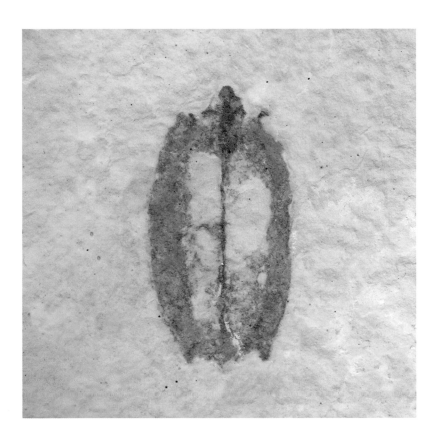

FIGURE 173

†*Odontocaryoidea* sp., an **ENDOCARP** from
the moonseed family Menispermaceae,
from the sandwich beds of FBM Locality L.
Specimen is 22 millimeters (0.9 inch) long
(BHI 5766).

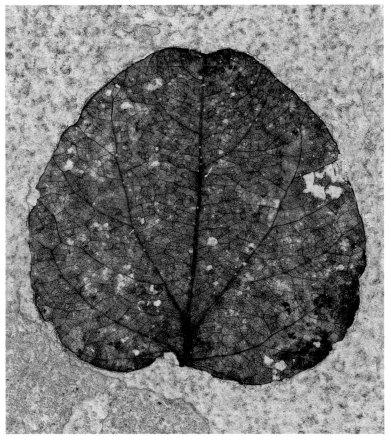

FIGURE 174

An undescribed species of leaf from the moonseed family Menispermaceae, from the 18-inch
layer of FBM Locality B. Specimen is 50 millimeters (2 inches) long (FOBU 10684).

Lotus, Lomatia, and Plane Trees (Order Proteales; Families Nelumbonaceae, Proteaceae, and Platanaceae)

The order Proteales is an ancient one that was diverse already over 100 million years ago in the mid-Cretaceous. It is well represented in the FBM by both aquatic and land species.

Lotus, or false water lilies (family Nelumbonaceae), are well represented in the FBM by lily pads, seed pods, and root systems (fig. 175). There is more than one species present. The most abundant species is the one illustrated here, with lily pads ranging from 100 to 200 millimeters (4 to 8 inches) in diameter. This kind is not yet identified in terms of species and may be new. Another undescribed species, so far only in private collections, is a very large form, more than 500 millimeters (20 inches) in diameter (e.g., fig. 176, *left*).

Nelumbo is a genus of aquatic plant with two living species native to the freshwaters of Asia and North America. The seed pod is very distinctive and looks like a shower head (fig. 175, *bottom left*). *Nelumbo* is the only genus in the family Nelumbonaceae. Lotus are closely related to a terrestrial tree family commonly referred to as the plane trees.

There is one fossil plant in the FBM that resembles the extant Australian genus *Lomatia* (fig. 176, *right*). *Lomatia* is a genus of evergreen flowering plant (family Proteaceae) with 12 living species (although the undescribed FBM species reported here also resembles the genus *Banksia*). Both *Lomatia* and *Banksia* are low-growing ground-cover plants or shrubs with superficially fern-like leaves. Today *Lomatia* is a particularly interesting plant because of its apparently great longevity. *Lomatia tasmanica*, found today in the rainforests of Tasmania, is incapable of reproducing sexually. It can only reproduce through clones: when a branch falls, that branch grows new roots. Each plant's lifespan is about 300 years, but it keeps growing and cloning itself over the centuries. Scientists believe that all *Lomatia tasmanica* living today are a single colony (effectively a single cloning plant). Based on dated fossils in the same region, scientists believe that this plant is between 43,000 and 135,000 years old! Today *Lomatia* and *Banksia* are found only in eastern Australia and western South America. The discovery of it in western North America suggests that it originally had a widespread distribution along

---

···⟩ FIGURE 175

Nelumbo sp., various parts of one or more undescribed species of floating lotus (water lilies) in the family Nelumbonaceae. *Top:* A leaf (lily pad), from FBM Locality G (FMNH PP43979). Longest diameter is 181 millimeters (7.1 inches). *Bottom left:* Seed pod from FBM Locality D, measuring 140 millimeters (5.5 inches) in diameter (FOBU 11744). *Bottom right:* Root system from FBM Locality A, measuring 80 millimeters (3.2 inches) in diameter (FMNH PP53578).

FIGURE 176

Left: A giant partial leaf of the lotus *Nelumbo* sp. (family Nelumbonaceae), showing the underside and stem attachment. Leaf measures 510 millimeters (1.7 feet) in diameter and is from the 18-inch layer of FBM Locality G; specimen is in a private collection. Photo courtesy of Rick Hebdon. Right: A small multi-lobed leaf resembling the Australian genus *Lomatia* (family Proteaceae) from the 18-inch layer of FBM Locality E, measuring 41 millimeters (1.6 inches) in length (FMNH PP45981).

the Pacific Rim. The leaves of *Lomatia* and *Banksia* are tough. Given the rarity of this plant in the FBM (two specimens to my knowledge) and the decomposed state of both specimens, I suspect this was not a plant that grew near the shore of Eocene Fossil Lake but was instead transported from the surrounding highlands via tributaries to the lake.

Plane trees, or sycamores (family Platanaceae), include the living and fossil sycamore genus *Platanus* and the extinct fossil genus †*Macginitiea*. Leaves and fruits from this family are present in the FBM (fig. 177). Specimens of †*Macginitiea* are extremely rare in the FBM and are mostly in private collections (e.g., fig. 178). This genus was widespread in western North America during the Paleocene and Eocene, and is much more common in the middle Eocene beds of Lake Uinta in Utah and Colorado. The FBM species commonly occurs as both three-lobed and five-lobed variants, and the depth of invagination between the lobes varies greatly.

Today the genus *Platanus* is native to subtropical and temperate regions of the Northern Hemisphere. Hybrids have been developed for cooler urban climates and are widely planted in cities worldwide. Plane trees are large trees and can reach 50 meters (164 feet) in height or more. They were probably part of the

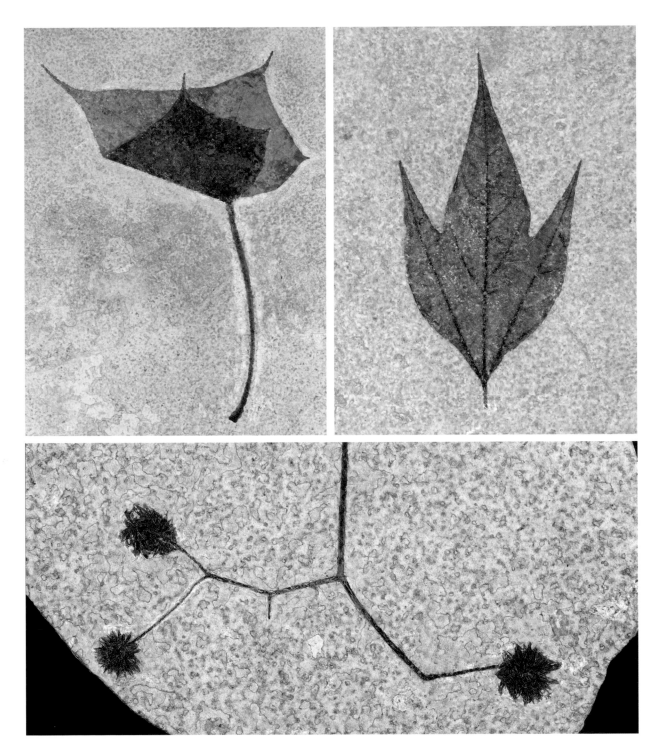

FIGURE 177

Undescribed leaves and fruits of the plane-tree family Platanaceae. From the FBM 18-inch layer. *Top left:* A large leaf (*Platanus*? sp.) with one side partly folded over. Specimen is from FBM Locality A and measures 235 millimeters (9.3 inches) in length (FMNH PP43976). *Top right:* A small leaf that is 120 millimeters (4.7 inches) in length, from FBM Locality A (FMNH PP43963). *Bottom:* A branch with three fruiting heads from FBM Locality A (FMNH PP43978); the three fruits range in size from 10 to 12 millimeters (0.4 to 0.5 inch) in diameter.

↑ FIGURE 178

†*Macginitiea wyomingensis* (Knowlton and Cockerell, 1919) Manchester, 1986, an extinct genus and species of the plane tree family Platanaceae. †*Macginitiea* is very rare in the FBM, but more common in the middle Eocene Lake Uinta deposits of Utah and Colorado. Scale bars in millimeters. Part and counterpart of a very well-preserved specimen of the five-lobed form from the 18-inch layer of FBM Locality E (UMNH PB235).

forest canopy in the highlands surrounding Eocene Fossil Lake. The scarcity of plane tree leaves in the FBM probably indicates that the trees were not growing near the lake and that the leaves were transported into the lake from the tributary streams and rivers from the highlands. The leaves and reproductive structures of Eocene plane trees are reviewed in Manchester (1986).

Sweetgum (Order Saxifragales, Family Altingiaceae)

The sweetgum *Liquidambar* (family Altingiaceae) occurs rarely in the FBM (fig. 179). *Liquidambar* is today represented by four living species restricted to Asia and North America. During the CENOZOIC, there were also species in Europe. The family has an extensive fossil record extending back into the Late Cretaceous. The order Saxifragales is a major group of flowering plants today containing over 2,500 living species. *Liquidambar* today forms part of the hardwood forest, and trees in the genus can reach heights of 40 meters (131 feet) or more. Based on its scarcity in the FBM, this was probably a tree in the highlands surrounding Fossil Lake, whose leaves washed into the lake in the tributary streams.

⇢ FIGURE 179

Liquidambar sp., a fine specimen of leaf from the sweetgum tree (family Altingiaceae). Leaf is 113 millimeters (4.5 inches) long and is from the sandwich beds of FBM Locality L (FOBU 13460).

Catalpa (Order Lamiales, Family Bignoniaceae)

Catalpa leaves are rare in the FBM (fig. 180). This genus of flowering plants is today native to warm temperate regions of East Asia, North America, and the Caribbean. It grows as a medium height tree, reaching about 20 meters (66 feet). Its leaves are very large. The scarcity of catalpa leaves in the FBM and the preference of living catalpa for a temperate climate suggest that these trees lived in the highlands surrounding Eocene Fossil Lake and were transported into the lake in the tributary streams feeding the lake.

FIGURE 180
Leaf from a catalpa tree (*Catalpa* sp.) in the family Bignoniaceae. Specimen is from the 18-inch layer of FBM Locality E, measuring 274 millimeters (10.8 inches) in length (FMNH PP55047).

EUDICOT FLOWERING PLANTS

Ivy (Order Apiales, Family Araliaceae)

There are some beautiful examples of the ivy family Araliaceae preserved in the FBM, including leaves and flowers (figs. 181–183). These remain undescribed. Today there are over 250 species of Araliaceae, living all over the world in a variety of habitats. Some are shrubs, some are vines, and some species are even small trees.

FIGURE 181

Leaf from the ivy family Araliaceae, measuring 200 millimeters (7.9 inches) in length, from the 18-inch layer of FBM Locality F (FOBU 13058).

FIGURE 182

Inflorescence from the ivy family Araliaceae, from the 18-inch layer of FBM Locality E (FMNH PP55052). *Left:* A branch measuring 575 millimeters (1.9 feet) in length from FBM Locality E. *Right:* Close-up of head.

FIGURE 183

A remarkable palmately compound leaf showing at least six leaflets in the ivy family Araliaceae, on a branch from the 18-inch layer of FBM Locality E. Note the variation in leaflet shape, indicating why shape alone is not a reliable character to identify isolated leaflet fossils. Entire specimen measures 300 millimeters (11.8 inches) from the top of highest leaf to the bottom of the stem. Private collection; photo courtesy of Carl and Shirley Ulrich.

Elderberry (Order Dipsacales, Family Adoxaceae)

There is a branch bearing a series of berries from the FBM in the Field Museum's collection tentatively identified as elderberry, family Adoxaceae (fig. 184). This family has about 20 living species today in temperate to subtropical regions in much of the world. There are no leaves from this family identified from the FBM so far, and identification of this specimen is provisional. Also the specimen illustrated does not show the distinctive seeds within the berries and therefore is somewhat inconclusive.

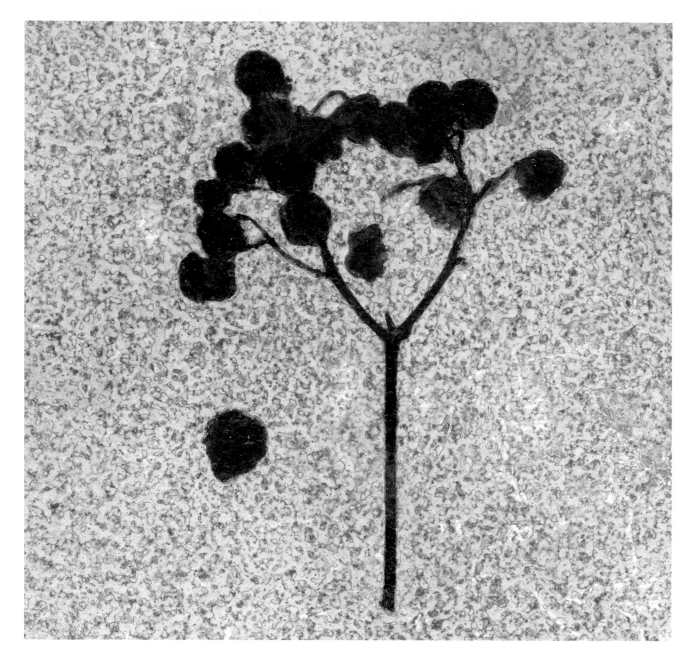

☦ FIGURE 184

A small grouping of elderberry? (family Adoxaceae) from the 18-inch layer of FBM Locality A (FMNH PP54661). Average diameter of each berry is 8 to 10 millimeters (0.3 to 0.4 inch).

⇢ FIGURE 185

Poplar leaves and CAPSULE (family Salicaceae) from the 18-inch layer of the FBM. *Top left: Populus* sp. leaf from FBM Locality A, measuring 98 millimeters (3.9 inches) in length (FMNH PP53554). *Bottom left: Populus* sp. capsule from FBM Locality E, with a length of 16 millimeters (0.6 inch) long (FMNH PP55048). *Right:* †*Populus wilmattae* Cockerell, 1925, from FBM Locality G, measuring 120 millimeters (4.7 inches) in length (BMNH V.62003).

Poplar and Spurge (Order Malpighiales, Families Salicaceae and Euphorbiaceae)

The order Malpighiales is one of the largest orders of flowering plants today, with 35 families and over 16,000 living species. Two families are fairly abundant in the FBM: the poplars and the spurges.

Poplars belong to the genus *Populus* in the family Salicaceae. *Populus* also contains aspens and cottonwoods. The extinct species †*Populus wilmattae* Cockerell (1925) occurs in the FBM, where it is found with a variety of leaf shapes. Leaves with very long-acuminate tips are characteristic of this species (fig. 185,

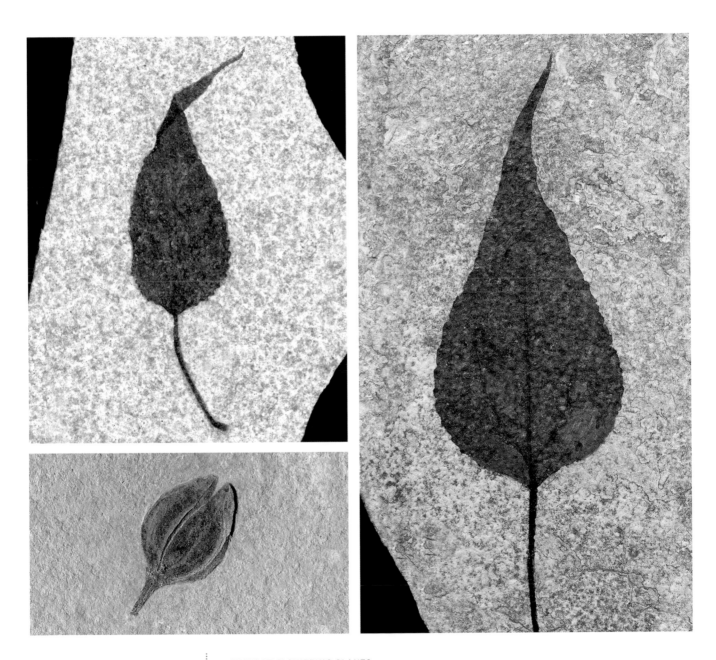

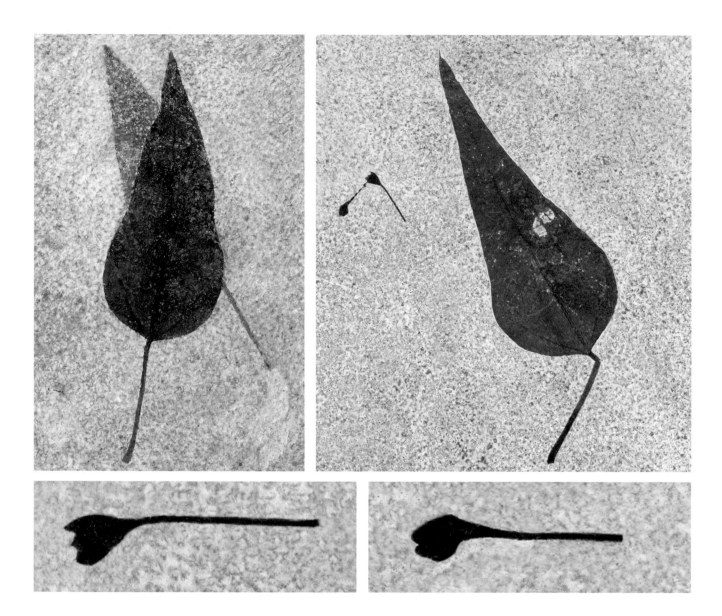

FIGURE 186

Leaves and flower buds of a species that Brown (1937) called †*Ficus mississippiensis* (Lesquereux, 1872), from the 18-inch layer of the FBM. This species does not belong in *Ficus*, and according to Manchester (pers. comm., 2011), it is better left as a species of Magnoliid or Rosid, but with uncertain family relationship. *Top left:* Specimen from FBM Locality B with two leaves (FMNH PP17998). The darker leaf is 123 millimeters (4.8 inches) long. *Top right:* Specimen from FBM Locality A on a slab with two small flowers (FMNH PP54657). Leaf is 165 millimeters (6.5 inches) long. *Bottom:* Two flower buds from above slab on right (digitally separated from each other). Left specimen is 25 millimeters (1 inch) long; right specimen is 19 millimeters (0.7 inch) long.

right). There are probably more than one species of poplar represented in the FBM, but the FBM forms still need to be studied in detail.

The genus *Populus* today contains about 30 living species and is native to most of the Northern Hemisphere. Living *Populus* reach heights of 50 meters (164 feet) or more today. The earliest known fossil poplars are from the Paleocene of western North America. Today poplars are often adapted to living in substrate along rivers (Karrenberg, Edwards, and Kollman 2002). This may explain their abundance in Fossil Lake. The tributary rivers and streams surrounding Fossil Lake were some of the main sources of leaves from inland trees. The foliage and fruits of Eocene poplars are reviewed in detail in Manchester, Judd, and Handley (2006).

One of the more common large leaves in the FBM is a kind referred to as †*Ficus mississippiensis* in Brown (1937) (also commonly misidentified as *Aleurites*). This species does not belong in the genera *Ficus* or *Aleurites*, and, according to Steve Manchester (pers. comm., 2011), it is in need of further study to determine what order, family, and genus it belongs in. One specimen of this leaf has two small flower buds associated with it (fig. 186, *top right* and *bottom*).

Legumes (Order Fabales, Family Fabaceae)

The order Fabales is represented in the FBM by several varieties of legumes (family Fabaceae). This family includes pea and bean plants and a variety of closely related forms. It is the third largest of all flowering plant families, with nearly 20,000 species living today. There are two named extinct species of legume identified in the FBM: †*Swartzia wardelli* MacGinitie, 1969, and †*Parvileguminophyllum coloradensis* (Knowlton, 1923).

†*Swartzia wardelli* is known only by a few specimens in the FBM (e.g., fig. 187A, *left*). The genus still survives today with about 200 living species native to tropical and subtropical regions of the Americas and Africa. Some species of *Swartzia* form shrubs and small trees, although some species in tropical regions reach heights of over 30 meters (100 feet).

†*Parvileguminophyllum* is another legume that occurs in the FBM (fig. 187A, *right*). A fine specimen at Fossil Butte National Monument closely resembles †*Parvileguminophyllum coloradensis* (Knowlton, 1923). This is an extinct genus containing a species previously described from the middle Eocene Parachute Creek Member of the Green River Formation in Colorado and Utah. Another undescribed legume was recently discovered in the FBM, represented by a beautiful specimen with very fine leaflets (fig. 187B). This specimen is currently under study.

FIGURE 187A

Legumes in the bean family Fabaceae. *Left:* †*Swartzia wardelli* MacGinitie, 1969, from the 18-inch layer of FBM Locality G. Specimen measures 40 millimeters (1.6 inches) in length (BMNH v.62002). *Right:* †*Parvileguminophyllum coloradensis* (Knowlton), 1923, from FBM Locality A, measuring 57 millimeters (2.2 inches) in length (FOBU 13377).

An undescribed legume with well-preserved leaflets from
FBM Locality D, near the K-spar tuff bed. The plant measures 255 millimeters (10 inches) in width (FMNH PP55447).

A large unidentified branch with large fruiting CAPSULEs in
various stages of disaggregation. *Left:* Large specimen is 485
millimeters (1.6 feet) long and is from the 18-inch layer of FBM
Locality A (FMNH PP55103). *Top right:* Close-up of a few parts of
a flower from the specimen at the left. *Bottom right:* Another
flower head, probably from the same plant species. Specimen is from FBM Locality L and measures 53 millimeters (2.1
inches) in height (FOBU 11573).

Unidentified Rosids

There are a number of other undescribed species of Rosids from the FBM that have yet to be assigned to an order or a family. One particularly interesting example is a branch with large bulbous fruiting CAPSULES in various stages of disaggregation (fig. 188, *left*). On this piece the upper left branch shows individual valves of disaggregated capsules, the upper right and lower left branches show capsules beginning to open apically to release seeds, and the lower right branch shows stalks from which capsules have become detached. This specimen is currently being studied by Steve Manchester.

The genus *Ficus* (family Moraceae) was also reported from the FBM in Brown (1937), but this identification is of doubtful validity (Steve Manchester, pers. comm., 2011).

Nettles and Elms (Order Rosales, Families Uticaceae and Ulmaceae, and Family Uncertain)

The Rosales is an order of flowering plants named for one of its nine families, the rose family. Although I was unable to identify any occurrences of the rose family (Rosaceae) itself in the FBM, two other families of the order are represented there: the nettle family and the elm family.

The nettle family, Uticaceae, is represented by a well-preserved leaf (fig. 189). Today the family is distributed worldwide with about 2,600 living species in a wide variety of habitats. These plants range from small herbs to shrubs and trees. The FBM leaf is large enough that it was either from a large shrub or a tree.

The elm (and zelkova) family, Ulmaceae, is represented in the FBM by †*Cedrelospermum nervosum* (Newberry, 1883). Fossils of this species include winged fruits and isolated leaves (fig. 190). Although these fruits and leaves are not found directly attached to each other in the FBM, Manchester (1989) described specimens of this species with a preserved branch of attached fruits and leaves from the middle Eocene Lake Uinta deposits of Utah. Branches with attached fruits also occur in the middle Eocene deposits of Messel, Germany. Thus, the association of these fruits to the leaves has been established elsewhere. The fruits are rare in the FBM, but the leaves are relatively common. Manchester (1989) observed that †*Cedrelospermum* is abundant in many fossil localities associated with volcanic ash, indicating that it may have been an early colonizer of open habitats. Plants in the family Ulmaceae are often wind-pollinated and today include about 150 living species in tropical to temperate regions of the world.

Various authors have misidentified specimens of †*Cedrelospermum* as *Zelkova* (e.g., Brown 1946; MacGinitie 1969). Manchester (1989) determined that there were no valid reports of *Zelkova* from anywhere in the Green River Formation. For a comprehensive review of the genus †*Cedrelospermum*, which was common both in North America and Europe, see Manchester (1989).

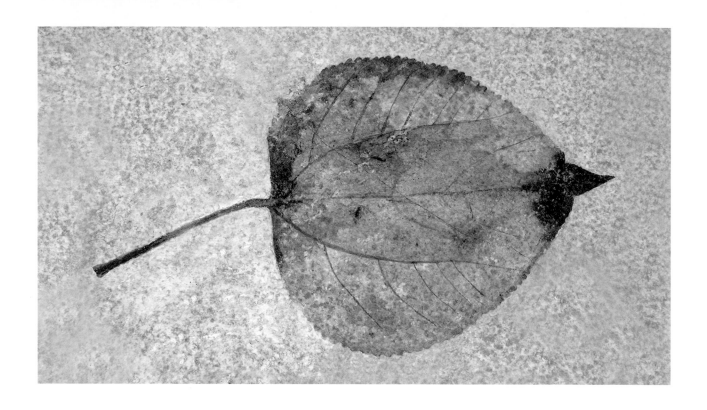

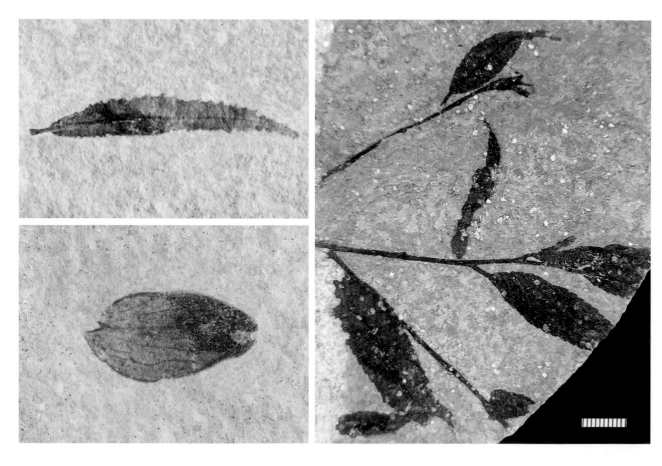

←··· FIGURE 189
A leaf of an undescribed species of nettle (family Urticaceae). Specimen measures 153 millimeters (6 inches) in length and is from the 18-inch layer of FBM Locality E (FMNH PP55049).

↑ FIGURE 191
A compound leaf of an unknown family probably from the order Rosales, from the 18-inch layer of FBM Locality A and measuring 105 millimeters (4.1 inches) in length (FMNH PP55095).

←··· FIGURE 190
†*Cedrelospermum nervosum* (Newberry, 1883) Manchester, 1989, an extinct member of the elm family Ulmaceae. *Left top:* Small leaf measuring 25 millimeters (1 inch) long from the 18-inch layer of FBM Locality E (FMNH PP45996). *Left bottom:* Winged fruit measuring 9 millimeters (0.4 inch) long from the 18-inch layer of FBM Locality A (FMNH PP46003). *Right:* Branch with leaves on a slab with many small white ostracods (†*Hemicyprinotus*), from the sandwich beds of FBM Locality K (USNM 276504). Scale is in millimeters.

Other fine specimens in the order Rosales continue to be discovered and await description (e.g., fig. 191).

Walnut and "Sweet Fern" (Order Fagales, Families Juglandaceae and Myricaceae)

Fagales is an order of plants containing some of the best known trees today, with over 1,500 living species. These species usually have hard wood, are wind-pollinated, and some have slim, elongate clusters of small flowers called catkins. In the FBM, there are at least two families of Fagales.

FIGURE 192
Leaves, branches, and catkins from the East Asian genus *Platycarya*, in the walnut family Juglandaceae. *Top: Platycarya* sp. from the 18-inch layer of FBM Locality A. Compound leaf measures 170 millimeters (6.7 inches) in length (FMNH PP53582). *Bottom left:* †*Platycarya castaneopsis* (Lesquereux, 1883) Wing and Hickey, 1984, from the 18-inch layer of FBM Locality A. Leaf measures 183 millimeters (7.2 inches) in length (FMNH PP43960). *Bottom right:* A pollen catkin of *Platycarya* sp. from the 18-inch layer of FBM Locality A, measuring 124 millimeters (4.9 inches) long (FMNH PP53572).

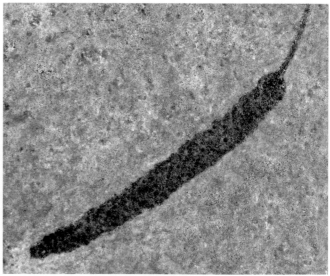

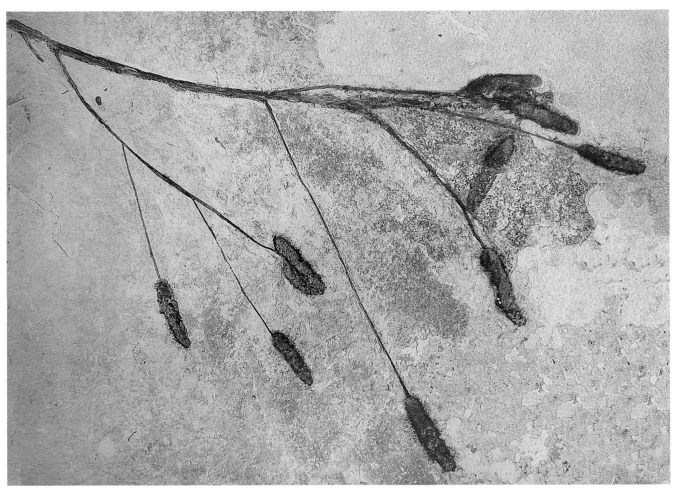

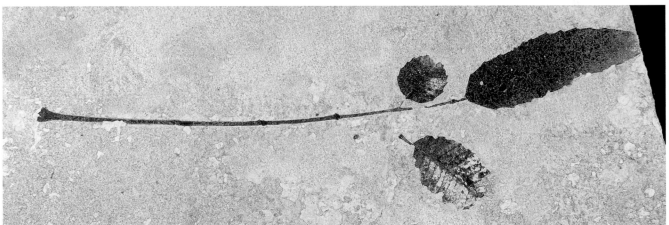

FIGURE 193

Platycarya sp. *Top:* Fruiting stage of an inflorescence from the 18-inch layer of FBM Locality E, measuring 408 millimeters (1.3 feet) in length; in a private collection. Photo courtesy of Danette McMurran. *Bottom:* A specimen showing compound leaves. Uncataloged specimen at the Burke Museum, University of Washington, measuring 440 millimeters (1.4 feet) in length, is from the sandwich beds of FBM Locality L.

FIGURE 194

Comptonia sp., commonly called the "sweet fern," even though it is a flowering plant and not a fern. It is in the family Myricaceae. *Left:* A branch with several leaves, from the 18-inch layer of FBM Locality A. Complete leaf on right of this branch measures about 65 millimeters (2.6 inches) in length (FOBU 11736). *Right:* Single isolated leaf from the 18-inch layer of FBM Locality A, measuring 37 millimeters (1.5 inches) in length (FMNH PP45916).

The walnut family, Juglandaceae, is represented in the FBM mainly by *Platycarya* (figs. 192, 193). This is a relatively common genus of the FBM flora. In addition to isolated LEAFLETS, there are many other types of fossils representing the genus *Platycarya* in the FBM, such as compound leaves, pollen catkins, and fruiting stages of inflorescence. Today *Platycarya* is represented by a single living species in eastern Asia, which grows as a tree to a height of about 45 meters (148 feet). It is a tree that inhabits warm temperate regions. This suggests that †*Platycarya castaneopsis* may have been a very common inhabitant of the highlands surrounding Fossil Lake. Eocene Juglandaceae were reviewed in detail in Manchester (1987).

Another Fagales genus present in the FBM is *Comptonia* (family Myricaceae). Species in this genus are commonly called "sweet fern," which is somewhat confusing because this is not a fern. This is not an uncommon element of the FBM flora and is sometimes found with multiple leaves associated on branch segments (fig. 194). Today *Comptonia* is survived by a single living species native to eastern North America. Fossil species show that this genus once had a much wider geographic distribution throughout much of the Northern Hemisphere.

Another Fagales species previously reported from the FBM was originally described in the oak genus *Quercus* as †*Q. castaneopsis* Lesquereux (1883). This species was later found by Wing and Hickey (1984) to not be an oak (family Fagaceae), but instead to belong in the genus *Platacarya* (family Juglandaceae).

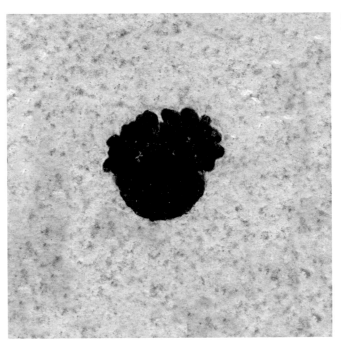 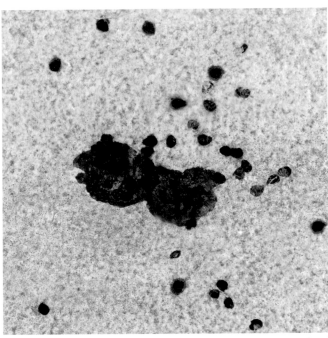

EUDICOT FLOWERING PLANTS

Guavas (Order Myrtales, Family Myrtaceae)

The myrtle family, Myrtaceae, is represented in the FBM by what appear to be seed pods of the genus *Psidium* (a guava). These seed pods are an abundant component of the FBM flora (fig. 195). Today there are about 100 living species in the genus native to tropical and subtropical regions of southeastern Asia, the Americas, and Africa. They are mostly shrubs and small trees. The fruits are important food sources for birds and mammals.

Sumac, Astronium, Soapberry, and Tree of Heaven (Order Sapindales; Families Anacardiaceae, Sapindaceae, and Simaroubaceae)

The order Sapindales today includes nine families containing almost 6,000 species. Three of these families are reported here from the FBM.

Sumac and astronium, both in the family Anacardiaceae, are well represented in the FBM. Sumac species are relatively common elements of the FBM paleoflora, represented there by at least two species: †*Rhus nigricans* and †*Rhus longipetiolata* (fig. 196). Today there are up to 250 living species included in the genus *Rhus*, in subtropical and temperate regions throughout the world. They grow as shrubs or small trees and propagate by birds and other small animals feeding on sumac fruits and spreading the seeds in their droppings. *Astronium* sp. is represented in the FBM by flowers (fig. 197, *top*). Several of these flowers

FIGURE 195
Psidium sp., a guava (family Myrtaceae); berries in two stages of seed eruption from the 18-inch layer of FBM Locality A. *Left:* Specimen is 26 millimeters (1 inch) high with seeds still encapsulated and on top of pod (FMNH PP54659). *Right:* Erupting berry (FMNH PP53580).

FIGURE 196

Sumac (family Anacardiaceae) from the FBM. *Top left:* †*Rhus nigricans* (Lesquereux, 1872) from the 18-inch layer of FBM Locality E. Middle stem measures 90 millimeters (3.5 inches) high (FOBU 47). *Top right:* †*Rhus longipetiolata* Lesquereux, 1883, from the 18-inch layer of FBM Locality A, measuring 57 millimeters (2.2 inches) in length (FMNH PP54664). *Bottom:* A leaf from another species of sumac, possibly *Rhus*. Specimen measures 100 millimeters (3.9 inches) in length and is from the 18-inch layer of FBM Locality A (FMNH PP43971).

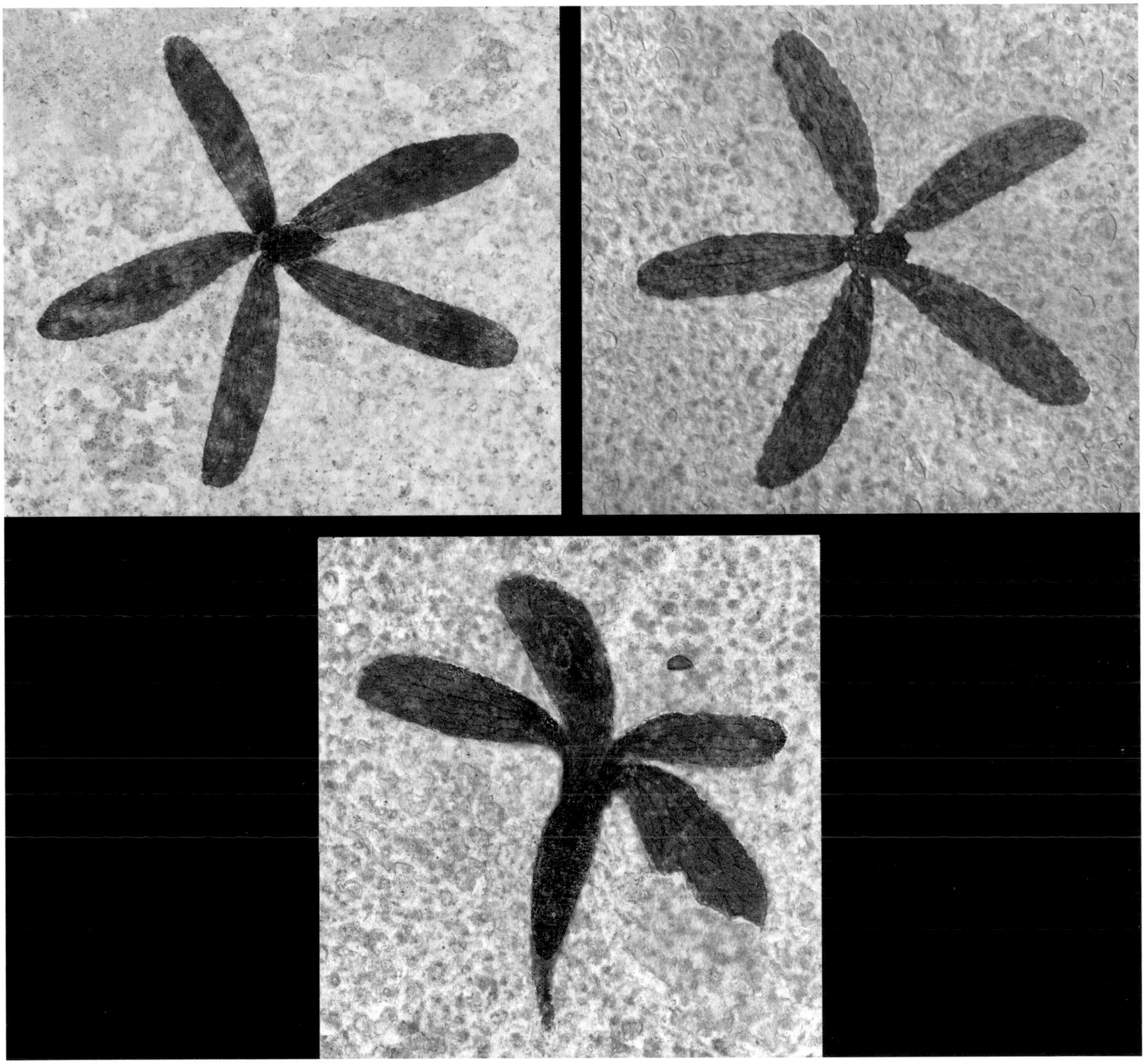

FIGURE 197

Top: Astronium flowers in the cashew and sumac family Anacardiaceae (*Astronium* sp.?). Specimen on the left measures 58 millimeters (2.3 inches) high and is from the 18-inch layer of FBM Locality A (FMNH PP54858). Specimen on the right measures 57 millimeters (2.2 inches) in diameter and is from the 18-inch layer of FBM Locality E. Private collection; photo courtesy of Bruce Baganz. *Bottom*: Another type of flower possibly belonging to the Anacardiaceae. Specimen is from the 18-inch layer of FBM Locality G and measures 40 millimeters (1.6 inches) in height (FMNH PP43970).

are known from the mid-lake quarries. Additional fruits and flowers in the FBM may also be assignable to the family Anacardiaceae (e.g., fig. 197, *bottom*).

The "lychee" or "soapberry" family Sapindaceae is represented in the FBM by several different leaf, seed, and fruit types, including *Koelreuteria* sp., *Allophylus* sp., *Sapindus* sp., *Dipteronia* sp., and *Cardiospermum* sp. Some of these generic assignments are provisional and awaiting further study.

There is a large fruit of a *Koelreuteria*-like plant known from the FBM (fig. 198), which may belong in that genus. Today *Koelreuteria* has three living tree species that are native to southern and eastern Asia.

†*Allophylus flexifolia* is a relatively common leaf in the FBM (fig. 199, *top*). The affinity of the FBM species to the extant *Allophylus* is still under study (Steve Manchester, pers. comm., 2011). The genus *Allophylus* today has over 100 living species that inhabit tropical and subtropical regions of Asia, the Americas, and

FIGURE 198
A large fruit of *Koelreuteria* sp. bearing at least four seeds (family Sapindaceae), from the 18-inch layer of FBM Locality B. Fruit measures 65 millimeters (2.6 inches) in length (FMNH PP55094).

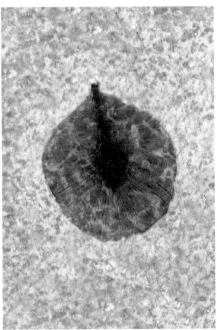

FIGURE 199

Various FBM species from the lychee family Sapindaceae. *Top*: †*Allophylus flexifolia* (Lesquereux, 1883), a large, well-preserved leaf from the 18-inch layer of FBM Locality A. Leaf measures 200 millimeters (7.9 inches) in length (FMNH PP53391). *Bottom left*: *Sapindus* sp., a soapberry plant from the 18-inch layer of FBM Locality C, measuring 82 millimeters (3.2 inches) long (FOBU 19). *Bottom middle*: A winged fruit of *Dipteronia* sp. from the 18-inch layer of FBM Locality A, measuring 19 millimeters (0.8 inch) in diameter (FMNH PP43950). *Bottom right*: A leaf from the balloon vine †*Cardiospermum coloradensis* (Knowlton, 1923), measuring 40 millimeters (1.6 inches) in length, from the 18-inch layer of FBM Locality K (FOBU 10725).

Africa. It usually has the form of a large shrub or small tree often found along stream banks and riverbanks. This suggests that the abundance of these leaves in the FBM may have been the result of the shrubs or trees growing along the banks of the Eocene tributaries to Fossil Lake.

Sapindus sp. is also known from the FBM (fig. 199, *bottom left*). The genus *Sapindus* is also known as the "soapberry" genus, because the fruit of its living species are used for making soap. Today there are about 12 living species in the genus. They are all shrubs or small trees that live in tropical to warm temperate regions of Asia and the Americas. Most species today are pollinated by birds or insects.

Dipteronia sp. is another member of the soapberry family that occurs in the FBM. It is represented there by several specimens of a seed that is completely encircled by a broad wing (fig. 199, *bottom middle*). Today there are two living species of *Dipteronia*, both of which live in China. There is one other fossil species, †*Dipteronia browni*, from the early Eocene to middle Eocene of Oregon, Washington, and British Columbia.

Another member of the soapberry family in the FBM is the balloon vine, †*Cardiospermum coloradensis* (fig. 199, *bottom right*). *Cardiospermum* today contains about 14 living species native to tropical regions of Asia, the Americas, and Africa. They are primarily herbaceous vines.

The winged seed †*Deviacer wolfei* Manchester, 1994, is also represented in the FBM (fig. 200). Manchester provisionally assigned this fossil species to the family Sapindaceae, but placement there is not conclusive.

One of the most common elements of the FBM flora is †*Ailanthus confucii*, or the "tree of heaven," in the family Simaroubaceae. This species is represented in the FBM almost exclusively by the very characteristically shaped winged seeds (fig. 201, *bottom*), although leaflets are also occasionally found (fig. 201, *top*). *Ailanthus* are the most common winged seeds in the FBM by far, known by hundreds of specimens. Today *Ailanthus* is survived by ten living species, and the

FIGURE 200
†*Deviacer wolfei* Manchester, 1994; a winged fruit of uncertain family, provisionally assigned to the family Sapindaceae in the original description. Specimen is from the 18-inch layer of FBM Locality C and is 40 millimeters (1.6 inches) long (FOBU 11716).

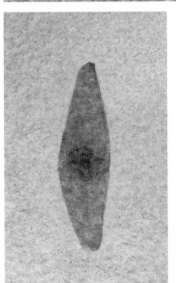
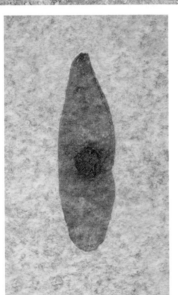
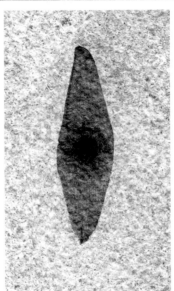
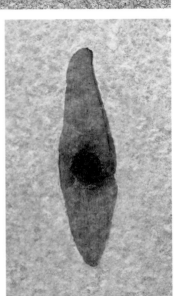

FIGURE 201

†*Ailanthus confucii* Unger, 1850, a close relative of the "tree of heaven" (family Simaroubaceae). *Top:* LEAFLET from the 18-inch layer of FBM Locality E, showing the characteristic rounded glandular tooth at its base. Specimen measures 135 millimeters (5.3 inches) in length. Note the insect damage at the APEX and within the leaf. Private collection; photo courtesy of Danette McMurran. *Bottom:* Winged seeds (fruits) from the 18-inch layer of FBM Locality A. This is one of the most common plant fossils in the FBM. Specimens from left to right are as follows: FMNH PP46002, 40 millimeters (1.6 inches) long; FMNH PP45915, 35 millimeters (1.4 inches) long; FMNH PP54660, 31 millimeters (1.2 inches) long; and FMNH PP55033, 30 millimeters (1.2 inches) long.

FIGURE 202

†*Chaneya tenuis* (Lesquereux, 1883), Wang and Manchester, 2000; a flower-like winged fruit conforming to the family Simaroubaceae. *Left:* Specimen measures 28 millimeters (1.1 inches) in diameter and is from the 18-inch layer of FBM Locality B (FOBU 11466). *Right:* Specimen measures 27 millimeters (1.1 inches) in diameter and is from the 18-inch layer of FBM Locality E (FMNH PP46524).

genus is native only to Asia, Australia, and New Guinea. Although present in the FBM, it eventually became extinct in North America, and its occurrence in North America and Europe today is as an invasive species introduced by man. *Ailanthus* is a fast-growing deciduous tree that grows to 45 meters (147 feet) in height. The fossil record of *Ailanthus* is reviewed in detail by Corbett and Manchester (2004).

Another FBM plant currently included in Simaroubaceae is †*Chaneya*. Wang and Manchester (2000) were the first to describe this genus of flower-like fruit, and they reported it from a number of Eocene and Oligocene localities in western North America, eastern Asia, and the late Miocene of Poland. Species in this genus have five **SEPALS** and a dark globular body in the center. It is known by several specimens from the FBM (e.g., fig. 202). Wang and Manchester did not definitively assign this genus to family, but suggested a possible affinity to Simaroubaceae. Earlier, Lesquereux (1883) had assigned such fossils to *Porana* (family Convolvulaceae), whereas MacGinitie thought they were *Astronium* (family Anacardiaceae). These fossils show multiple ovaries per calyx, which fits some Simaroubaceae but excludes Convolvulaceae and Anacardiaceae. It is not known what FBM leaf type goes with this genus, and specimens with associated fruits and flowers on the same branch are needed to solve this. For further discussion and description of †*Chaneya*, see Wang and Manchester (2000) and Manchester and Zastawniak (2007).

Plants in the Order Malvales (Family Malvaceae and Family Unknown)

The Mallow family (Malvaceae) is represented in the FBM by †*Florissantia speirii* (Lesquereux, 1883), which is known from this locality by only a single flower in the collection at the University of Michigan (fig. 203, *right*). Although this genus

  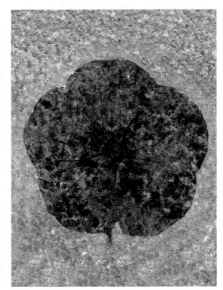

FIGURE 203

Various plants in the order Malvales.
Left and middle: †*"Sterculia" coloradensis,*
Brown, 1929. Specimen FMNH PP53549
measures 177 millimeters (7 inches) long,
and specimen FMNH PP43944 measures
106 millimeters (4.2 inches) long. Both
specimens are from the 18-inch layer of
FBM Locality A. *Right:* †*Florissantia speirii*
(Lesquereux, 1883), an extinct genus of
flower in the Mallow family, Malvaceae.
Specimen is from the 18-inch layer and
measures 32 millimeters (1.3 inches) in
diameter (UMMP 16853).

is extinct, its affinities are well established based on more completely preserved specimens with details of ovary, style, stamens, and pollen from the John Day Formation of Oregon. Like †*Chaneya*, this species was originally attributed to the genus *Porana*. †*Florissantia* fruits and flowers were described in detail in Manchester (1992).

Trilobed leaves that were once thought to represent the Chocolate group of the Mallow family have traditionally been identified as †*"Sterculia" coloradensis,* but inclusion of this species in the extant genus *Sterculia* (and the family Malvaceae) is probably inaccurate (S. Manchester, pers. comm., 2010). This species is a common component of the FBM flora as well as in other Green River Formation localities (e.g., fig. 203, *left* and *middle*).

Today the order Malvales has about 6,000 species grouped within nine families. Most species inhabit tropical or subtropical regions, where they are widely distributed. They also have a limited distribution in temperate regions.

Plants of Unknown Affinity

There are many plants in the FBM whose identities remain elusive. Of course, some of these are the thousands of leaves found in the FBM without clear vein preservation. But some yet unclassified plants are well preserved but simply lack any recognizably diagnostic characters to place them in an existing family or order. Some of these well-preserved but unassigned forms are discussed below.

There is a peculiar winged fruit that occurs in the FBM that was given the species name †*Lagokarpos lacustris* by McMurran and Manchester (2010). This winged fruit is not uncommon in the FBM, where it is sometimes called "rabbit ears" because of its peculiar shape (fig. 205). Although McMurran and Manchester named this species based on type material from the FBM, they were unable to classify this extinct species to family or order. They were also not able to find any of these winged fruits associated with leaves. Perhaps if it is one day found attached to a branch with leaves, it will be more accurately classified. Other five winged fruits that are well represented in the FBM remain undescribed (fig. 206).

EUDICOT FLOWERING PLANTS

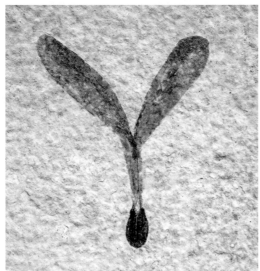

⇡ FIGURE 205

†*Lagokarpos lacustris* McMurran and Manchester, 2010, winged fruits of unknown family and order relationship. Additional material with seeds, flowers, and leaves attached to each other is needed to resolve the relationships of this seed. *Left:* HOLOTYPE specimen showing well-preserved wing vena-tion and globose fruit body; it measures 39 millimeters (1.5 inches) in length (FOBU 9835). *Middle:* Specimen with fruit starting to split into halves. Length (measured along curve of longest wing) is 39 millimeters (1.5 inches) (FMNH PP53568). *Right:* PARATYPE with fruit further split. Length (measured along curve of longest wing) is 39 millimeters (1.5 inches) (FMNH PP43951).

⇠⋯ FIGURE 204

Unidentified leaves, leaflets, seeds, and branches, all from the 18-inch layer of FBM Locality A. There is much plant material from the FBM that remains unidentified. Hopefully as time goes on, more attention will focus on this rich source of Eocene plant diversity. *Left:* Speci-men measuring 220 millimeters (8.7 inches) high (FMNH PP54658). *Top right:* Specimen measuring 200 millimeters (7.9 inches) long (FMNH PP45982). *Middle right:* Specimen measuring 90 millimeters (3.5 inches) long (FMNH PP53547). *Bot-tom right:* Specimen measuring 70 millimeters (2.8 inches) in length (FMNH PP43959).

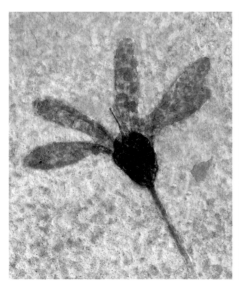

⇡ FIGURE 206

An unidentified five-winged fruit. *Left:* Specimen from the 18-inch layer of FBM Locality A, measuring 35 millimeters (1.4 inches) in height (FMNH PP43957). *Right:* Another specimen also showing a smaller bud. Specimen from the 18-inch layer of FBM Locality A, measuring 41 millimeters (1.6 inches) in height (FMNH PP46525).

FIGURE 207
Various unidentified flowers from the
FBM. *Left:* Unidentified flowers on
branches from the 18-inch layer of FBM
Locality A. Specimen measures 80 mil-
limeters (3.2 inches) in height (FMNH
PP53540). *Right:* A small unidentified
flower from the 18-inch layer of FBM
Locality A, measuring 21 millimeters (0.8
inch) in diameter (FMNH PP53387).

In addition, there are numerous leaves and flowers that remain unclassified but are well-enough preserved so that further study may allow more detailed classification (e.g., figs. 204, 207). And as always in science, as our knowledge of plant phylogeny improves with time, some of the classified plants above may warrant reclassification in different families or orders. That is both progress and the continuing challenge.

What Is Most Obviously Missing from the FBM Plant Flora Compared to the Modern Flora?

One major modern plant group that is missing in the FBM is the grass family Poaceae. Grasses are today some of the most diverse of all plants, with about 10,000 living species, but no grasses have been reported from the FBM. Fossil grasses are known as far back as the early Eocene, but the 52-million-year-old Fossil Lake may have been before grasses had undergone much diversification. Also, grasses and other herbs are rarely preserved in lacustrine sediments compared to deciduous-leaved trees and shrubs. Grasses do not shed leaves readily, so the absence of grasses in the FBM may simply be due to lack of preservation and fossilization of grasses that existed around Fossil Lake (i.e., TAPHONOMY). Recently some Late Cretaceous dinosaur coprolites were discovered that contained PHYTOLITHS resembling grass phytoliths, although exact classification of these is still under study. Succulents are also absent from the FBM, but like grasses this may also be a factor of taphonomy.

Trace Fossils

Trace fossils (sometimes also called "ichnofossils") demonstrate by their numbers that Eocene Fossil Lake was a very biologically active ecosystem. This type of fossil is a trace or geological record of biological activity in the past rather than the animals and plants themselves, and include trackways, burrows, feeding marks, and fossilized dung referred to as "coprolites."

The vast majority of trace fossils in the FBM are well-preserved coprolites, which come in all different shapes and sizes (figs. 208, 209). With millions of fishes and other aquatic vertebrates, there was a lot of fecal matter being produced on a daily basis within the Eocene lake, and a lot of it fossilized. This is unusual because most fossilized freshwater fish localities have few, if any, coprolites (Edwards 1976). It is also unusual because preservation of coprolites in the FBM is very three-dimensional. It may seem counterintuitive, that something as soft as fecal matter would be less compressed during the fossilization process than vertebrate skeletons. The reason for such three-dimensionality of coprolites is a series of

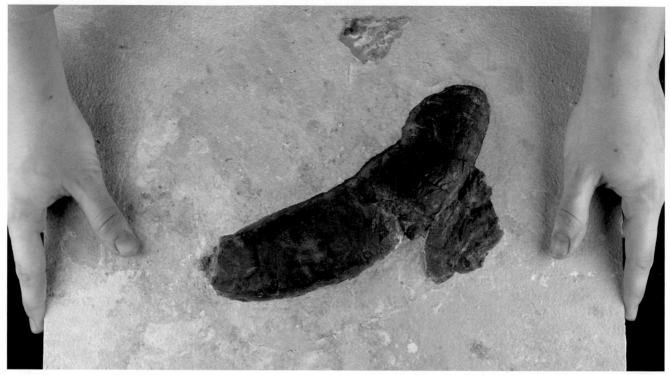

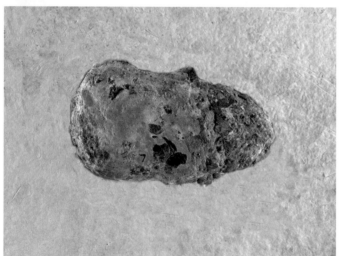

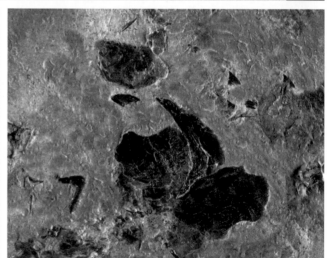

FIGURE 208

Coprolites from large animals. *Top:* Big coprolite most probably from a crocodile, given its size. Specimen is from the 18-inch layer of FBM Locality A and measures 230 millimeters (9.1 inches) in length (FMNH PR3042). *Bottom left:* Large coprolite containing many †*Knightia* bones, probably from a small crocodilian or large turtle. Specimen is from the 18-inch layer of FBM Locality A and measures 76 millimeters (3 inches) in length (FMNH PR3041). *Bottom right:* Close-up of the †*Knightia* bones from specimen illustrated in bottom left.

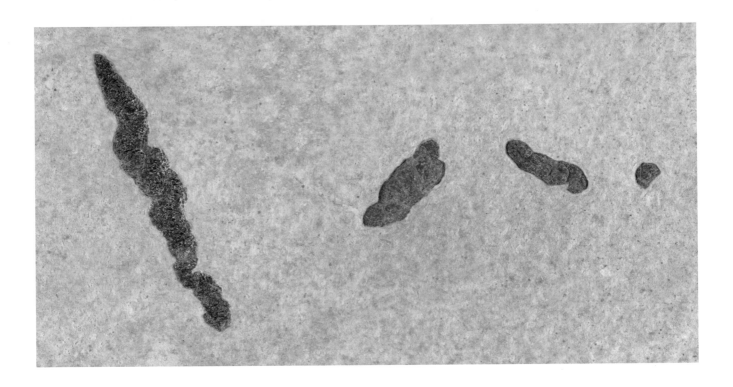

FIGURE 209
Coprolites from small animals in a variety
of shapes. Most probably from fishes.
Long, rope-shaped coprolite on the left
measures 37 millimeters (1.5 inches) and
is from the 18-inch layer of FBM Locality
A (FMNH PF15381).

chemical reactions that occur between the fecal matter, *E. coli* (common digestive
bacteria), and the sedimentary environment, making the LITHIFICATION of fecal
matter an unusually rapid process. In fact, lithification of fecal matter can take as
little as a few weeks in some cases (Chin et al. 2003).

There are many distinct types of coprolites in the FBM. Some are too large
to have come from fishes or other small organisms that inhabited Fossil Lake, so
by process of elimination (no pun intended) the large crocodiles of Fossil Lake
must have produced them (fig. 208, *top*). Shape and contents are other indicators. A plentiful supply of medium-size drop-shaped coprolites containing fish
bones and scales (fig. 208, *bottom*) were likely from smaller crocodilians or large
turtles that were feeding on fishes. In general, they indicate that the favorite
food fishes were the herring †*Knightia* and small individuals of the †paraclupeid
†*Diplomystus*.

The smaller coprolites in the FBM are probably the work of fishes and come
in a variety of shapes and sizes including drop-, bead-, patty-, and rope-like
shapes (e.g., fig. 209). The rope-shaped coprolites are among the most common
in the FBM, and microscopic examination of them reveal no bone or shell fragments, internal structures, or fossilized microorganisms. These types of coprolites are also occasionally found *in situ* within body fossils of †*Knightia*. Thus
these rope-like coprolites are probably mostly produced by †*Knightia eocaena*
and represent digested plankton such as green algae. Edwards (1976) examined

the chemical composition of these and found them to now be composed primarily of the mineral apatite. He also noted that fossil localities that produce rope-shaped coprolites are extremely rare in the fossil record.

Paleontologists who specialize in the study of coprolites are called paleoscatologists. In the mid-1970s, a graduate student named Paul Edwards from the University of Nebraska began to study the FBM coprolites in detail. Tragically, he was killed in an accident in 1976, and his one paper on the FBM coprolites (Edwards 1976) was published posthumously. Hopefully one day another student or professional paleoscatologist will pick up where Edwards left off in the FBM. There is an amazing diversity and abundance of well-preserved coprolites in the FBM, and most of it awaits detailed study.

Other types of trace fossil present in the FBM include trackways, drag marks, burrows, and feeding marks. Trackways or trails are extremely rare in the FBM and consist primarily of trails left by swimming fishes. Martin, Vasquez-Prokopec, and Page (2010) were first to describe a type of bottom-feeding trail known by several specimens from the FBM that they attributed to the gonorynchid fish †*Notogoneus* (e.g., fig. 210). Such feeding trails constitute evidence that the mid-

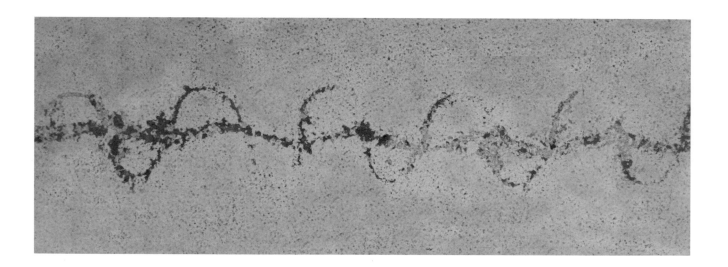

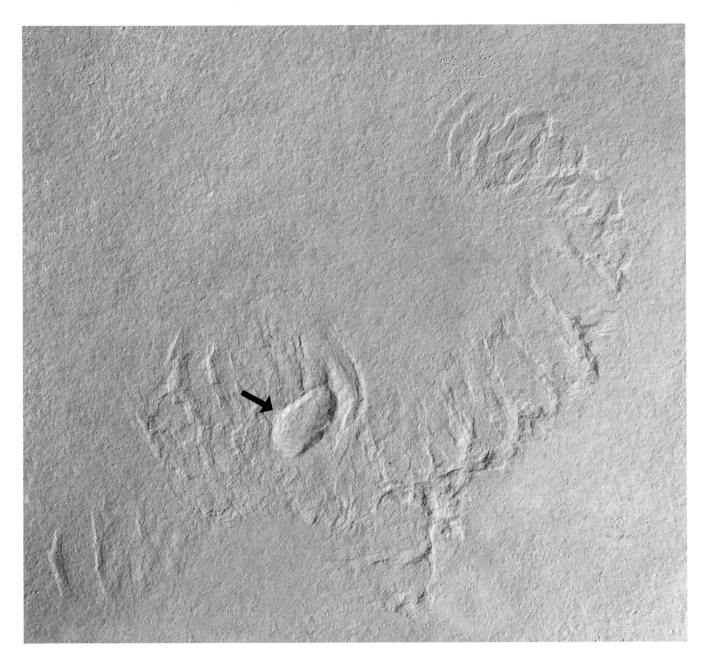

←⋯ FIGURE 210
Trace fossil. Feeding trail of a bottom-feeding fish
(probably made by †*Notogoneus osculus*), from the
18-inch layer of FBM Locality B. *Top:* Slab measuring 430 millimeters (1.4 feet) in length (FOBU 11619).
Bottom: Reconstruction of a †*Notogoneus osculus* as
it might have looked making the above trail.

⬆ FIGURE 211
Trace fossil of a bivalve from the sandwich beds of
FBM Locality M. Arrow points to possible bivalve
that apparently left the trail. This was probably an
instance of a small current or perhaps a predator
moving this piece along the bottom. Specimen
is FMNH PE60951, and the shell that the arrow is
pointing to measures 21 millimeters (0.8 inch)
along the longest axis.

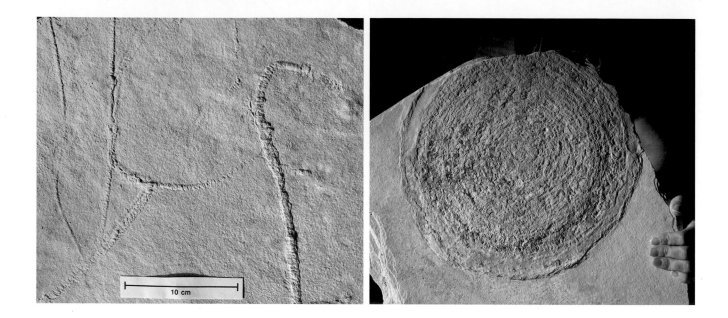

FIGURE 212
Other trace fossils from FBM Locality A. *Left:* Burrows of an aquatic invertebrate, from the 18-inch layer (FMNH PE61027). Such burrow fossils are extremely rare in the 18-inch layer of the FBM. *Right:* Large concentric structure probably resulting from escape of gas from the lake-bottom sediments. Specimen is from the sandwich beds (FMNH PE61044).

lake bottom waters of Fossil Lake were not always ANOXIC (although between water turnovers and for most of the time in the lake they likely were; page 358). Other trails of uncertain origin are also present in the FBM. Drag marks are one of the most common trace fossils in the FBM. Most have yet to be identified to origin, but some include the organism that caused them. One such piece in the Field Museum's collection is a slab from the sandwich beds showing drag marks associated with a bivalve (fig. 211). Picture a slight current caused by a large swimming animal above, moving a clam shell around the lake bottom. There are also many types of drag marks in the 18-inch layer of the FBM whose origins are unknown. Burrows, or holes dug by small animals such as worms, are extremely rare within the FBM, especially from the 18-inch layer and the sandwich bed localities listed here (appendix A). They can occasionally be found at the very margins of Fossil Lake, possibly indicating that the depth of the lake dropped off rapidly with distance from the shore. They were reported from the FBM by Buchheim, Cushman, and Biaggi (2012, 172). Feeding marks are abundant trace fossils in the FBM and are found mostly on fossil leaves (e.g., see fig. 36). In addition, some are found on vertebrates fossils representing predation (fig. 104) and on the limestone bedding plane representing feeding marks or burrows from bottom-feeding fish or invertebrates (figs. 210 and 212, *left*).

Reading the Pages of
Deep History

So how do we interpret this extinct community, so beautifully preserved within the FBM? The diverse assemblage of fossils presents us with **EMPIRICAL** data with which to reconstruct deep history. It raises a number of questions as well.

The Paleoecology of Fossil Lake: What Was Life Like 52 Million Years Ago in Southwestern Wyoming?

The fossils in the FBM make up a collage of 52-million-year-old "snapshots" documenting a biologically diverse freshwater lake surrounded by subtropical lowlands and more distant temperate upland regions. The plants and animals were transitional species between the age of the dinosaurs and modern day. The early Eocene included the warmest global temperatures of the Cenozoic era (Wilf 2000); and the warm, humid climate of Fossil Lake was similar to the present climate of the Gulf coast and southern Atlantic regions of the United States, with an annual rainfall of 30 to 40 inches and nearly frostless winters (Bradley 1929, 1948; MacGinitie 1969). That ecosystem and ecology contrast sharply with the cool mountain-desert climate that exists in the region today (fig. 213). There were active volcanoes about 120 miles north of the lake that had several major eruptions during the lake's existence and left thin ash layers within the FBM. Occasionally these eruptions caused massive forest fires and catastrophic mass kills in the lake, as evidenced by occasional beds that are heavily blackened with carbon and charcoal occurring just above or below fish mass-mortality layers.

A natural balance of prey and predator existed in the lake for millennia as part of a complex food chain. Algae such as *Pediastrum* sp. were the PRIMARY PRODUCERS that converted sunlight and carbon dioxide into sugar and served as the primary food source of the lake. PRIMARY CONSUMERS included vast schools of the filter-feeding fish †*Knightia eocaena*. This species existed in the lake by the billions and fed on primary producers such as algae and other microorganisms. It was the most common of all the fish species in the lake, and it reproduced in great numbers, providing one of the most abundant food sources for higher links in the food chain. SECONDARY CONSUMERS included predaceous fishes (†*Crossopholis*, *Lepisosteus*, *Amia*, *Atractosteus*, †*Phareodus*, †*Diplomystus*, †*Priscacara*, and †*Mioplosus*), as well as frigatebirds, bats, trionychid turtles, aquatic lizards, small crocodilians, the otter-like †pantolestid, and many other animals. †*Knightia* was the primary link in the food chain that fueled this level of the ecosystem. And at the top there were TERTIARY CONSUMERS such as the large crocodiles, alligators, monitor lizards, and giant trionychid turtles that preyed on the secondary consumers. On shore the primary producers were land plants and shallow-water plants that were fed upon by herbivores ranging from insects to the small horse †*Protorohippus venticolus*. Insects were, in turn, fed upon by secondary consumers like birds, bats, fishes, and insectivorous mammals; and perhaps †*Protorohippus* occasionally fell prey to the large crocodiles in the lake.

Near the shore there were aquatic organisms such as water lilies, floating ferns, ceratophyllum, and many kinds of swimming insect larvae and nymphs. Along the shoreline were cattails, horsetail, elephant ear plants, ferns, sumac, balloon vines, and palms. An abundance of very large well-preserved palm fronds in the FBM indicate that large groves of palm trees must have grown close to the water's edge. Dragonflies and damselflies filled the air along the shoreline while clouds of march flies and small biting gnats swarmed over the water. The nearshore aquatic plants helped provide nursery grounds for the many schools of baby fishes that dwelt there. The lake was filled by water flowing down from the uplands, including a major river to the northeast. This resulted in some river inhabitants showing up in the northern nearshore FBM quarries from time to time, like the pickerel †*Esox kronneri*, the Eocene mooneye †*Hiodon falcatus*, the trout-perch †*Amphiplaga brachyptera*, clams, the otter-like †*Palaeosinopa*, as well as freshwater shrimp and crayfish.

FIGURE 213

Fossil Lake, then and now. *Top:* Reconstruction of near-eastern shore as it may have looked 52 million years ago. The two dawn horses (†*Protorohippus*) drink at the water's edge while being viewed by a crocodile (†*Borealosuchus*), and out in the lake frigatebirds (†*Limnofregata*) dive to the surface to snatch small herring (†*Knightia*) for dinner. Palms (†*Sabalites*) and elephant ear plants (*Colocasia*) line the shore near the lake's edge, with climbing ferns (*Lygodium*) on some of the palm trunks. Cattail (*Typha*) and water lotus (*Nelumbo*) grow in the water near shore. In the far distance, part of the surrounding highlands is visible. *Bottom:* What remains of Fossil Lake today: the limestone of the FBM at the tops of isolated buttes with eroded valleys in between. The remaining record of Fossil Lake is encapsulated in the sedimentary rock layers being mined there.

The river also carried cooler-climate vegetation into the lake. The more temperate, cooler-climate plants in the distant highlands—including conifer, plane trees, and sweetgum—shed their leaves and fruits into streams and rivers flowing down the mountains.

Around the lake were small groups of the three-toed "dawn horse," †*Protorohippus venticolus*, which grazed on vegetation. The palms and other trees were filled with a diverse array of birds, insects, and small carnivorous mammals. The large true carnivores of today of the mammalian order Carnivora had not yet evolved, and those mammalian forms occupying that niche were very different than those of today. One such form was a small carnivorous tree-climbing mammal, the first to have developed a prehensile tail allowing it an agile life in the trees (fig. 144). It swung freely through the tree canopy with its long tail and graceful arms and legs, feeding on small animals along the way. The top predators of the time continued to be reptilian, such as the large crocodiles and monitor lizards, as well as the giant 2-meter-tall flightless bird †*Gastornis*. The daytime skies contained frigatebirds that swooped down from the sky to feed on fishes and other small vertebrates in the lake. Rails, rollers, and birds that resembled a cross between ducks and flamingos were common shorebirds, and many species of parrots inhabited the trees surrounding the lake. It is hard today to think of Wyoming as a hotspot for parrot diversity! There were colonies of bats near the north end of the lake, where most of the 35 or so fossil bat specimens have been discovered. Today, bats use echolocation as a way to navigate and find food in the dark. One of the bat species that lived near Fossil Lake, †*Onychonycteris finneyi*, appears to have lacked the ability to echolocate and may have been a day flier also inhabiting nearshore trees.

From palm trees to crocodiles, the FBM fossils provide a quasi-photographic record of a lost world. There are more than just a few isolated plant and animal species preserved here; there is a complex and vibrant ecosystem locked in stone.

Death and Mass Mortality in Eocene Fossil Lake: What Killed Everything?

The spectacular abundance of fossils in the FBM and their often life-like preservation raises the question: What killed these organisms? There are many possible explanations. Many catastrophic deaths were due to mass-mortality events that killed thousands or even hundreds of thousands of individuals at a time. Other deaths were clearly due to individual causes such as predation. Then there are specimens in which the cause of death is still a mystery.

Ecological catastrophes are not rare in nature. In fact, throughout Earth's history, and even today, catastrophic events involving mass mortalities in aquatic environments are common. But it is rare that documentations of geologically ancient catastrophes are as well preserved as they are in the FBM. Ecological catastrophes occur at many different scales, and they could be classified as one of two basic types: *global mass extinction* or *regional mass mortality*. Global mass-extinction events are rare phenomena that involve the extinction of 50 percent or more of all species on the planet and can take hundreds or even thousands of years from start to finish. There have been at least five of these in the last 540 million years, and a good example is the one at the end of the Cretaceous, discussed on page 2. The FBM captures a picture of the early recovery of the North American BIOTA, roughly midway between the post-Cretaceous global mass-extinction event and today.

The FBM mortalities were not global in nature. Instead, the many mortality layers of the FBM represent regional mass mortalities. Regional mass mortalities are relatively common ecological events. These events are localized rather than global in scale and take place quickly, usually in a matter of hours, days, or weeks. Today, these type of mortality events commonly occur in aquatic environments and result from factors ranging from an overly warm summer day in a shallow bay to an oil spill in the ocean. There are several places where the regional rapid die-off of millions of fishes and aquatic organisms occurs repeatedly, even annually (e.g., Scott and Crossman 1973, 126). Smith (1949) described modern examples of bays where the bottom was covered with a solid mass of fish carcasses. Although localized die-offs in nature are not uncommon, conditions that allow fossilized preservation of these mass mortalities are very rare.

Several scientific explanations have been offered for the localized mass mortalities that occurred within the FBM. It is important to remember that the FBM is not a single mass mortality. The FBM spanned at least many thousands of years, and within it there are some layers with millions or billions of fossils, and others with very few fossils. Many different regional mass mortalities occurred throughout the duration of Eocene Fossil Lake. Within the mid-lake quarries, these are represented by several different zones, including many consisting almost exclusively of †*Knightia eocaena*. The †*Knightia* "death layers" represent enormous schools that were quickly killed, and some layers contain up to hundreds of bodies per square meter. There is one †*Knightia eocaena* mass-mortality zone that occurs on parts of Fossil Ridge at the base of the 18-inch layer (e.g., fig. 214, *top*), one that occurs near the K-spar tuff layer (fig. 74), and several others in between. Within the 18-inch layer, there are also several mass mortalities of the percoid fish †*Cockerellites liops*, the most prominent of which is a layer just above a volcanic ash (bentonitic) layer about one-third of the way down in the 18-inch layer of the mid-lake deposits (e.g., fig. 93). There is also a mass-mortality layer of small specimens of the percoid †*Mioplosus* near the K-spar tuff zone in FBM Locality A (fig. 83, *bottom*). All of these species-specific mass mortalities represent mass kills of enormous schools of fishes over a short period of time, sometimes covering areas of thousands of square meters or more. This brings us back to the question: What caused the mass mortalities in the FBM?

FIGURE 214

Mass mortality of the herring family, Clupeidae, past and present. *Top:* Mass mortality of †*Knightia eocaena* from the base of the 18-inch layer on Fossil Butte. At this particular site, this layer has a density of several hundred fishes per square meter, with each fish averaging about 100 millimeters (3.9 inches) in length. Slab is AMNH 13101. *Bottom:* A mass mortality of the cludeid *Alosa pseudoharangus* (also called the alewife) that was typical during the summer in Lake Michigan during the 1960s. This kill, like the fossil illustrated above, has a density of several hundred fishes per square meter with each fish also averaging about 100 millimeters (3.9 inches). Photo taken on a Chicago beach, but mass kill also covered part of the lake bottom. Clupeids have had a long history of sensitivity to abrupt fluctuations in environmental conditions.

Explanations for individual mass mortalities vary. The †*Cockerellites liops* mass mortality just above the ash bed in the 18-inch layer seems easy to explain: the intense volcanic activity must have changed the water chemistry to cause localized die-offs within Fossil Lake, and at that time there were large schools of †*Cockerellites liops* in the area affected. Volcanic activity may also have been responsible for mass mortalities of insects such as the march fly zone in the capping layer below the 18-inch layer (fig. 34, *bottom*), although march flies are so prolific that a large swarm caught in a storm over the lake would also have provided a mortality layer.

Reasons for mass mortalities at other levels, consisting primarily of the clupeid herring †*Knightia eocaena*, were probably related either to changing temperature or changing water chemistry. Today living clupeids are extremely susceptible to changing water conditions. From the 1960s through the 1980s, Lake Michigan had summer mass mortalities of the clupeid *Alosa pseudoharangus*, which involved millions of carcasses washing up on shore and millions more sinking to the bottom of the lake (e.g., fig. 214, *bottom*). Clupeids are not only very susceptible to mass die-offs; they reproduce faster than most other fish species and can produce billions of eggs to quickly repopulate lakes with the new generations of clupeids. A single living clupeid can today lay as many as 200,000 eggs at one time. It is not therefore surprising that †*Knightia eocaena* was such a common element of the FBM mass mortalities.

Other factors besides volcanic activity could have caused lethal changes of water chemistry in parts of Eocene Fossil Lake. One such factor could have been excessive ALGAL BLOOMS. Explosive growth of green algae would have eventually crashed due to overpopulation, and the decaying organic material from such crashes could have depleted the water's oxygen content to lethal levels. Similarly, overgrowth of cyanobacteria ("blue-green algae") and plankton in regions of the lake may also have caused mass mortalities. PLANKTONIC ORGANISMS are usually beneficial to the fish community, for all primary fish feed directly or indirectly on these tiny organisms. However, under certain conditions, one or a few species of these organisms multiply with abnormal rapidity. The discoloration of the water caused by such rapid multiplication or by swarming of microscopic organisms is called water bloom (Brongersma-Sanders 1957). Cyanobacteria can generate a strong poison in the water during water blooms (Fitch et al. 1934; Shelubsky 1951). Komarovsky (1951) described heavy mortalities in fish ponds in the Jordan Valley of Israel caused by water blooms and cyanobacteria. McGrew (1975) suggested that minor annual blooms of cyanobacteria frequently caused limited mortalities throughout the FBM, and that an occasional "super bloom" or water bloom was responsible for some of the mass-mortality layers. This theory also fits because cyanobacteria are known to cause precipitation of $CaCO_3$ (the thin white laminations in fig. 6A, *bottom left*), which would facilitate the burial of the specimens and impede decomposition.

Another possible reason for mass mortalities in the FBM is stratified water turnover. As we have seen, Fossil Lake appears to have been subtropical, and for this reason, it was probably stratified, or thermally "layered," from time to time. In subtropical areas during the coldest part of the year, the water temperature of a lake is basically homogenous. The wind blowing across the water's surface sets up a slow circulation throughout the lake, and oxygenated surface waters are carried to all parts of the lake. With the arrival of summer, the surface water warms up rapidly, and as it warms, its density and viscosity decrease until a distinct upper layer of relatively warm, light, less viscous water is established (fig. 215, *left*). This layer, the epilimnion, rests on a deeper, colder, denser body of water known as the hypolimnion, with a relatively thin transitional zone, or thermocline, between the two. In the stratified lake, only the light, warm water of the epilimnion mixes with the oxygenated surface water. In the hypolimnion, the decay of organic matter continues to use up the remaining dissolved oxygen and become concentrated with hydrogen sulfide until it is lethal to most organisms. The lake remains stratified until periodic storms or seasonal turnover mix the layers of the lake (Bradley 1948). In Fossil Lake, the periodic water turnovers may have caused localized mass mortalities as fishes came in contact with the lethal waters of the hypolimnion rising to the lake's surface in some areas.

In addition to mass mortalities, there are also individual deaths that can be attributed to specific causes. For example, the turtle in figure 104 appears to have been bitten by a crocodilian. It initially escaped being eaten by the predator, but it eventually died, possibly from an infection or other bite-related problem. Some fishes appear to have choked to death on prey too big to be swallowed (e.g., figs. 71, 90). Other specimens show indications of predation, although it is unclear whether they were preyed upon before or after death. For example, there are a number of complete, beautifully articulated bird skeletons missing the head (fig. 142B, *right*), which suggest that the head was removed either by a predator causing its death or by a carrion feeder while it was floating in the lake. Then there are perplexing cases, like the small three-toed horse in figure 152. This beautifully articulated, complete skeleton from FBM Locality D invites the question: How did this horse make it to the center of the lake, sink, and get buried in sediment

FIGURE 215

Schematic cross-section model of a subtropical lake during stratification (*left*) and during mixing (*right*). During periods of stratification, the lake's upper warmer and lighter waters form an "epilimnion" that does not circulate with the bottom waters. The bottom waters (the "hypolimnion") are separated from the epilimnion by a thermocline (between the two dashed lines), and the loss of oxygen and buildup of toxic gases eventually make the hypolimnion uninhabitable for most aquatic animals that continue to inhabit the epilimnion above.

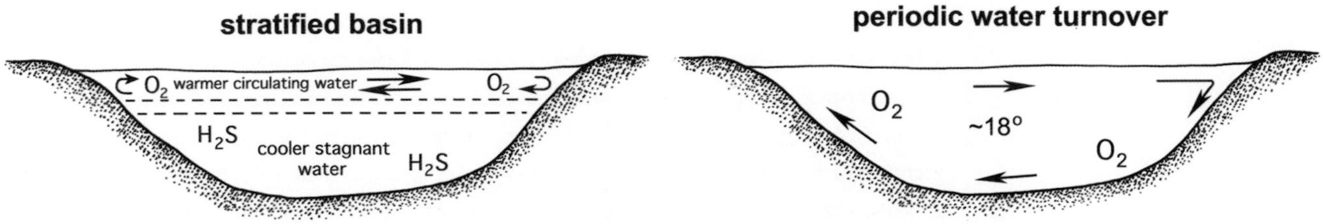

stratified basin

O_2 warmer circulating water O_2

H_2S cooler stagnant water H_2S

periodic water turnover

O_2 ~18° O_2

before it was torn apart by scavengers? Perhaps it was attempting to swim across the lake and drowned before it could make it. So far, this is the only horse known from the FBM.

As is the case in modern lakes today, the causes of death in Fossil Lake were varied. Even the periodic mass mortalities seen there would be expected for such a complex ecosystem over centuries of time.

Nature's Portraits of Deep History: What Preserved Everything?

It is not the apparent scale of mortality in Fossil Lake that is so remarkable. Over centuries of time, many billions of organisms will live and die in any large, bio-diverse lake. What is truly exceptional is that so many of the dead organisms were well preserved as fossils. Bones and other tissues resulting from mortalities do not usually remain very long after death in lakes today. The huge mass mortalities of the clupeid *Alosa* that occurred in Lake Michigan in the 1960s through the 1980s have largely disappeared from the lake bottom. If you dig down into the lake sediments where the *Alosa* mortalities occurred year after year, you are unlikely to find fish bones. Why? Because it takes very special and unusual environmental conditions for dead animals to become fossilized. Lake Michigan has neither the sedimentation rate nor adequate protection of carcasses from scavengers for the dead fishes to easily become fossilized. The FBM, on the other hand, is an example where all the right conditions were present: an aquatic environment with repeated catastrophes killing inhabitants of the lake, river tributaries dumping leaves and other organisms from surrounding areas into the lake, water chemistry that did not dissolve the bones and shells, environmental and sedimentological conditions that allowed the burial and fine preservation of the organisms as fossils, and protection of the rich fossil-bearing limestone from erosion over the last 52 million years so it could one day be excavated by paleontologists and other fossil quarriers.

The study of decaying organisms and fossilization is called TAPHONOMY. Because the FBM contains such a rich fossil record, it is occasionally the subject of taphonomic studies. There is a wide range of decay damage on the FBM fossils, ranging from perfect articulation of bones (skeletons with all the bones still assembled roughly in anatomical position) to "blown" skeletons (fig. 216). McGrew (1975) noted that the FBM has an unusually high percentage of vertebrates fossilized as well-articulated individuals. He excavated a large sample of the FBM 18-inch layer from FBM Locality E, from which he extracted 385 fishes. He found that 68 percent of them were perfectly or near-perfectly articulated. Only 22 percent of the specimens had any significant disarticulation. Sullivan et al. (2012) found similar results based on a sample size of 1,133 fishes showing 70 percent with perfect or near-perfect articulation. McGrew attributed the differences in preservation to seasonal algae blooms, which would cause both large fish kills

Stage 1 disarticulation

Stage 2 disarticulation

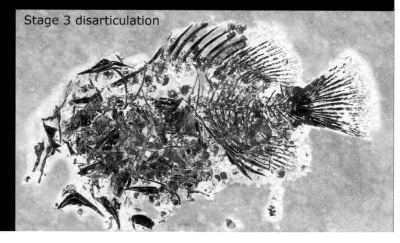

Stage 3 disarticulation

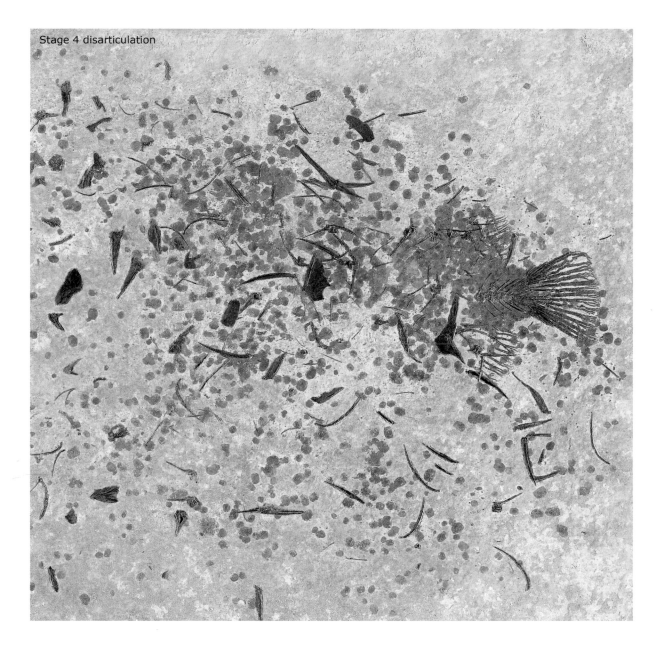

Stage 4 disarticulation

← ↑ FIGURE 216

Stages of skeleton articulation-disarticulation, probably correlated with the length of time that the dead fish remained on the bottom prior to burial. The quicker the burial, the better the articulation later as a fossil. All specimens are †*Cockerellites liops* from the 18-inch layer of FBM Locality A, and all with total lengths of approximately 130 millimeters (5.1 inches). *Stage 1 disarticulation:* Perfectly articulated specimen, representing the quickest burial after death (FMNH PF12107). *Stage 2 disarticulation:* Specimen with some slight loss of scales, pelvic girdle displaced under head, and a few other slight bone displacements (FMNH PF12103). *Stage 3 disarticulation:* Specimen with much more disarticulation of the head and abdominal region (FMNH PF12071). *Stage 4 disarticulation:* Specimen completely "blown" anterior to caudal region (FMNH PF12074).

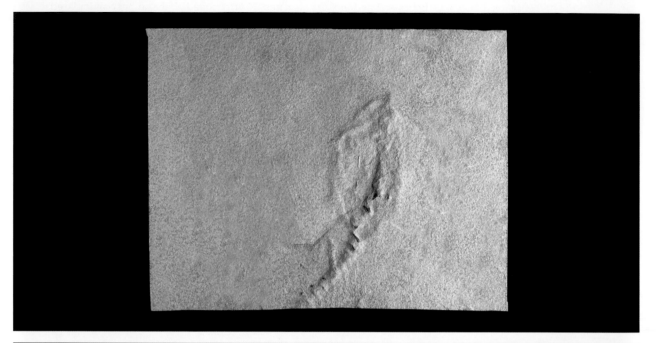

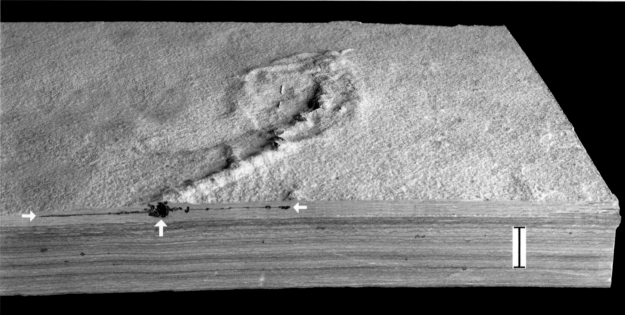

FIGURE 217

The fishes and other vertebrate animals fossilized in the FBM are extremely compressed within the horizontal bedding plane. To illustrate this, a large unprepared †*Mioplosus labracoides* within a slab of the FBM 18-inch layer was sawn in half. *Top:* Top view of a slab with anterior half of fish showing as bumps under the surface, head pointing up and slightly to right. *Bottom:* Slab tipped to show the edge with fish in cross section (between small white arrows). Larger arrow points to section through vertebral column. Note that the entire fish is sandwiched between two of the thin laminations of the rock (laminations also discussed on pages 11–13). The thickness of the fish's body where there are no bones is only a fraction of a millimeter. The total length of the fish was 310 millimeters (1 foot). Scale bar equals 20 millimeters (0.8 inch).

and increased sedimentation rates. He proposed that the best-articulated FBM skeletons were those that died at the time of the algal bloom and were quickly buried. Although it is highly speculative to link fine preservation with algal blooms, I agree that it is probably somehow tied to periods of increased sedimentation. However, what caused these increases in sedimentation may have varied through the history of Fossil Lake. Changes in lake water chemistry resulting from storms or seasonal water turnovers in the lake, as well as algal blooms, may have contributed. During the stratified periods the **ANOXIC**, possibly H_2S-rich hypolimnion would also have protected dead animal and plant remains on the lake bottom from decomposition. More recent studies (Hellawell and Orr 2012) suggest that microbial mats on the lake bottom may have adhered to sunk carcasses and held them together until they were eventually buried.

There are many different types of fossilization in the Green River Formation, but the fossils of the FBM are almost all permineralized. Permineralization is a process in which mineral-rich water forms crystals within the tiny spaces and pores of the skeletons, down to the cellular level. In the FBM, the crystals are calcite, and microscopic examination of FBM bones shows that the walls of the original bone cells are still preserved, often filled with white or colorless calcite. There was plenty of calcite ($CaCO_3$) in the water of Fossil Lake, because the tributaries filling the lake came from rivers and streams running down calcite-rich mountains made largely of limestone. Therefore the pH of Fossil Lake was probably high. There was also some carbonization of fossils in the FBM, particularly in the mid-lake region, resulting in fossil bones that are black or dark brown in color. This process results from the loss of volatile elements of the original organism and heat and pressure from compression of the fossil. Carbonization leaves a black carbon pigment, which gives many fossils their dark color. The fossils in the FBM are highly compressed within single fine laminations not pierced by the fossil (figs. 217 and 5, *bottom right*). The exact mechanism of such precise compression between sedimentary lamina has yet to be fully explained.

The conditions for preservation in Fossil Lake were obviously exceptional during its existence. The amazing diversity, abundance, and life-like preservation of the FBM fossils are rivaled by very few other fossil localities in the world.

Development of a Post-Cretaceous Biota: Where Did All the Plants and Animals Come From?

Even though the animals and plants in and around Fossil Lake lived in the early Eocene, the biogeographic relationships of the community there strongly reflect a much earlier part of Earth history, when the Asian continent was connected to western North America. **BIOGEOGRAPHY** is the study of geographic distribution patterns of animals and plants. This branch of science not only tells us where organisms live; it can also reveal past differences in Earth's surface structure. The theory of continental drift, for example, was first hinted at by distributions of plants and animals that did not fit with the current configuration of continents. Decades ago, scientists discovered that many groups of Brazilian plants and animals had their closest relatives in Africa rather than in other parts of the Americas. Such biogeographic patterns suggested that land bridges between continents once existed that are no longer present. Later, independent geological evidence was found to support a notion of moving continents, and today scientists accept continental drift (or plate tectonics) as part of Earth history. Earth's biota co-evolved with Earth's geography.

In the FBM, the evolutionary relationships of the fossilized plants and animals reflect a geographic configuration of the North American continent that is very different from the present day. MacGinitie (1969) reported this biogeographic pattern when he described the Green River flora of middle Eocene Lake Uinta in Utah and Colorado. In his study, he states that "the composition of the western [North American] fossil flora was more Asiatic than European until late in the Tertiary." Similarly, some of my earlier works (Grande 1985b, 1989, 1994b) reported that the teleost fish fauna of the Green River Formation shows a strong trans-Pacific relationship, as does the paddlefish family Polyodontidae. Many other FBM plants and animals discussed in this book—such as *Koelreuteria*, *Ailanthus*, †*Bahndwivici*, and others—also show the connection between western North America and Asia. The strong Asian relationships of the FBM can be explained by two main factors. First, there was a constant interchange of plants and animals between Asia and western North America over the land connection that persisted for millions of years during the Late Cretaceous (fig. 1). Thus, much of the FBM biota may have ultimately come from Asia. Second, there was a complete isolation of western North America from eastern North America by the Western Interior Seaway through much of the Cretaceous and later by intense mountain-building activity during the Early Cenozoic. Presumably, the separation from eastern North America acted to insulate western North American plants and animals from European influences for many millions of years.

Concluding Remarks

Today the species that once flourished in and around Fossil Lake are long extinct. Even the environment of the Fossil Lake region could be thought of as extinct. The warm moist subtropical system eventually gave way to the high mountain desert and cool climate that exist there today. Did the region's community disappear abruptly at the end of the early Eocene, or did it gradually disappear over the course of millions of years? We do not know. The known geological record is not complete enough to tell us the exact duration of Fossil Lake's decline in days, months, or even centuries. Some of the FBM species belonged to major groups that appear to have become completely extinct at or shortly before the end of the Green River Lake System (e.g., the †paraclupeid and †asineopsid families of fishes, the †baenid turtles, the †pantolesteid and †apatemyid families of mammals, and several families of birds including †Lithornidae, †Gallinuloididae, †Presbyornithidae, †Messelornithidae, †Foratidae, †Fluvioviridavidae, †Sandcoleidae, and †Zygodactylidae). Most of the families and orders present in the FBM are still extant and have continued to SPECIATE and diversify into many species in the present-day BIOTA.

Although the Fossil Lake community no longer exists as a living entity, it left a visual record of its existence like few such ancient communities ever have. The FBM provides a vivid physical record of the early evolutionary development of the modern North American biota. It is a demonstration of how life eventually fills a void. The end of the Cretaceous saw a global extinction event that eliminated over 70 percent of all species living at the time. This mass extinction provided a variety of relatively open ecological niches for the rapid diversification of new plant and animal species. And it is not just individual species that evolved. The interactions among the biota, the environment, and Earth's surface all evolved interconnectedly. The multifaceted evolution of life on this planet is a complex and wondrous thing, and exceptional fossil assemblages from the FBM are like snapshots from deep time, locked in stone.

Postscript

The preface of this book included a few comments about the importance of education and influential teachers in attracting young students, such as my former self, to pursue a scientific career. Therefore, I thought it might be appropriate to end with a postscript on my own efforts to pass something down to the next generation of paleontologists.

Taking a lesson from my old professor Dr. Sloan, I have tried to remain active in teaching, even though my museum job has never required it. I have taught a number of university and graduate courses over the years through the University of Chicago and the University of Illinois at Chicago. But one of the most satisfying courses to me is the one I have taught annually for the last nine years through the Graham School of the University of Chicago. This course targets primarily advanced high school students, and I am convinced that high school is one of the most critical periods in the development of a student's outlook on the future. This summer course, called "Stones and Bones," attracts some of the most motivated students I have ever had the pleasure of working with. It is

The "Stones and Bones" class of 2005 along with museum staff and volunteers on location at Lewis Ranch, Fossil, Wyoming, in July 2005.

a field-paleontology course, with two weeks of classroom instruction in Chicago and two weeks in Fossil Basin, Wyoming, digging fossils for the Field Museum. The students get to work on an actual field excavation in one of the world's most productive fossil sites, and the Field Museum (not to mention my own research program) benefits from the fruits of their labor. Some of the pieces illustrated in this book were discovered during these classes, and many of my students have gone on to universities and eventually to graduate school in paleontology. The students come from all over the world for this course, some from as far away as Saudi Arabia or Beijing. But once the students have all come to Chicago and started the class together, it takes them very little time to form a cohesive team, and a bond often forms among the students that is kept for years after the course ends. I love teaching the class, the students enjoy the experience, and everybody involved benefits. Maybe one day, one of the students will write a sequel to this volume. . . .

Lance Grande
The Field Museum, Chicago

Acknowledgments

Several fossils illustrated in this book were donated to the Field Museum of Natural History in Chicago between 1983 and 2012, or to the American Museum of Natural History in New York while I was there 1979 to 1983. For these very generous donations of important fossils to research museums, I thank Robert Kronner, James E. Tynsky, Thomas Lindgren, Dean Sherman, Bruce P. and Olive W. Baganz, Gerome Montgomery, Rick Hebdon, Greg Laco, David C. Rilling, Robert and Bonnie Finney, Carl and Shirley Ulrich, Dan Judd, Thomas Maloney, and Richard Jackson.

I would like to thank Marlene Donnelly, Lori Grove, and Jenna Rieker for helping me construct some of the drawn figures in this book. For reviewing various parts of the first draft of this book, I thank Arvid Aase and Paul Buchheim (general geology and history); Shannon Hackett, John Bates, Nate Smith, Julia Clarke, and Dan Ksepka (birds); Dan Summers, Jim Louderman, Corrie Moreau, Al Newton, Margaret Thayer, and Petra Sierwald (insects and spiders); Pat Holroyd, Walter Joyce, and Howard Hutchison (turtles); Steve

Manchester, Ian Glasspool, and Scott Wing (plants); Peter Crane (plants, and most of the book in its entirety); John Flynn and Ken Rose (mammals); Erin De-Witt, Elaine Zeiger, Steve Gieser, and Mark Alvey (general). Many of the bird determinations were made thanks largely to work by Julia Clarke and Dan Ksepka, and a grant from the National Science Foundation to Julia Clarke, Lance Grande, Derek Briggs, and Richard O. Prum (NSF EAR-0719943).

For discussions about various identifications, localities, and other matters covered in this book, I thank Arvid Aase, Paul Buchheim, Michael Smith, Dave McGinnis, Rudiger Bieler, David Grimaldi, Neal Evenhuis, Vladimir Blagoderov, Thomas Wesener, Ken Rose, Wighart Koenigswald, Carl Ulrich, Shirley Ulrich, James E. Tynsky, Rick Hebdon, Richard Lewis, Betty Lewis, Tom Lindgren, Gerome Montgomery, Chris Brochu, John Flynn, Michelle Spaulding, Howard Hutchison, Pat Holroyd, and Ian Glasspool.

For letting me study specimens in their care or for collection management, I thank Arvid Aase, John Maisey, Jacques Gautier, Mark Florence, Jessica Cundiff, Daniel Brinkman, Tristan Birkemeier, Storrs Olson, Pete Larson, Burkhard Pohl, Bill Simpson, Ian Glasspool, Paul Mayer, and Michael Cassiliano. For beautifully skillful preparation work on some of the specimens illustrated here, I thank Connie Van Beek, Bill Simpson, Jim Holstein, Lisa Herzog, Irene Broede, Akiko Shinya, and Mary Peters. Thanks to all the people who assisted me with fieldwork and/or helped run my "Stones and Bones" field course in the FBM over the years, particularly Jim Holstein, Akiko Shinya, Connie VanBeek, Mike Eklund, Brian Morrill, Stephen Gieser, Lisa Bergwall, and all of the Stones and Bones students. Their monumental efforts and enthusiasm in Wyoming make the fieldwork a great joy to undertake. I also thank the Negaunee Foundation and the National Science Foundation for support of this work and other scientific endeavors of the Field Museum over the years.

I thank Jim E. Tynsky for letting me and my various field crews and student groups work on his quarry lease in FBM Locality A over the years, and to the Lewis family for letting us camp on their land while were quarrying. I also thank the late Rick Jackson for his generosity in letting me work in his state quarry (FBM Locality G) in the early 1980s. His is another generous soul whose contribution should not be forgotten.

And, finally, thanks to Dee, just for being Dee.

Appendix A

Topographic map coordinates for the major FBM quarries are given below. FBM Locality codes A–N are keyed to the maps in figures 218–220 and the diagrams in figures 4 and 7. This section is meant to document locations of the major commercial FBM quarries that have been active during the last 60 years. Most of these are located either in what was once the mid-lake region of Fossil Lake (referred to here as "mid-lake" quarries) or what was once along the nearshore eastern side of Fossil Lake (referred to here as "nearshore" quarries). The earliest quarries (dating back more than a century) were mid-lake quarries located on Fossil Butte and Fossil Ridge (not given letter codes here, but see maps in figures 218 and 219). Today active quarries also include the nearshore localities (e.g., the highly productive Thompson Ranch localities H and M).

Most of the mid-lake quarries have traditionally focused on mining the "18-inch layer" (also referred to as "F-1," or the "blackfish layer"), but since the 1980s, some of the mid-lake quarries have also expanded their excavations into beds above the 18-inch layer

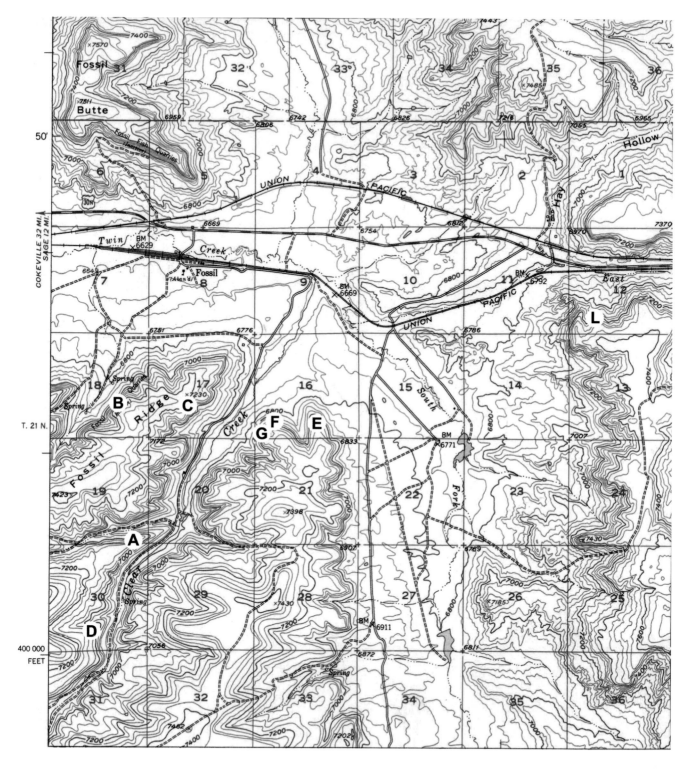

FIGURE 218

General locations of some major FBM quarries discussed in this book. FBM Localities A–G are mid-lake quarries, and FBM Locality L is in a region near what was the eastern shore of Fossil Lake. These locations are described further on pages 375–78. Source of base map is part of USGS Kemmerer Quadrangle, 15 minute series (topographic). Note that map figs. 218 and 219 overlap each other, and map fig. 220 is an area directly below the area of map fig. 218.

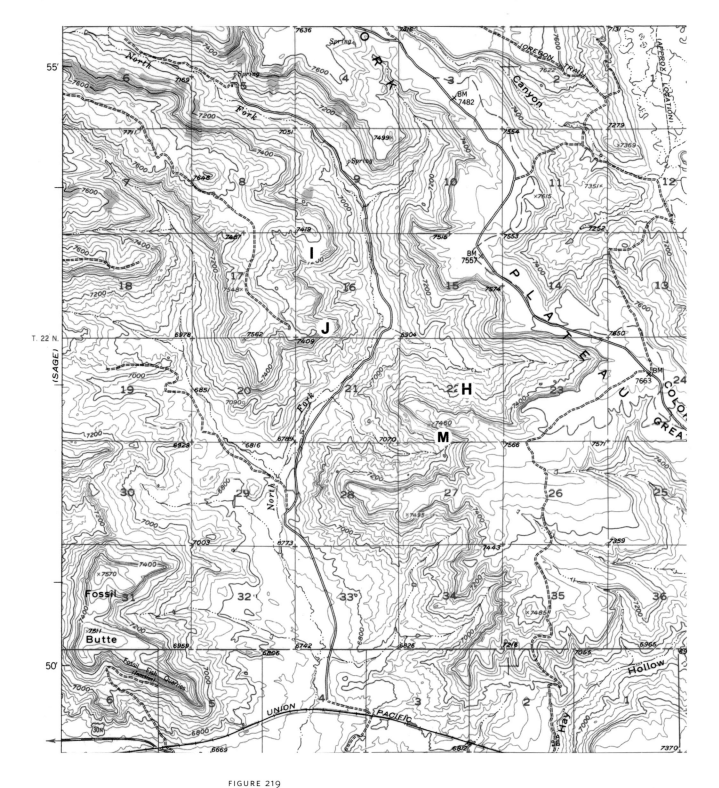

FIGURE 219

General locations of some major FBM quarries near what was the northeastern shore of Eocene Fossil Lake. These locations are described further on pages 377–78. Source of base map is part of USGS Kemmerer Quadrangle, 15 minute series (topographic). Note that map figs. 218 and 219 overlap each other, and map fig. 220 is an area directly below the area of map fig. 218.

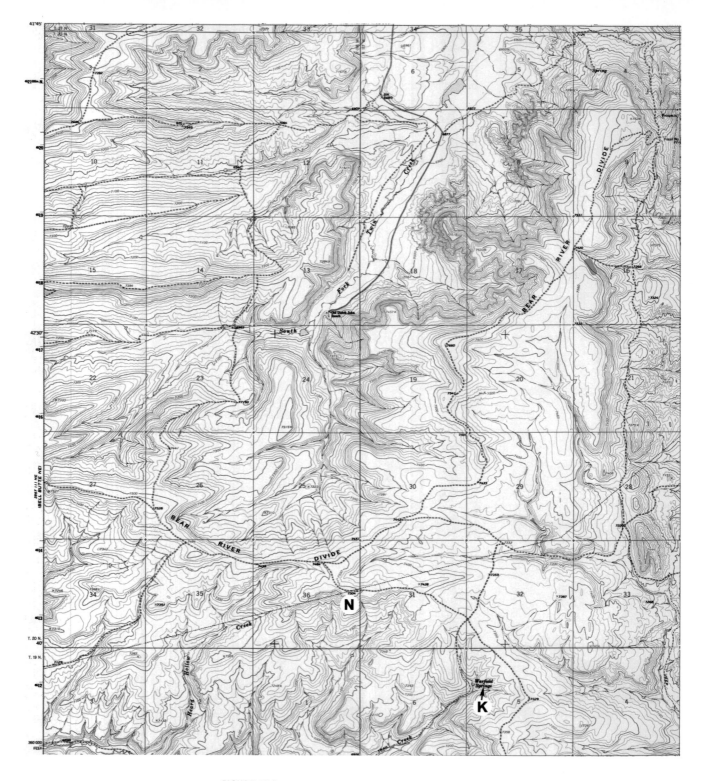

FIGURE 220

General location of FBM Locality K, and a quarry in the Angelo Member of Fossil Basin discussed in this book. These sites are described further on page 379. Source of base map is part of USGS Warfield Springs Quadrangle, 7.5 minute series (topographic). This map shows an area directly below map fig. 218.

near the K-spar tuff. More recently these quarries have also expanded into the sandwich layer just below the 18-inch layer (see figure 4 for stratigraphic positions of these layers). Most of the nearshore quarries have traditionally focused on the "sandwich beds" (also referred to as "F-2"). The sandwich beds at Thompson Ranch (FBM Localities H and M) are also often referred to as the "split-fish layer," but this common name is also used for other layers within the FBM and not a consistent stratigraphic zone indicator. "Sandwich" is a better term because is refers to a specific stratigraphic horizon. Although the sandwich bed equivalent can be accurately identified in the mid-lake quarries, I have been unable to identify the 18-inch layer equivalent in the nearshore quarries.

Some of the quarries on private and state land allow tourists on their site to dig fossils for a set price per day. The quarry operator will confiscate any truly rare pieces found, but the tourist will be allowed to keep the more common fossils they find. Information on quarry tours is available in the town of Kemmerer from the Chamber of Commerce or from the Fossil Butte National Monument. Digging for fossils elsewhere in the FBM is illegal unless on private land with the permission of the owner.

FBM Locality A—Lewis Ranch site #1 (figs. 11–14, 213, *bottom*). This mid-lake quarry is located in SE¼, SE¼, Sec. 19, T.21N., R.117W., and NE¼, NE¼, Sec. 30, T.21N., R.117W., Kemmerer 15 minute Quadrangle (USGS). This quarry was first opened and leased by James E. Tynsky in 1984, and he was still working the site when this book was written. (See discussion of the Tynsky family operations on pages 27–29.) Most of the Field Museum quarrying activity has been in this quarry. Fossil mining in this locality began primarily in the 18-inch layer, but starting in 1989 fossil mining expanded into fossiliferous layers well above the 18-inch layer near the K-spar tuff bed (see fig. 4). These upper beds contain mass mortalities of small †*Knightia eocaena* (e.g., fig. 74), at least one mass-mortality layer of small †*Mioplosus labracoides* (e.g., fig. 83, *bottom*) and other mass-mortality zones of mixed vertebrate species and mollusks (frontispiece). Starting in 2009, the sandwich beds just below the 18-inch layer were also opened up to mining at this locality. The sandwich layers in the mid-lake localities are more organic rich with darker-colored fossils and laminations than in the nearshore Thompson Ranch quarries (e.g., FBM Locality H).

FBM Locality B—Lewis Ranch site #2. A mid-lake quarry, located in NW¼, SE¼, Sec. 18, Sec. 19, T.21N., R117W., Kemmerer 15 minute Quadrangle (USGS). Fossils from this quarry have been mined primarily from the 18-inch layer, but starting in 1990 mining of the fossiliferous beds near the K-spar tuff layer also began. Locality B is also known as the "Smith Hollow" quarry and is one of the oldest commercial quarries within the Fossil Butte Member deposits. This quarry was worked by Lee Craig more than a century ago (see pages 23–25). Beginning in the 1970s and 1980s, it was also worked by Wallace Ulrich (son of Carl and Shirley

Ulrich—see FBM Locality E, below). In the mid-1980s, it was mined by Thomas Lindgren, Scott Walter, Anthony Lindgren, and others, and today, in 2010, it is mined extensively by the Green River Stone Company and Jerome Montgomery.

FBM Locality C—Lewis Ranch site #3. A mid-lake quarry, located in NW¼, SE¼, Sec. 17, T.21N., R.117W., Kemmerer 15 minute Quadrangle (USGS). This quarry mines the 18-inch layer and the upper beds near the K-spar tuff layer. Opened in 1990 when it was leased to Ronald Mjos, Cliff Miles, Clark Miles, and Jeff Parker. Later it was worked by Dennis Kingery, and then by Jim A. Tynsky (now deceased) and his son Duane, who continues to work the quarry at the time of this printing.

FBM Locality D—Lewis Ranch site #4. A mid-lake quarry, located in SW¼, NE¼, Sec. 30, T.21N., R.117W., Kemmerer 15 minute Quadrangle (USGS). Excavation currently takes place in two different quarries about 300 meters (985 feet) apart in a number of different stratigraphic horizons within the FBM, including the 18-inch layer, the sandwich layer, and layers above the 18-inch layer near the K-spar tuff bed. It was originally opened in 1989 by Ron Mjos and partners, who shortly after moved to FBM Locality C. It has since been worked at various times by quarriers Dennis Kingery, Jim E. Tynsky, and Stewart Grieves.

FBM Locality E—Wyoming State commercial site #1; leased to Carl J. and Shirley Ulrich and also worked at various times by their son Wallace (see also Locality B above). A mid-lake quarry, located in SW¼, SE¼, Sec. 16, T.21N., R.117W., Kemmerer 15 minute Quadrangle (USGS). The Ulrich 18-inch layer quarry is the oldest of the state commercial quarries and has been operating there since the late 1950s. For nearly 20 years (late 1950s to 1978), the Ulrich quarry was the only full-time state-leased commercial quarry in the 18-inch layer using modern collecting techniques. Consequently, most 18-inch layer specimens collected between 1960 and 1978 came from this quarry. Before working Locality E, the Ulriches worked various other deposits, including sites on Fossil Butte, since the late 1940s. See discussion of the Ulrich family quarry operations on pages 25–28.

FBM Locality F—Wyoming State commercial site #2. A mid-lake quarry, located in N½, SE¼, SW¼, SW¼, Sec. 16, Sec. 16, T.21N., R.117W., Kemmerer 15 minute Quadrangle (USGS). Over the last 15 years, this site has been leased and worked by a number of quarrying operations including W. G. Somers and Jack A. Scott (jointly), William Boundy, and Richard Dayvault. So far, only the 18-inch layer has been significantly mined at this locality. Currently, this site and the contiguous FBM Locality G are being worked as a single large quarry by Rick Hebdon and Company. See discussion of the Hebdon family quarry operations on page 29.

FBM Locality G—Wyoming State commercial site #3. A mid-lake quarry, located in S½, SW¼ , SW¼, SW¼, Sec. 16, T.21N., R.117W., Kemmerer 15 minute Quadrangle (USGS). So far, only the 18-inch layer has been extensively mined at this locality. This quarry was originally leased and opened up by Richard W. Jackson in 1978. Later quarriers leasing the site included Ronald Mjos, Cliff Miles, and Brigham Young University. As of the date of this publication, it was being leased by Rick Hebdon (see Locality F), and Localities F and G have been joined into a single quarry. This locality has produced the largest fish specimen known so far from the Green River Formation, a specimen of †*Lepisosteus atrox* about 2 meters (7 feet) in length deposited at UW (see Grande 2010, fig. 387B).

FBM Locality H—Thompson Ranch locality #1 (fig. 15). A nearshore quarry, located in NE¼, SW¼, Sec. 22, T.22N., R.117W., Kemmerer 15 minute Quadrangle (USGS) (mistakenly listed as "T20N" in Grande 1989). This was the first commercial quarry to extensively mine the sandwich beds and was originally referred to as a "split-fish" quarry. Because the term "split-fish" has been used for several different stratigraphic horizons, I use the more specific terms "sandwich bed" or "sandwich layer" here (also see fig. 4). Locality H is the oldest and has been one of the most productive of the sandwich bed quarries. Previous to 1986, Locality H was leased and worked by three generations of the Tynsky family between 1969 and 1985. The quarry was first run by Sylvester Tynsky and his sons Robert and James A. Tynsky through the mid-1970s. From about 1975 to 1985, the quarry was run by Robert Tynsky's son, James E. Tynsky. James E. Tynsky currently runs FBM Locality A (see above). The quarry has been leased since 1986 to Richard (Rick) Hebdon. The fossils of the sandwich beds of the Thompson Ranch localities (FBM Localities H, I, J, and M) all lie within the northeastern region of Fossil Lake. They are usually a different color than those of the more southern sandwich bed outcrops (FBM Locality L, a nearshore locality from the mid-eastern side of Fossil Lake, and FBM Locality K, a nearshore locality from the southeastern side of Fossil Lake). The fossils of the Thompson Ranch region are usually lighter in color, have an orange to orange-brown hue, and do not have the prominent dark bands or laminations that the 18-inch layer has (e.g., compare figs. 6A, *upper right*; 37; 44; 66, *top*; 68, *bottom*; 78; 97; 103; and 142A to figs. 5A, *upper left*; 21; 42; 50; 62; 66, *bottom*; 68, *top*; 102; and 130). Bird feathers, insects, and plants are much less likely to be visibly preserved in the Thompson Ranch localities than in the other FBM localities. Sometimes they are preserved as colorless impressions that will show up under black light (e.g., a large bird wing with flight feathers preserved that are invisible under normal light but visible under black light; specimen FMNH PA729).

FBM Locality I—Wyoming State commercial site #4. A nearshore quarry, located S½, NW¼, NW¼, Sec. 16, T.22N., R.117W., Kemmerer 15 minute Quadrangle (USGS). Opened in the late 1970s and leased to James A. Tynsky. This is a very productive sandwich bed quarry, similar to the nearby Thompson Ranch localities.

FBM Locality J—Wyoming State commercial site #5. A nearshore quarry, located in SE¼, SW¼, Sec. 16, T.22N., R.117W., Kemmerer 15 minute Quadrangle (USGS). Opened in 1978 by James E. and Karen Tynsky. This is a very productive sandwich bed quarry, similar to the nearby Thompson Ranch localities.

FBM Locality K—Warfield Springs locality (on Hebdon Ranch); currently owned and operated by Rick Hebdon. A nearshore quarry, located in N½, SW¼, NW¼, Sec. 5, T.19N., R.117W., Warfield Creek 7.5 minute Quadrangle (USGS). This quarry focuses on the sandwich beds and has been mined commercially since about 1975, when it was initially worked by Richard Hebdon and his father, Virl. This is the southernmost commercial quarry of the FBM, and fossils are a much darker color than they are in the Thompson Ranch sandwich beds. The bones are usually black or dark brown, and the fish scales, bird feathers, insects, and plants are dark brown in color, much like the fossils of the 18-inch layer of the mid-lake quarries. The limestone is usually very soft and bright white in color, except for a few thin intermittent layers. The deposits appear to represent a region of Fossil Lake that had deltaic condition.

FBM Locality L—Western nearshore private-land quarries, west of FBM Locality A. A nearshore quarry, located in the SW¼ of Sec. 12, T.21N., R117W., Kemmerer 15 minute Quadrangle (USGS). Locality currently with two separate commercial quarries; one belonging to Pete Severns and the other to Charles Nunn. These quarries mainly work the sandwich beds. The fossils and sometimes the laminations of the sandwich beds in this quarry tend to be more darkly colored than those at the Thompson Ranch and associated localities (e.g., FBM Localities H, I, J, and M). The bones are usually dark brown or black, and the fish scales, bird feathers, insects, and plants are dark brown in color, much like the fossils of the 18-inch layer of the mid-lake quarries.

FBM Locality M—Thompson Ranch Quarry #2. A nearshore quarry, located in the SW¼ of Sec. 22, T.22N., R117W., Kemmerer 15 minute Quadrangle (USGS). As of 2010 locality managed and leased by Bob and Bonnie Finney. Previously this quarry was also worked by Bryan Wade and then by Robert Kronner, who both leased the property. Later Charlie Nunn took over the lease prior to the Finneys purchasing the property. These quarriers all focused on the sandwich beds.

FBM Locality N—The "powerline quarry" nesting site of the bird †*Presbyornis* that contains thousands of isolated bones from this genus along with fragments of eggshell (Leggitt, Buchheim, and Biaggi 1998). Although this specific quarry does not target the Fossil Butte Member, it is located in the Angelo Member of Fossil Basin just above the Fossil Butte Member. Because this locality lies directly above the FBM and another †*Presbyornis* nesting site occurs in the lower unit of the Fossil Butte Member, it is reasonable to assume that the shore deposits of the FBM would also contain †*Presbyornis* nesting sites. These shore deposits have yet to be discovered or may have been eroded away over time. The powerline nesting site is located in the SE¼ of Sec. 36, T.20N., R. 118W., Warfield Creek 7.5 minute Quadrangle (USGS). This is the only quarry in the locality list here that does not focus on the FBM.

Appendix B

The following table is a list of valid fish species recognized here from the FBM, together with the authors and original description citations for existing names. The families to which these species are thought to belong are given in the left-hand column.

Families present in the FBM	Genera present in the FBM	Reported FBM fish species recognized in this book
†Heliobatidae	†*Heliobatis* Marsh, 1877	†*H. radians* Marsh, 1877
†Asterotrygonidae, new family	†*Asterotrygon* Carvalho, Maisey, and Grande, 2004	†*A. maloneyi* Carvalho, Maisey, and Grande, 2004
Polyodontidae	†*Crossopholis* Cope, 1883	†*C. magnicaudatus* Cope, 1883
Lepisosteidae	*Lepisosteus* Lacépède, 1803	†*L. bemisi* Grande, 2010
	Atractosteus Rafinesque, 1820	†*A. simplex* (Leidy, 1873) †*A. atrox* (Leidy, 1873)
	†*Masillosteus* Micklich and Klappert, 2001	†*M. janei* Grande, 2010
Amiidae	*Amia* Linnaeus, 1776	†*A. pattersoni* Grande and Bemis, 1998
	†*Cyclurus* Agassiz, 1844	†*C. gurleyi* (Romer and Fryxell, 1928)
Hiodontidae	*Hiodon* Lesueur, 1818	†*H. falcatus* (Grande, 1979)
Osteoglossidae	†*Phareodus* Leidy, 1873	†*P. encaustus* (Cope, 1871) †*P. testis* (Cope, 1877)
†Paraclupeidae	†*Diplomystus* Cope, 1877	†*D. dentatus* Cope, 1877
Clupeidae	†*Knightia* Jordan, 1907	†*K. eocaena* Jordan, 1907 †*K. alta* (Leidy, 1873)
Gonorynchidae	†*Notogoneus* Cope, 1885	†*N. osculus* Cope, 1885
Esocidae	*Esox* Linnaeus, 1758	†*E. kronneri* Grande, 1999
†Asineopidae	†*Asineops* Cope, 1870	†*A. squamifrons* Cope, 1870
Percopsidae	†*Amphiplaga* Cope, 1877	†*A. brachyptera* Cope, 1877
Latidae	†*Mioplosus* Cope, 1877	†*M. labracoides* Cope, 1877
Moronidae	†*Priscacara* Cope, 1877	†*P. serrata* Cope, 1877 †*P. n. sp. A* (undescribed) †*P. n. sp. B* (undescribed)
	†*Cockerellites* Jordan, 1923	†*C. liops* (Cope, 1877)
	†*Hypsiprisca* new genus	†*H. hypsacantha* (Cope, 1886) †*H. n. sp* (undescribed)

Appendix C

The following table is a list of distinct bird species from the FBM recognized here, together with the authors and original description citations for existing names. The families (or most closely related living families in some cases) to which these species are thought to belong are given in the left column. An asterisk (*) indicates that the identification is only provisional and further study or better material is needed to verify.

Families present in the FBM	Genera present in the FBM	Reported FBM bird species recognized in this book
†Lithornithidae	†*Pseudocrypturus* Houde, 1988	†*P. cercanaxius* Houde, 1988
	new genus (undescribed)	new genus and species (undescribed)
†Gallinuloididae	†*Gallinuloides* Eastman, 1900	†*G. wyomingensis* Eastman, 1900
†Presbyornithidae	†*Presbyornis* Wetmore, 1926	†*P.* new species? (undescribed)
*"anseriform-like"	new genus (undescribed)	new genus and species (undescribed)
*†Gastornithidae?	†*Gastornis?* Herbert, 1855	inconclusive identification
†Fluvioviridavidae	†*Fluvioviridavis* Mayr and Daniels, 2001	†*F. platyrhamphus* Mayr and Daniels, 2001
†Eocypselidae	†*Eocypselus* Harrison, 1984;	new species A (undescribed)
	new genus (undescribed)	new genus and species B (undescribed)
Steatornithidae	†*Prefica* Olson, 1987	†*P. nivea* Olson, 1987
*Eurypygidae?	new genus (undescribed)	new genus and species (undescribed)
Frigatidae	†*Limnofregata* Olson, 1977	†*L. azygosternon* Olson, 1977
		†*L. hasegawai* Olson and Matsuoka, 2005
Threskiornithidae	new genus (undescribed)	new genus and species (undescribed)
†Foratidae	†*Foro* Olson, 1992	†*F. panarium* Olson, 1992
†Messelornithidae	†*Messelornis* Hesse, 1992	†*M. nearctica* Hesse, 1992
†Salmilidae	new genus (undescribed)	new genus and species (undescribed)
*"rail-like"	new genus (undescribed)	new genus and species (undescribed)
Galbulidae?	†*Neanis* Shufeldt, 1913	†*N. kistneri* (Feduccia, 1973)
*Messelirrisoridae	new genus (undescribed)	new genus and species (undescribed)
†Primobucconidae	†*Primobucco* Broadkorb, 1970	†*P. mcgrewi* Broadkorb, 1970
Leptosomidae	†*Plesiocanthus* Weidig, 2006	†*P. wyomingensis* Weidig, 2006
		†*P. major* Weidig, 2006
†Sandcoleidae	†*Anneavis* Houde and Olson, 1992	†*A. anneae* Houde and Olson, 1992
Coliidae	†*Celericolius* Ksepka and Clarke, 2010a	†*C. acriala* Ksepka and Clarke, 2010a
*"falconiform-like"	new genus (undescribed)	new genus and species (undescribed)
†Messelasturidae	†*Tynskya* Mayr, 2000	†*T. eocaena* Mayr, 2000
†Halcyornithidae	†*Cyrilavis* Ksepka et al., 2011	†*C. colburnorum* Ksepka et al., 2011
		†*C. olsoni* (Feduccia and Martin, 1976)
†Quercypsittidae	†*Avolatavis* Ksepka and Clarke, 2012	†*A. tenens* Ksepka and Clarke, 2012
†Zygodactylidae	†*Eozygodactylus* Weidig, 2010	†*E. americanus* Weidig, 2010
		†*E.* new species (undescribed)
*"*Morsoravis*-like"	new genus (undescribed)	new genus and species (undescribed)
*"long-beak"	new genus (undescribed)	new genus and species (undescribed)

Appendix D

There are many methods employed by commercial fossil quarriers to enhance the appearance of the fossils they excavate and sell. Many of the quarriers sell FBM fossils as art objects or souvenirs rather than as scientific specimens. RESTORATION is the most common form of enhancement, but INLAYING (sometimes also called "insetting") is also used. Dealers are usually open and honest about what forms of enhancement they use on the pieces they sell, but it is a wise policy to ask about this before acquiring an important or expensive piece. All of the fossils in this book are natural, except for a few cases where it is otherwise indicated in the figure caption.

Commercial restoration of fossils from the FBM can range from enhancement of fin tips and coloring of holes in body scales (fig. 221, *bottom*), to composite reconstruction of stingrays and other fossils using different individuals (sometimes called "Franken-fossils") or fake heads inset on headless bird skeletons. This restoration also varies in quality from difficult to detect, to blatantly

FIGURE 221

Comparison of fast-track commercial preparation to detailed museum preparation. Commercial quarriers generally cannot afford to put as much time into preparation of common species as museums often do. For example, here are two specimens of †*Mioplosus labracoides* from near the K-spar tuff layer in FBM Locality A. Both fishes are the same size (140 millimeters [5.5 inches] in total length) and were on the same slab of rock in the same state of preservation. The one on the top was prepared as a research specimen under a microscope over a period of six weeks by a museum preparator at the Field Museum (FMNH PA783). The one on the bottom was prepared as a commercial souvenir using a fast-track method with an engraving tool over a period of less than two hours by a commercial preparatory (FMNH PF15374). Note the damage, which is partly obscured by a painted-on coat of brown coloring. No color or restoration of any kind was added to the top specimen.

obvious. But restoration can always be detected by close examination, with microscopes, X-rays, or illumination with black light.

If a person is acquiring an FBM fossil based on its aesthetic appeal as an art object, the restoration often does not matter. Or if it is simply a common species of fish such as a †*Knightia* acquired as a souvenir, it also matters little. Restored pieces are usually relatively cheap on the commercial market. But for scientists, researchers, and museums, it is important to have original pieces consisting of a single individual that is well-enough preserved for accurate anatomical study.

Restoration is a substitute for time in preparation of commercial specimens. A commercial quarrier might only spend an hour preparing a specimen that would take a museum a month to prepare. Of course, there is a qualitative difference, but for most commercial quarriers, the point of diminishing return comes more quickly than for scientific institutions, particularly for the pieces with low commercial value. Time is money for the commercial operators, and they cannot afford to pay thousands of dollars for preparation of a fish fossil that they will sell for between $2 and $100. They may fill orders of 5 to 10,000 small fishes to a single buyer for a few dollars per fish. Preparation time on such pieces must be near zero to make a profit. The rarer pieces are a different story, though, and most commercial operators have come to realize this. When very valuable pieces are discovered by commercial quarriers (e.g., fig. 152), they realize it is in their best interest to sometimes contract higher-skilled preparators to spend enough time to do the preparation. Museums have resources to put much more time into the preparation of all fossils, particularly the more common species, because they do not have to profit from selling the pieces they prepare. The restoration of fossils by commercial quarriers is not an attempt to be dishonest. They are generally clear about which pieces have been restored, what has been restored, and which pieces are original.

Some commercial fossil quarriers employ a process called inlaying to enhance the aesthetics of fossils, particularly in the nearshore localities of Thompson Ranch. Sometimes this is used when a fossil fish is found too close to the edge of a slab. The fish is cut out and carefully set into the center of a blank piece of rock. This can sometimes be difficult to detect with the naked eye; but it can be easily detected by the use of an X-ray machine, which often shows the edges of the inset pieces (e.g., fig. 131), or by looking at the corner join between the inset slab and the larger slab it is set into. Many large slabs from FBM Locality M that have beautifully positioned multiple fishes surrounding a stingray or a perfectly positioned mix of various fishes marketed as a "stone aquarium" are pieces with aesthetically strategic inlays. These can be very attractive but do not necessarily represent natural assemblages of fishes.

Appendix E

This book is very different from my *Bulletin 63* publication (Grande 1980, 1984), which covered the entire Green River Formation. This book focuses only on the Fossil Butte Member (FBM), which is paleontologically the most important member of the Green River Formation. The FBM is only about 12 to 22 meters (39 to 72 feet) thick depending on where in Fossil Basin the section is measured. Confining the paleontological scope of this volume in this way constricts its focus in both time (to about 52 million years before present) and space (to early Eocene Fossil Lake in what is now southwestern Wyoming). This narrower focus allows me to present the fossils as more of a contemporaneous community with an interconnected set of ecosystems from the past. It attempts to examine the complex web of life that existed in Eocene Fossil Lake 52 million years ago and to link it to our modern-day biota. The main Lake Gosiute fossil localities (Laney Member) are several million years younger than the Fossil Butte Member, and some of the Lake Uinta fossil localities are several million years older (e.g., Flagstaff Member). The FBM is currently mined at many localities within Fossil Basin. References to FBM Locality A through FBM Locality N throughout this book are explained and plotted on maps in appendix A.

A dagger symbol (†) precedes the names of all extinct taxonomic groups (groups known only by fossil species or individuals). As far as is known, all FBM species are extinct today (with the possible exception of one plant species), as are many of the FBM genera, a few of the FBM families, and some of the FBM orders. Words set in an **ALL-CAPITALS** font are defined in a glossary at the back of this book. Institutional abbreviations followed by catalog numbers are given in the figure captions for all fossils illustrated in this book wherever possible. A key to the institutional abbreviations follows on pages 393–94. Most of the fossils illustrated here are in the world's largest collection of FBM material, the collection of the Field Museum of Natural History, Chicago.

Appendix F

Sources for phylogenetic trees structure used in this book (although not necessarily taxonomic ranking of group names) are as follows:

Arthropods (Edgecombe 2010; Regier et al. 2010)
Insects (Grimaldi and Engel 2005)
Fishes (Nelson 2006; Grande 2010)
Sarcopterygia (Liem et al. 2001)
Turtles (Sterli 2010)
Birds (Hackett et al. 2008)
Mammals (Rose 2006)
Bats (Simmons et al. 2008)
Plants (Stevens 2008)

Institutional Abbreviations
Used in This Book

The key to the institutional acronyms in this book is as follows:

AMNH   The American Museum of Natural History, New York

BHI   Black Hills Institute, Hill City, South Dakota

BMNH   The Natural History Museum, London

BMS   Buffalo Museum of Science, Buffalo, New York

BSP   Bayerische Staatssammlung für Paläontologie und Historische Geologie, München, Germany

FMNH   The Field Museum of Natural History, Chicago

GAAM   Great American Alligator Museum, New Orleans, Louisiana

MCZ   Harvard Museum of Comparative Zoology, Cambridge, Massachusetts

NAMAL   North American Museum of Ancient Life, Lehi, Utah

PU	Princeton University Collection, now mostly at Yale University, New Haven, Connecticut
SMA	Saurier Museum, Aathal, Switzerland
SMF	Senckenberg Museum, Frankfurt, Germany
SMNK	Staatliches Museum fur Naturkunde Karlsruhe, Germany
TMP	Royal Tyrrell Museum of Paleontology, Drumheller, Alberta, Canada
UAM	Alabama Museum of Natural History, University of Alabama, Tuscaloosa, Alabama
UMMP	University of Michigan Museum of Paleontology, Ann Arbor, Michigan
UMNH	Natural History Museum of Utah, Salt Lake City
USNM	National Museum of National History (Smithsonian Institution), Washington, DC
UW	University of Wyoming, Laramie
WDC	Wyoming Dinosaur Center, Thermopolis, Wyoming
WSGS	Wyoming State Geological Survey, Laramie
WW	Walla Walla University, Department of Biological Sciences, College Place, Washington

Glossary

ABDOMINAL REGION.   Belly region.

ABSOLUTE DATES.   A term for geologic dates that have been determined
by some direct measurement such as RADIOMETRIC DATING, as opposed to relative dates determined through correlation of fossil
assemblages or geological anomalies. Some scientists prefer the
terms *chronometric* or *calendar* dating, because the word "absolute" implies an unwarranted certainty and precision. Absolute dating provides a computed numerical age in contrast with RELATIVE
DATES, which provide only an order of events.

ACTINOPTERYGIANS.   A group also called the "ray-finned fishes," which
today contains over 90 percent of all living fish species and has
fossils dating back over 400 million years before present. They are
characterized by unique modifications of the pectoral and pelvic
fins.

ALGAL BLOOMS.   An extremely rapid increase in the population of algae in
an aquatic system.

AMNIOTIC EGG.   Also called "cleidoic egg." An egg with a protective shell
and other features that allow eggs to be laid on land. Its origin is

thought to have been a critical point in vertebrate evolution that allowed verte-brates to leave the aquatic environment and diversify in the terrestrial environment.

ANADROMOUS. Said of an adult fish that lives in a marine environment but migrates into a freshwater environment to breed.

ANADRONY. Migration from salt water to freshwater as adults.

ANOXIC. Depleted of oxygen.

APEX. With reference to leaves and LEAFLETS, the apex is the tip and the upper 5 to 15 percent of the distal end.

ARBOREAL. Living in trees.

ASPIRATION SPECIMENS. Fossil fishes with other animals preserved in their mouth or stomach.

AUTOTROPHS. *See* PRIMARY PRODUCERS.

AVAILABILITY. Term applied to a taxonomic name that has been created following the ap-propriate international rules of nomenclature. If the name also has PRIORITY, then it is the VALID name.

BEDDING PLANE. A layer of SEDIMENTARY ROCK or sediment (STRATA) representing sur-faces of exposure between sedimentary depositional events. In Fossil Lake, these represent pervious surfaces of the lake bottom and are horizontal surfaces along which the strata separate or "split."

BINOMIAL. The two-part name of an organism, comprising a generic name and specific name.

BIOGEOGRAPHY. The study of geographic distributions of plants, animals, and other organisms.

BIOTA. The total collection of organisms of a geographic region or a time period. In the context of a classification system, "biota" is the group category that includes all life on Earth (all kingdom-level groups).

CALCIMICRITE. A type of limestone formed by recrystallization of lime mud. It is com-posed of calcareous particles ($CaCO_3$).

CAPSULE. In botany, a fruit that opens at maturity to release seeds.

CELLULOSE. An organic compound that forms the primary cell walls of green plants. It is the most common organic compound on Earth, forming about one-third of all plant matter.

CENOZOIC. The geologic time period lasting from about 65 million years ago to the pres-ent day, beginning in the Paleocene.

CENTRA (singular form: CENTRUM). The main body of a vertebra, often disc-like in shape, that usually supports dorsal arches and ventral ribs.

CHICXULUB ASTEROID. An asteroid estimated to have been at least 10 kilometers (6 miles) in diameter that impacted Earth at the end of the Cretaceous (65 million years ago), ultimately leading to the extinction of over 70 percent of species living at that time. The impact crater, one of the largest on Earth, is in the Yucatán re-gion and measures more than 180 kilometers (110 miles) in diameter.

CHITIN. A derivative of glucose that forms the shell or EXOSKELETON of arthropods, as well as the cell walls of fungi.

CO-EVOLUTIONARY PROCESSES. Processes that involve one species changing in response to another species changing. For example, many species of flowering plants have co-evolved with the species of insects that pollinate them.

CRETACEOUS.   A geologic period that ranged from about 145 to 65 million years before present.

CROP.   In a bird's digestive system, a pouch near the gullet or throat that can be used to store food temporarily.

CROWN GROUP.   A group that includes all living species of the group, plus all extinct descendants back to the common ancestor of all the living species. *See* STEM GROUP. The stem group/crown group distinction is not accepted (or at least recognized) by all systematists today.

CURSORIAL.   Adapted specifically for running.

DARWIN.   Charles Darwin (1809–1882) was a naturalist who proposed the idea of natural selection as a process to explain the evolution of new species, in his groundbreaking book *The Origin of Species* (1859). No glossary definition can do justice to Darwin's rich history and complexity.

DECAPOD.   An aquatic malacostracan arthropod with 10 feet or legs.

DESCRIBED.   Said of a taxon with a valid scientific name.

DESCRIPTION.   A set of published features used to describe a species or higher taxon.

DETRITIVORES.   Organisms that obtain nutrients by consuming detritus (decomposing plant and animal parts as well as organic fecal matter).

DEWATER.   Part of the process where aquatic sediments are turned into SEDIMENTARY ROCK by water being driven out of the sediments by great pressure from the weight of overlying sediments

DEWATERED.   *See* DEWATER.

DEWEY DECIMAL SYSTEM.   A system of classification used by libraries to organize books in a way that allows efficient retrieval of specific titles.

DIAGNOSIS.   A set of descriptive features that serve to distinguish a TAXON from all other closely related taxa.

DICOT.   Short for "dicotyledon." One of two subdivisions of flowering plants, in which the embryonic seedlings have two leaves. Dicots usually have RETICULATE or branched veins rather than parallel. *See also* MONOCOT.

DIURNAL.   Active by day, less active at night. *See also* NOCTURNAL.

ECHOLOCATION.   A biological sonar system used to navigate and detect objects using an elaborate combination of sound and hearing rather than sight. Used by bats, whales, and a few other animal groups.

ECOLOGICAL TIME.   As defined here, sequential events directly observable by humans. We see the sun rise, and we see it set on a given day. In contrast, we did not see the end of the Cretaceous.

ELYTRA.   The modified, hardened forewings of certain insects such as beetles, sometimes commonly called the "shard" or "body shell."

EMPIRICAL.   Gained by means of direct observation or experience.

ENDOCARP.   The "pit" or hard inner portion of edible fruits, such as the pit of a cherry, peach, or olive.

ENDOSPERM.   The tissue produced inside the seeds of most flowering plants.

ENTOMOLOGIST.   A scientist who studies insects.

EPITHET.   The second part of a BINOMIAL, the species name (e.g., the *sapiens* part of *Homo sapiens*).

EUSOCIAL. The highest level of social organization, as exists with ants, bees, and wasps. Among mammals, only mole rats exhibit this level of social behavior.

EXOSKELETON. External skeleton that supports and protects insects and many other invertebrates, as compared with the endoskeleton, the internal skeleton of vertebrates.

EXTANT. Still in existence. Said of taxa that are not extinct.

FECUNDITY. The ability to reproduce offspring.

FILTER FEEDER. An aquatic animal that feeds by filtering small organisms out of the water with net-like structures in the mouth or throat.

FLUVIAL. Processes associated with rivers or streams and the sediments or SEDIMENTARY ROCK created by them.

FORMATION. The fundamental unit of STRATIGRAPHY consisting of layers, or STRATA, with similar LITHOLOGY, history, and other properties. Often divided into subunits called MEMBERS.

FOSSORIAL. Burrowers.

FUSIFORM. Spindle-shaped, wide in the middle, and tapered at both ends, like an American football.

GALL. Thickening of leaf tissue caused by irritation.

GASTROLITH. "Stomach-stone" or "gizzard-stone." A stone retained in the muscular gizzard used to grind food in animals lacking grinding teeth, or molars, as in birds, crocodilians, and dinosaurs.

GEOLOGIC TIME. A system of sequencing events of deep history in chronological order used by geologists, paleontologists, and other earth scientists to describe the history of Earth over billions of years, including time well before ECOLOGICAL TIME. Dates of geologic time are usually extrapolations of RADIOMETRIC DATING techniques, INDEX FOSSILS, and STRATIGRAPHY.

HIERARCHICAL ARRANGEMENT. A system of groups ranked one above the other; subsets within sets.

HOLOTYPE, TYPE SPECIMEN. The unique specimen designated as the official "name holding" specimen in any new species description. The holotype is often used as representative of a species' characteristics.

INDETERMINATE GROWTH. Growth that is not terminated during the life span of an individual, as opposed to "determinate growth," which stops once a genetically predetermined structure has completely formed.

INDEX FOSSILS. Fossils used to define and identify geologic periods of time. They work on the premise that a peculiar assemblage of fossils that occurs in a radiometrically dated locality can be used to calculate the age of other localities that have not been radiometrically dated.

INLAYING. A process of inserting additional pieces into a slab. This is most often done by commercial fossil dealers, particularly for rare (and valuable) specimens like birds (e.g., fig. 131). Sometimes this is merely insetting pieces that split off on the counter-slab during excavation, sometimes it is a process of imbedding extra fossils into a slab for aesthetic reasons, and sometimes it is a process that restores missing parts by taking them from different specimens and adding them ("Franken-fossils").

JUNIOR SYNONYM.   In zoological nomenclature, a name for a species that is invalid because it does not have PRIORITY.

KEROGEN.   A mixture of organic matter and chemical compounds found in SEDIMENTARY ROCK.

LACUSTRINE.   From a lake.

LAGOON.   A shallow body of water connected to a larger body of water. Most often the term is used to describe shallow marine environments, but in LACUSTRINE systems, it can refer to nearshore shallow embayments.

LAGOONAL.   Of or pertaining to a LAGOON.

LAMINATED.   In very fine layers.

LARVAE (plural form of larva).   A distinct juvenile form that many animals transition through before metamorphosis into adults.

LEAFLET.   A part of a compound leaf that resembles a whole leaf, but instead of having a true stem, it has an extension of the leaflet's central vein that attaches to the side of a stem together with other leaflets.

LIMNOLOGY.   The study of inland water systems.

LINNAEAN SYSTEM.   The hierarchical classification created by Carolus Linnaeus (1707–1778) as set forth in his *Systema naturae* (1735).

LINNAEUS.   Carolus Linnaeus, Latinized name of Carl von Linné (1707–1778). Swedish botanist, zoologist, and physician who laid the foundation for the modern system of naming and classifying biological organisms. Although he was himself a believer of SPECIAL CREATIONISM, he also laid the groundwork for a unified theory of evolution by putting organisms in a hierarchical classification based on shared characteristics.

LITHIFIED.   Turned to rock.

LITHOLOGY.   The general physical characteristics of rocks on a rock formation.

MACROFOSSILS.   Fossils that are large enough to be visible without a microscope.

MATRIX.   Rock that contains fossils.

MEMBER.   A unit of stratigraphy that is a subdivision of a FORMATION.

MERISTIC.   In ichthyology, relating to counting features of fish, such as vertebrae, fin rays, or teeth.

MICROFOSSILS.   Fossils that require a microscope to see.

MOLARIFORM.   Said of teeth that are shaped like molars, usually used for crushing and grinding.

MONOCOT.   Short for "monocotyledon." One of two subdivisions of flowering plants, in which the embryonic seedlings have only a single leaf. Monocots usually have veins that are parallel rather than branched or reticulate. *See also* DICOT.

MONOTYPIC.   With regard to NOMENCLATURE, a genus or higher TAXON that contains only a single species.

NEOTYPE.   A specimen later selected to be the type after a HOLOTYPE has been lost.

NOCTURNAL.   Active at night, less active by day. Nocturnal animals generally have highly developed senses of hearing and smell. *See also* DIURNAL.

NOMENCLATURE.   Group names and the system of principles, terms, and procedures related to naming.

NON-MONOPHYLETIC. A taxonomic group that is artificial or unnatural because it does not include all the descendants of a common ancestor (e.g., "reptiles" without birds or "Osteichthys" without tetrapods).

NYMPH. An immature growth stage of an insect that is more advanced than a larva in that it resembles the adult stage. *See* LARVAE.

OCCLUSAL. The surface of an upper- or lower-jaw tooth that makes contact or near contact with the corresponding surface of a tooth in the lower or upper jaw.

OMNIVORE. Any species that eats both plants and animals.

ONTOGENIC. Pertaining to ONTOGENY.

ONTOGENY. The origin and early growth and development of an organism, from embryo to adult.

OVERBURDEN. Overlying layers of dirt and rock that must be removed to get to the desired fossil layers.

PALEOECOLOGY. The study of extinct ecosystems in the fossil record, focusing on ecological features such as life cycles, environment, and interactions.

PALEOENTOMOLOGIST. A scientist who studies fossil insects.

PALEOGENE. The geologic time period at the outset of the Cenozoic, between the Cretaceous and Neogene, comprising the Paleocene, Eocene, and Oligocene.

PALAEOGNATHOUS BIRDS. A relatively primitive group of birds that evolved well before the end of the Cretaceous and whose living species include ostriches, kiwi, emu, cassowary, tinamous, rheas, and moas. The group (Paleognathae, sometimes also called Palaeognathae) also contains a number of fossil groups including the extinct †lithornithids. Palaeognathous birds are characterized by a number of characters in the palate, sternum, and pelvis, and they form a sister group to a group containing all other living birds.

PALMATE VENATION. Vein arrangement in a leaf with the veins radiating outward from the base of the leaf like fingers spread out from the palm of a hand

PARASITIC. A type of relationship between different organisms where one (the parasite) benefits at the expense of the other (the host).

PARASITOID. An organism that spends a significant portion of its life attached to or within a single host in a parasitic relationship that ultimately kills or sterilizes its host.

PARATYPE. An additional referred specimen used with the HOLOTYPE in the original species description.

PHOTOSYNTHETIC. Said of organisms that are able to convert carbon dioxide into sugars using energy from sunlight.

PHYLOGENETIC RELATIONSHIPS. Relative closeness among species on a PHYLOGENETIC TREE (i.e., which species or other TAXON is the closest relative to given species).

PHYLOGENETIC TREE. A tree that expresses phylogenetic (or evolutionary) relationships by the order of branching. See p. 50.

PHYLOGENY. Evolutionary interrelationships of species and higher groups.

PHYTOLITHS. A "plant stone"; a rigid microscopic body that occurs inside many plants. Silica phytoliths are found in grasses, various grains, and some trees.

PISCIVORY. Fish eating.

PLANKTONIC ORGANISMS. Small or microscopic organisms, including algae and protozoans, that float or drift in great numbers in bodies of water.

POSTCRANIAL. Referring to the parts of the skeleton excluding the head.

PREHENSILE. Said of an animal's appendage (a monkey's tail, an elephant's trunk) that is adapted to grasp or hold objects such as tree branches.

PRIMARY CONSUMERS. Organisms that eat PRIMARY PRODUCERS (also called herbivores).

PRIMARY FRESHWATER FAMILIES. Fishes normally confined to freshwater for their entire life cycle.

PRIMARY PRODUCERS. Organisms at the bottom of the food chain that make their own food from sunlight or deep sea thermal vents (also called autotrophs). These organisms serve as food for PRIMARY CONSUMERS.

PRIORITY. The concept that the first name applied to a TAXON is the name that will be used.

PROTO-FEATHERED. Having PROTO-FEATHERS.

PROTO-FEATHERS. Primitive feathers lacking some characteristics of modern feathers.

PROTRUSIBLE JAWS. A structural arrangement of the jaw bones, muscles, and other tissues that enable the fish to extend (protrude) its mouth forward and withdraw it at will. This aids in food capture from a water column or the water bottom.

RACHIS. In botany, a main axis, referring to the main stem or stalk.

RADIOMETRIC DATING. A technique used to date rocks, usually based on measuring a naturally occurring radioactive isotope and its decay products, using known decay rates. It is the principal source of information about the ABSOLUTE DATES of rocks and other geological features, including the age of Earth itself.

RELATIVE DATES. Geologic dates that have been determined by ordering the relative order of past events, without necessarily determining their ABSOLUTE DATE. Also dates determined through correlation with other STRATA based on similar fossils or geological anomalies.

RESTORATION. The addition of color on missing parts of a fossil to improve or enhance its appearance.

RETICULATE. Resembling a net or network.

SECONDARY CONSUMERS. Organisms that eat PRIMARY CONSUMERS in the food chain.

SEDIMENTARY ROCK. Rock derived from sedimentation of material at Earth's surface, particularly in aquatic environments.

SEDIMENTOLOGY. The study of physical and chemical properties of SEDIMENTARY ROCK and the processes involved in its formation.

SEPALS. In an angiosperm flower, the modified leaves just below its petals that protect the flower before it opens and often resemble the petals of the open flower.

SISTER GROUP. In phylogenetic systematics, a group that is more closely related to a group than any other. A monophyletic group.

SOCIAL INSECTS. Insects that live in colonies and have division of labor, group integration, and overlap of generations (e.g., ants, bees, and termites).

SPECIAL CREATIONISM. A theological assertion that the origin of the universe and all life within it sprang into being by divine decree, following the book of Genesis. This is usually interpreted as having taken place just over 6,000 years ago in the span of six 24-hour days.

SPECIATE. *See* SPECIATION.

SPECIATION. The process of new biological species arising.

STEINKERN. The fossilized shape of a hollow organic structure such as the inside of a mollusk shell, formed when sediment consolidated within the structure remains after the structure itself disintegrates or dissolves.

STEM GROUP. All species not part of the CROWN GROUP. By definition, all stem group species are extinct. The stem group/crown group distinction is not accepted (or at least recognized) by all systematists today. *See* CROWN GROUP.

STRATA. Distinct layers of rock or sediments.

STRATIFICATION. Separation into distinct layers.

STRATIGRAPHY. The study of geologic STRATA, or layers.

STROMATOLITE. A layered accretionary structure formed in shallow waters of sedimentary grains by cyanobacteria (formerly known as "blue-green algae").

SUCTION FEEDING. A method of ingesting prey organisms in fluid by suctioning them into the predator's mouth.

SUPERSATURATED. A solution that contains so much of a dissolved substance that changes in temperature or pressure result in solid material precipitating out of it.

SYMBIOTIC. A close, prolonged association between two or more different organisms, sometimes of mutual benefit or dependence.

SYMPATRIC. Occupying the same geographic area without interbreeding.

SYSTEMATICS. The study of diversification and its organization.

TAPHONOMIC. Having to do with TAPHONOMY.

TAPHONOMY. The study of decaying organisms over time and how they become fossilized (if they do).

TAXA. *See* TAXON.

TAXON, pl. TAXA. A taxonomic group. An official name for a classified group (phylum, order, family, genus, species) of organisms.

TAXONOMIC. Of or pertaining to TAXONOMY or TAXA.

TAXONOMIC WASTEBASKET. Some group names have been found to represent non-natural (i.e., NON-MONOPHYLETIC) groups but have nevertheless been maintained for the sake of convenience. Well-known examples are "fishes" when the group excludes tetrapods (because bony fishes are more closely related to TETRAPODS than they are to sharks) and "reptiles" when the group excludes birds (because certain reptiles, such as crocodiles and dinosaurs, are more closely related to birds than they are to snakes or turtles). Wastebasket groups are of little use to evolutionary studies.

TAXONOMIST. A scientist who studies TAXONOMY.

TAXONOMY. The study of TAXA and classification.

TERTIARY. The geologic time period lasting from about 65 million to about 2.6 million years ago, from the start of the Paleocene to the end of the Pliocene. Not used much anymore and largely replaced by the term CENOZOIC.

TERTIARY CONSUMERS. Organisms that eat SECONDARY CONSUMERS in the food chain.

TETRAPODS. Vertebrate animals that have four limbs or have descended from ancestors that had four limbs. Tetrapods today include mammals, birds, reptiles, turtles, and amphibians.

TYPE SECTION.   That sequence of STRATA identified as the original sequence described from the original location or area; the standard against which the same sequence is identified in other areas.

TYPE SPECIES.   The species that is designated by the author to typify a new genus in the original description of that genus.

TYPE SPECIMEN.   One (sometimes more than one) specimen that is designated in a formal description of a new species to which the scientific name is formally attached. The primary type is called the HOLOTYPE. Type specimens are generally deposited in major museum collections.

UNDERSTORY.   The part of the forest that grows below the canopy.

UNDESCRIBED.   In TAXONOMY, a species or other TAXON that has not yet been officially described and named.

UNDIAGNOSABLE.   With no apparent uniquely defining characteristics. Indistinguishable from at least one other TAXON.

UPWELLING.   An aquatic phenomenon that involves wind-driven motion of dense, cooler nutrient-rich water from below to the surface of a body of water.

VALID.   Said of a taxonomic name that satisfies the criteria of AVAILABILITY and PRIORITY.

VARVES.   Annual laminations in sedimentary rocks.

VASCULAR PLANTS.   Those plants that have uniquely specialized tissues for conducting water, minerals, and PHOTOSYNTHETIC materials through the plant.

WASTEBASKET GROUP.   See TAXONOMIC WASTEBASKET.

ZOOPLANKTON.   Tiny organisms other than plants that drift in the water column of oceans, seas, and freshwater, and are usually too small to see with the naked eye. They differ from phytoplankton, which make their own energy through photosynthesis or other chemical reactions.

ZYGODACTYL.   Having two toes projecting forward and two projecting backward, as in parrots, woodpeckers, cuckoos, and some owls.

References Cited

Agassiz, L. 1833–44. *Recherches sur les Poissons Fossiles.* 5 volumes plus supplement. Published by the author, printed by Petitpierre: Neuchâtel. Dates for publication of individual parts are given by W. H. Brown in Woodward and Sherborn (1890, 5: xxv–xxix).

Baer, J. L., 1969. "Paleoecology of Cyclic Sediments of the Lower Green River Formation, Central Utah." *Brigham Young University Geology Studies* 16: 1.

Berry, E. W. 1925. "A New Salvinia from the Eocene." *Torreya* 25: 116–18.

Bertelli, S., B. E. K. Lindow, G. J. Dyke, and L. M. Chiappes. 2010. "A Well-Preserved Bird from the Early Eocene Fur Formation of Denmark." *Palaeontology* 53: 507–31.

Bowen, G. J., A. L. Daniels, and B. B. Bowen. 2008. "Paleoenvironmental Isotope Geochemistry and Paragenesis of Lacustrine and Palustrine Carbonates, Flagstaff Formation, Central Utah, U.S.A." *Journal of Sedimentary Research* 78: 162–74.

Bradley, L. 1987. "Lee Craig—Controversial Fossil Man." Typewritten manuscript in the archives of Fossil Butte National Monument.

Bradley, W. H. 1929. "The Varves and Climate of the Green River Epoch." *U.S. Geological Survey Professional Paper* 158E: 87–110.

———. 1931. "Origin and Microfossils of the Oil Shale of the Green River Formation of Colorado and Utah." *U.S. Geological Survey Professional Paper* 168: 1–58.

———. 1948. "Limnology and the Eocene Lakes of the Rock Mountain Region." *Geological Society of America, Bulletin* 59: 635–48.

———. 1964. "Aquatic Fungi from the Green River Formation of Wyoming." *American Journal of Science* 262: 413–16.

Broadkorb, P. 1970. "An Eocene Puffbird from Wyoming." *Contributions to Geology* 9: 13–15.

Brochu, C. A. 1997. "A Review of 'Leidyosuchus' (Crocodyliformes, Eusuchia) from the Cretaceous through Eocene of North America." *Journal of Vertebrate Paleontology* 17 (4): 679–97.

———. 2010. "A New Alligatorid from the Lower Eocene Green River Formation of Wyoming and the Origin of Caimans." *Journal of Vertebrate Paleontology* 30 (4): 1109–26.

Brongersma-Sanders, M., 1957. "Mass Mortality in the Sea." *Geological Society of America, Memoirs,* 67: 941–1010.

Brown, R. W. 1929. "Additions to the Flora of the Green River Formation." *United States Geological Survey Professional Paper* 154: 279–99.

———. 1934. "The Recognizable Species of the Green River Flora." *United States Geological Survey Professional Paper* 185-C.

———. 1937. "Additions to Some Fossil Floras of the Western United States." *U.S. Geological Survey Professional Paper* 186-J: 163–207.

———. 1946. "Alterations in Some Fossil and Living Floras." *Journal of the Washington Academy of Sciences.* 36: 344–55.

Buchheim, H. P. 1994a. "Eocene Fossil Lake, Green River Formation, Wyoming: A History of Fluctuating Salinity." In "Sedimentology and Geochemistry of Modern and Ancient Saline Lakes," edited by R. Renaut and W. Last. *Society for Sedimentary Geology Special Publication* 50: 239–47.

———. 1994b. "Paleoenvironments, Lithofacies and Varves of the Fossil Butte Member of the Eocene Green River Formation, Southwestern Wyoming." *Contributions to Geology* 30: 3–14.

Buchheim, H. P., R. A. Cushman Jr., and R. E. Biaggi. 2012. "Stratigraphic Revision of the Green River Formation in Fossil Basin, Wyoming: Overfilled to Underfilled Lake Evolution." *Rocky Mountain Geology* 46: 165–81.

———. 1998. "Eocene Fossil Lake: The Green River Formation of Fossil Basin, Southwestern Wyoming." In "Modern and Ancient Lacustrine Depositional Systems," edited by J. Pittman and A. Carroll. *Utah Geological Association Guidebook* 26: 1–17.

Buchheim, H. P., and R. R. Surdam. 1981. "Paleoenvironments and Fossil Fishes of the Laney Member, Green River Formation, Wyoming." In *Communities of the Past*, edited by J. Gray, A. J. Boucot, and W. B. N. Berry. Stroudsburg, PA: Hutchinson Ross.

Calixto-Albarrán, I., and José-Luis Osorno. 2000. "The Diet of the Magnificent Frigatebird during Chick Rearing." *Condor* 102: 569–76.

Carvalho, M. R. de, J. G. Maisey, and L. Grande. 2004. "Freshwater Stingrays of the Green River Formation of Wyoming (Early Eocene), with the Description of a New Genus and Species and an Analysis of Its Phylogenetic Relationships (Chondrichthyes: Myliobatiformes)." Bulletin of the American Museum of Natural History 284: 1–136.

Cashion, W. B. 1967. "Geology and Fuel Resources of the Green River Formation, Southeastern Uinta Basin, Utah and Colorado." *U.S. Geological Survey, Professional Paper* 548: 1–48.

Cavender, T. M. 1986. "Review of the Fossil History of North American Freshwater Fishes." In *The Zoogeography of North American Freshwater Fishes*, edited by C. H. Hocutt and E. O. Wiley. New York: John Wiley and Sons.

———. 1998. "Development of the North American Tertiary Freshwater Fish Fauna with a Look at Parallel Trends Found in the European Record." *Italian Journal of Zoology* 65: 149–61.

Chang, M.-M., N. Wang, and F.-X. Wu. 2010. "Discovery of †*Cyclurus* (Amiinae, Amiidae, Amiiformes, Pisces) from China." *Vertebrata PalAsiatica* 48: 85–100.

Chin, K., D. A. Eberth, M. A. Schweitzer, T. A. Rando, W. J. Sloboda, and J. R. Horner. 2003. "Remarkable Preservation of Undigested Muscle Tissue within a Late Cretaceous Tyrannosaurid Coprolite from Alberta Canada" *Society for Sedimentary Geology, Research Letters*, 286–94.

Clarke, J. A., D. T. Ksepka, N. A. Smith, and M. A. Norell. 2009. "Combined Phylogenetic Analysis of a New North American Fossil Species Confirms Widespread Eocene Distribution for Stem Rollers (Aves, Coracii)." *Zoological Journal of the Linnean Society* 157: 586–611.

Clarke, J. A., C. P. Tambussi, J. I. Noriega, G. M. Erickson, and R. A. Ketcham. 2005. "Definitive Fossil Evidence for the Extant Avian Radiation in the Cretaceous." *Nature* 433: 305–8.

Cockerell, T. D. A. 1925. "Plant and Insect Fossils from the Green River Eocene of Colorado." *Proceedings of the U.S. National Museum* 66: 1–13.

Conrad, J. L. 2006. "An Eocene Shinisaurid (Reptilia, Suamata) from Wyoming, U.S." *Journal of Vertebrate Paleontology* 26: 113–26.

Conrad, J. L., O. Rieppel, and L. Grande. 2007. "A Green River (Eocene) Polychrotid (Squamata: Reptilia) and a Re-examination of Iguanian Systematics." *Journal of Paleontology* 81: 1365–73.

———. 2008. "Re-assessment of Varanid Evolution Based on New Data from *Siniwa ensidens* Leidy, 1870 (Squamata, Reptilia)." *American Museum Novitates* 3630: 1–15.

Cope, E. D. 1870. "Observations on the Fishes of the Tertiary Shales of Green River, Wyoming Territory." *Proceedings of the American Philosophical Society* 11: 380–84.

———. 1871. "On the Fishes of the Tertiary Shales of Green River, Wyoming Territory." *U.S. Geological and Geographical Survey of the Territories Annual Report* 4.

———. 1872. "Description of Some New Vertebrata from the Bridger Group of the Eocene [*Axestus byssinus*]." *Proceedings of the American Philosophical Society* 12: 462.

———. 1873. "Some Extinct Turtles from the Eocene Strata of Wyoming [*Trionyx heteroglyptus*]." *Proceedings of the Academy of Natural Sciences of Philadelphia* 277–79.

———. 1877. "A Contribution to the Knowledge of the Ichthyological Fauna of the Green River Shales." *Bulletin of the U.S. Geological and Geographical Survey of the Territories* 3, art. 34: 807–19.

———. 1879. "A Sting Ray from the Green River Shales of Wyoming." *American Naturalist* 13: 333.

———. 1881. "On the Vertebrata of the Wind River Eocene Beds of Wyoming." *Bulletin of the U.S. Geological and Geographical Survey of the Territories* 6: 183–202.

———. 1883. "A New Chondrostean from the Eocene." *American Naturalist* 17: 1152–53.

———. 1885a. "Eocene Paddle-fish and Gonorhynchidae." *American Naturalist* 19: 1090–91.

———. 1885b. *Vertebrata of the Tertiary Formations of the West.* Book 1. United States Geological Survey of the Territories. [Although this publication bears two dates (1884 on p. i, 1883 on p. ix), it was not issued until 1885 (Hay 1901).]

———. 1886. "On Two New Forms of Polyodont and Gonorhynchid Fishes from the Eocene of the Rocky Mountains." *Memoirs of the National Academy of Sciences* 3, 17th memoir, 161–65.

Corbett, S. L., and S. R. Manchester. 2004. "Phytogeography and Fossil History of *Ailanthus* (Simaroubaceae)." *International Journal Of Plant Sciences* 165 (4): 671–90.

Currano, E. D., P. Wilf, S. L. Wing, C. C. Labandeira, E. C. Lovelock, and D. L. Royer. 2008. "Sharply Increased Insect Herbivory during the Paleocene-Eocene Thermal Maximum." *Proceedings of the National Academy of Sciences* 105: 1960–64.

Cushman, R. A., Jr., 1983. "Palynology and Paleoecology of the Fossil Butte Member of the Eocene Green River Formation in Fossil Basin, Lincoln County, Wyoming." Master's thesis, Loma Linda University, Riverside, CA.

Dayvault, R. D., L. A. Codington, D. Kohls, W. D. Hawes, and P. M. Ott, 1995. "Fossil Insects and Spiders from Three Locations in the Green River Formation of the Piceance Creek

Basin, Colorado." In *The Green River Formation in Piceance Creek and Eastern Uinta Basins*, edited by W. R. Averett, 97–115. Grand Junction, CO: Grand Junction Geological Society.

Eastman, C. R. 1900. "New Fossil Bird and Fish Remains from the Middle Eocene of Wyoming." *Geological Magazine (London)* 7: 54–58.

Edgecombe, G. D. 2010. "Arthropod Phylogeny: An Overview from the Perspectives of Morphology, Molecular Data and the Fossil Record." *Arthropod Structure & Development* 39: 74–87.

Edwards, P. 1976. "Fish Coprolites from Fossil Butte, Wyoming." *Contributions to Geology, University of Wyoming* 14 (2): 115–17.

Estes, R. 1976. "Middle Paleocene Lower Vertebrates from the Tongue River Formation, Southeastern Montana." *Journal of Paleontology* 50 (3): 500–520.

Eugster, H. P., and L. A. Hardie. 1975. "Sedimentation in an Ancient Playa-lake Complex: The Wilkins Peak Member of the Green River Formation of Wyoming." *Bulletin of the Geological Society of America* 86 (3): 319–34.

Feduccia, A. 1973. "A New Eocene Zygodactyl Bird." *Journal of Paleontology* 47 (3): 501–3.

———. 1977. "*Neanis schucherti* Restudied: Another Eocene Piciform Bird." Collected Papers in Avian Paleontology Honoring the 90th Birthday of Alexander Wetmore, *Smithsonian Contributions to Paleobiology* 27: 95–99.

———. 1999. *The Origin and Evolution of Birds*. 2nd ed.. New Haven, CT: Yale University Press.

Feduccia, A., and L. D. Martin. 1976. "The Eocene Zygodactyl Birds of North America (Aves: Piciformes)." *Smithsonian Contributions to Paleobiology* 27: 101–10.

Feldmann, R. M., L. Grande, C. Birkhimer, J. T. Hannibal, and D. L. McCoy. 1981. "Decapod Fauna of the Green River Formation (Eocene) of Wyoming." *Journal of Paleontology* 55: 788–99.

Fitch, C. P., L. M. Bishop, W. L. Boyd, F. L. A. Gordner, C. T. Rogers, and J. E. Tilden. 1934. "'Waterbloom' as a Cause of Poisoning in Domestic Animals." *Cornell Veterinarian* 24: 30–39.

Fouch, T. D. 1976. "Revision of the Lower Part of the Tertiary System in the Central and Western Uinta Basin, Utah." *U.S. Geological Survey Bulletin* 1405-C.

France, A. 1881. *Le Crime de Sylvestre Bonnard*. [*The Crime of Sylvestre Bonnard*, English translation 2007]. New York: Mondial.

Gaffney, E. S. 1972. "The Systematics of the North American Family Baenidae (Reptilia, Cryptodira)." *American Museum of Natural History Bulletin* 147: 241–320.

Garner, P. 1997. "Green River Blues." *Creation* 19: 18–19.

Gaudant, J. 1981. "Contribution de la paléoichthyologie continentale à la reconstitution des paléoenvironnements cénozoïques d'Europe occidentale: Approche systématique, paléoécologique, paléogéographique et paléoclimatologique." PhD diss., Université Pierre et Marie Curie, Paris.

Gilmore, C. W. 1938. "Fossil Snakes of North America." *Geological Society of America, Special Paper* 9: 1–96.

Grande, L. 1979. "*Eohiodon falcatus*, a New Species of Hiodontid (Pisces) from the Late Early Eocene Green River Formation of Wyoming." *Journal of Paleontology* 53: 103–11.

———. 1980. "The Paleontology of the Green River Formation with a Review of the Fish Fauna." 1st ed. *Geological Survey of Wyoming, Bulletin* 63: 1–334.

———. 1982a. "A Revision of the Fossil Genus *Diplomystus* with Comments on the Interrelationships of Clupeomorph Fishes." *American Museum Novitates* 2728: 1–34.

———. 1982b. "A Revision of the Fossil Genus *Knightia*, with a Description of a New Genus from the Green River Formation (Teleostei, Clupeidae)." *American Museum Novitates* 2731: 1–22.

———. 1984. "Paleontology of the Green River Formation, with a Review of the Fish Fauna." 2nd ed. *Geological Survey of Wyoming, Bulletin* 63: 1–334. (1st ed., 1980.)

———. 1985a. "Recent and Fossil Clupeomorph Fishes with Materials for Revision of the Sub-groups of Clupeoids." *Bulletin, American Museum of Natural History* 181: 231–372.

———. 1985b. "The Use of Paleontology in Systematics and Biogeography, and a Time Control Refinement for Historical Biogeography." *Paleobiology* 11: 1–11.

———. 1987. "Redescription of †*Hypsidoris farsonensis* (Teleostei: Siluriformes) with a Reassessment of Its Phylogenetic Relationships." *Journal of Vertebrate Paleontology* 7: 24–54.

———. 1989. "The Eocene Green River Lake System, Fossil Lake, and the History of the North American Fish Fauna." In *Mesozoic/Cenozoic Vertebrate Paleontology: Classic Localities, Contemporary Approaches*, edited by J. Flynn. 28th International Geological Congress Fieldtrip Guidebook T322, American Geophysical Union.

———. 1994a. "Repeating Patterns in Nature, Predictability, and 'Impact' in Science." In *Interpreting the Hierarchy of Nature: From Systematic Patterns to Evolutionary Process Theories*, edited by L. Grande and O. Rieppel, 61–84. San Diego, CA: Academic Press.

———. 1994b. "Studies of Paleoenvironments and Historical Biogeography in the Fossil Butte and Laney Members of the Green River Formation." *Contributions to Geology, University of Wyoming* 30 (1): 15–32.

———. 1999. "The First *Esox* (Esocidae: Teleostei) from the Eocene Green River Formation, and a Brief Review of Esocid Fishes." *Journal of Vertebrate Paleontology* 19 (2): 271–92.

———. 2001. "An Updated Review of the Fish Fauna from the Green River Formation, the World's Most Productive Lagerstätten." In *Eocene Vertebrates: Unusual Occurrences and Rarely Sampled Habitats*, edited by G. Gunnell, 1–38. New York: Plenum.

———. 2010. "An Empirical Synthetic Pattern Study of Gars (Lepisosteiformes) and Closely Related Species, Based Mostly on Skeletal Anatomy: The Resurrection of Holostei." *American Society of Ichthyologists and Herpetologists Special Publication* 6: 1–871; Supplementary Issue of *Copeia* 10(2A).

Grande, L., and W. Bemis. 1991. "Osteology and Phylogenetic Relationships of Fossil and Recent Paddlefishes (Polyodontidae) with Comments on the Interrelationships of Acipenseriformes." *Society of Vertebrate Paleontology Memoir* 1: i–viii, 1–121; supplement to *Journal of Vertebrate Paleontology* 11 (1).

———. 1998. "A Comprehensive Phylogenetic Study of Amiid Fishes (Amiidae) Based on Comparative Skeletal Anatomy: An Empirical Search for Interconnected Patterns of Natural History." *Society of Vertebrate Paleontology Memoir* 4: i–x, 1–690; supplement to *Journal of Vertebrate Paleontology* 18 (1).

Grande, L., and H. P. Buchheim. 1994. "Paleontological and Sedimentological Variation in Early Eocene Fossil Lake." *Contributions to Geology* 30: 33–56.

Grande, L., and J. T. Eastman, and T. M. Cavender. 1982. "*Amyzon gosiutensis*, a New Catostomid Fish from the Green River Formation." *Copeia*, 523–32.

Grande, L., and T. Grande. 2008. "Redescription of the Type Species for the Genus †*Notogoneus* (Teleostei: Gonorynchidae) Based on New, Well-Preserved Material." *Memoir, Journal of Paleontology* 70: 1–31.

Grande, L., F. Jin, Y. Yabumoto, and W. E. Bemis. 2002. "*Protopsephurus liui*, a Well-Preserved Paddlefish (Acipenseriformes: Polyondontidae) from the Lower Cretaceous of China." *Journal of Vertebrate Paleontology* 22: 209–37.

Grande, L., and J. G. L. Lundberg. 1988. "Revision and Redescription of the Genus †*Astephus* (Siluriformes: Ictaluridae) with a Discussion of Its Phylogenetic Relationships." *Journal of Vertebrate Paleontology* 8: 139–71.

Grimaldi, D., and M. S. Engel. 2005. *Evolution of the Insects*. Cambridge: Cambridge University Press.

Gruber, G., and N. Micklich., eds. 2007. *Messel: Treasures of the Eocene*. Darmstadt: Wissenschaftliche Buchgesellschaft.

Hackett, S. J., R. T. Kimball, S. Reddy, R. C. K. Bowie, E. L. Braun, M. J. Braun, J. L. Chojnowski, W. A. Cox, K.-L. Han, J. Harshman, C. J. Huddleston, B. D. Marks, K. J. Miglia, W. S. Moore, F. H. Sheldon, D. W. Steadman, C. C. Witt, and T. Yuri. 2008. "A Phylogenic

Study of Birds Reveal Their Evolutionary History." *Science* 320: 1763–67.

Harrison, C. J. O. 1984. "A Revision of Fossil Swifts (Vertebrata, Aves, Suborder Apodi), with Descriptions of Three New Genera and Two New Species." *Meded. Werkgr. Tert. Kwart. Geol.* 21: 157–77.

Hay, O. P. 1908. "The Fossil Turtles of North America." *Carnegie Institution of Washington* 75: 1–568.

Hebert, E., 1855. "Note sur le tibis du Gastornis pariensis [*sic*]." *C R Academy of Sciences* 40: 579–82.

Heer, O. 1861. "Beiträge zur nähern Kenntnis der sachsischthüringischen Braunkohlen Flora." *Abhandlungen des Naturwissenschaftlichen Vereines für Sachsen und Thüringen in Halle* 2: 407–38.

Hellawell, J., and P. J. Orr. 2012. "Deciphering Taphonomic Process in the Eocene Green River Formation of Wyoming." *Palaeobiodiversity and Palaeoenvironments* 92: 353–65.

Herendeen, P. S., D. H. Les, and D. L. Dilcher. 1990. "Fossil Ceratophyllum (Ceratophyllaceae) from the Lower Tertiary of North America." *American Journal of Botany* 77: 7–16.

Hesse, A. 1988. "Die Messelornithidae: Eine neue Familie der Kranichartigen (Aves: Gruiformes: Rhynocheti) aus dem Tertiar Europas und Nordamerikas." *Journal of Ornithology* 129: 83–85.

———. 1992. "A New Species of *Messelornis* (Aves: Gruiformes: Messelornithidae) from the Middle Eocene Green River Formation." In *Papers in Avian Paleontology Honoring Pierce Brodkorb,* edited by K. E. Campbell. *Science Series of the Natural, History Museum of Los Angeles County* 36: 171–78.

Hilton, E. J., and L. Grande. 2008. "Fossil Mooneyes (Teleostei: Hiodontiformes, Hiodontidae) from the Eocene of Western North America, with a Reassessment of their Taxonomy." In *Fishes and the Break-up of Pangaea*, edited by L. Calvin, A. Longbottom, and M. Richter, 221–251. Geological Society, London, Special Publications 295.

Houde, P. W. 1988. "Paleognathous Birds from the Early Tertiary of the Northern Hemisphere." *Publications of the Nuttall Ornithological Club* [Cambridge, MA] 22: 1–148.

Houde, P. W., and S. L. Olson. 1981. "Paleognathous Carinate Birds from the Early Tertiary of North America." *Science* 214: 1236–37.

———. 1989. "Small Arboreal Nonpasserine Birds from the Early Tertiary of North America." In *Acta XIX congressus internationalis ornithologici*, edited by H. Ouellet. Ottawa: University of Ottawa Press.

———. 1992. A Radiation of Coly-like Birds from the Eocene of North America (Aves: Sandcoleiformes New Order)." In *Papers in Avian Paleontology Honoring Pierce Brodkorb*, edited by K. E. Campbell Jr. Natural History Museum of Los Angeles County, Science Series 36: 137–60.

Hutchison, J. H. 1998. "Turtles across the Paleocene/Eocene Epoch Boundary in West Central North America." In *Late Paleocene-Early Eocene Climatic and Biotic Events in the Marine and Terrestrial Records*, edited by M. P. Aubry, S. G. Lucas, and W. A. Berggren, 401–8. Princeton, NJ: Princeton University Press.

Jackson, R. W. 1980. *The Fish of Fossil Lake: The Story of Fossil Butte National Monument.* Dinosaur Nature Association, Dinosaur National Park, Colorado.

Jepsen, G. L., 1930. "New Vertebrate Fossils from the Lower Eocene of the Bighorn Basin, Wyoming." *Proceedings of the American Philosophical Society* 69: 117–31.

———. 1966. "Early Eocene Bat from Wyoming." *Science* 154: 1333–39.

———. 1970. "Bat Origins and Evolution." In *Biology of Bats*, vol. 1, edited by W. A. Wimsatt. New York: Academic Press.

Jordan, D. S. 1905. *A Guide to the Study of Fishes.* New York: Henry Holt.

———. 1907. "The Fossil Fishes of California; with Supplementary Notes on Other Species of Extinct Fishes." *Bulletin of the Department of Geology, University of California* 5: 95–145.

———. 1923. "A Classification of Fishes Including Families and Genera as Far as Known." *Stanford University Publications, University Series, Biological Series* 3: 77–243.

Karrenberg, S., P. J. Edwards, and J. Kollmann. 2002. "The Life History of Salicaceae Living in the Active Zone of Floodplains." *Freshwater Biology* 47: 733–48.

Knowlton, F. H. 1923. "Revision of the Flora of the Green River Formation." *United States Geological Survey Professional Paper* 131: 133–82.

Knowlton, F. H., and T. D. A. Cockerell, 1919. "A Catalogue of Mesozoic and Cenozoic Plants of North America." U.S. Geological Survey Bulletin 696.

Koenigswald, W. v., K. D. Rose, L. Grande, and R. D. Martin. 2005a. "Die Lebensweise Eozäner Säugetiere (Pantolestidae und Apatemyidae) aus Messel (Europa) im Vergleich zu Neuen Skelettfunden aus dem Fossil Butte Member von Wyoming (Nordamerika)." *Geologisches Jahrbuch Hessen* 132: 43–54.

———. 2005b. "First Apatemyid Skeleton from the Lower Eocene Fossil Butte Member, Wyoming (USA), Compared to the European Apatemyid from Messel, Germany." *Palaeontographica* Abt. A, Band 272: 149–69.

Komarovsky, B. 1951. Some Characteristic Water-blooms in Lake Tiberias and Fish Ponds in the Jordan Valley." *Verhandlungen der Internationalen Vereinigung fur Theoretische und Angewandte Limnologie,* 11: 219–23.

Ksepka, D. T. 2009. "Broken Gears in the Avian Molecular Clock: New Phylogenetic Analyses Support Stem Galliform Status for *Gallinuloides wyomingensis* and Rellid Affinities for *Amitabha urbsinterdictensis*." *Cladistics* 25: 173–97.

Ksepka, D. T., and J. A. Clarke. 2009. "Affinities of *Palaeospiza bella* and the Phylogeny and Biogeography of Mousebirds [Coliiformes]." *Auk* 126: 245–59.

———. 2010a. "New Fossil Mousebird (Aves: Coliiformes) with Feather Preservation Provides Key Insight into the Ecological Diversity of an Eocene North American Avifauna." *Zoological Journal of the Linnaean Society* 160: 685–706.

———. 2010b. "*Primobucco mcgrewi* (Aves: Coracii) from the Eocene Green River Formation: New Anatomical Data and the Earliest Definitive Record of Stem Rollers." *Journal of Vertebrate Paleontology* 30: 215–25.

———. 2012. "A New Stem Parrot from the Green River Formation and the Complex Evolution of the Grasping Foot in Pan-Psittaciformes." *Journal of Vertebrate Paleontology* 32: 395–406.

Ksepka, D. T., J. A. Clarke, and L. Grande. 2011. "Stem Parrots (Aves, Halcyornithidae) from the Green River Formation and a Combined Phylogeny of Pan-Psittaciformes." *Journal of Paleontology* 86: 835–52.

Ksepka, D. T., J. A. Clarke, S. J. Sterling, and L. Grande. Submitted. "First Fossil Evidence of Wing Shape in a Stem Relative of Swifts and Hummingbirds (Aves, Panapodiformes)."

Labandeira, C. C., K. R. Johnson, and P. Wilf. 2002. "Impact of the Terminal Cretaceous Event on Plant-Insect Associations." *Proceedings of the National Academy of Sciences of the United States of America* 99 (4): 2061–66.

Lacépède, B. G. E. 1803. *Histoire naturelle des poissons 5*. Paris: Chez Plassan, Imprimeur-Libraire.

La Rocque, A. 196. *Molluscan Faunas of the Flagstaff Formation of Central Utah*. Geological Society of America.

Leggitt, V. L., H. P. Buchheim, and R. E. Biaggi. 1998. "The Stratigraphic of Three Presbyornis Nesting Sites: Eocene Fossil Lake, Lincoln County Wyoming." *National Park Service, Geologic Resources Division Technical Report* 98/1: 61–68.

Leidy, J. 1856. "Notice of Some Remains of Fishes Discovered by Dr. John E. Evans." *Proceedings of the Academy of Natural Sciences, Philadelphia*.

———. 1869. "Notice of Some Extinct Vertebrates from Wyoming and Dakota." *Proceedings of the Academy of Natural Sciences, Philadelphia*.

———. 1870a. "Description of *Emys jeanesi*, *E. haydeni*, *Baena arenosa*, and *Saniwa ensidens*." *Proceedings of the Academy of Natural Sciences, Philadelphia*.

———. 1870b. "Remarks on *Poicilopleuron valens*, *Clidastes intermedius*, *Leiodon proriger*,

Baptemys wyomingensis, and *Emys stevensonianus*." *Proceedings of the Academy of Natural Sciences, Philadelphia*.

———. 1871. "Remarks on Fossil Vertebrates from Wyoming [*Baena undata*]." *Proceedings of the Academy of Natural Sciences, Philadelphia*.

———. 1872. "On a New Genus of Extinct Turtles [*Chisternon*]." *Proceedings of the Academy of Natural Sciences, Philadelphia*.

———. 1873. "Contributions to the Extinct Vertebrate Fauna of the Western Territories." *Report of the U.S. Geological and Geographical Survey of the Territories* 1: 14–358.

Lesquereux, L. 1872. *U.S. Geological and Geographical Survey of the Territories. Annual Report for 1871*, 289–90.

———. 1873. "Enumeration and Description of Fossil Plants from the Western Territories Formations." F. V., *U.S. Geological Survey of the Territories, Annual Report* 6: 372–427.

———. 1883. "Contributions to the Fossil Flora of the Western Territories. III. The Cretaceous and Tertiary Floras." *U.S. Geological Survey of the Territories, Annual Report* 8.

Lesueur, C. A. 1818. "Descriptions of Several New Species of North American Fishes (Continued)." *Journal of the Academy of Natural Sciences, Philadelphia* 1: 359–69.

Li, G.-Q., L. Grande, and M. V. H. Wilson. 1997. "The Species of *Phareodus* (Teleostei: Osteoglossidae) from the Eocene of North America and Their Phylogenetic Relationships." *Journal of Vertebrate Paleontology* 17: 487–505.

Liem, K. F., W. E. Bemis, W. F. Walker, and L. Grande. 2001. *Functional Anatomy of the Vertebrates: An Evolutionary Perspective*. 3rd ed. Fort Worth: Saunders College Publishers.

Linnaeus, C. 1758. *Systema naturae*. Editio X. [*Systema naturae per regna tria naturae, secundum classes, ordines, genera, species, cum characteribus, differentiis, synonymis, locis.* Tomus I. Editio decima, reformata.] Holmiae. [Nantes and Pisces in Tom. 1, pp. 230–338. Date fixed by ICZN, Code Article 3.]

Loewen, M. A., and H. P. Buchheim. 1998. "Paleontology and Paleoecology of the Culminating Phase of Eocene Fossil Lake, Fossil Butte National Monument, Wyoming." *National Park Service, Geologic Resources Division Technical Report* 98/1: 73–80.

Lucas, F. A. 1900. "Characters and Relations of *Gallinuloides,* a Fossil Gallinaceous Bird from the Green River Shales of Wyoming." *Bulletin of the Museum of Comparative Zoology* 36.

MacGinitie, H. D., 1969. "The Eocene Green River Flora of Northwestern Colorado and Northeastern Utah." *University of California Publications in Geological Sciences* 83: 1–140.

Manchester, S. R. 1986. "Vegetative and Reproductive Morphology of an Extinct Plane Tree (Platanaceae) from the Eocene of Western North America." *Botanical Gazette* 147 (2): 200–226.

———. 1987. "The Fossil History of the Juglandaceae." *Monographs in Systematic Botany from the Missouri Botanical Garden* 21: 1–137.

———. 1989. "Attached Reproductive and Vegetative Remains of the Extinct American-European Genus *Cedrelospermum* (Ulmaceae) from the Early Tertiary of Utah and Colorado." *American Journal of Botany* 76: 256–76.

———. 1992. "Flowers, Fruits, and Pollen of *Florissantia*, an Extinct Malvalean Genus from the Eocene and Oligocene of Western North America." *American Journal of Botany* 79: 996–1008.

———. 1994. "Fruits and Nuts of the Middle Eocene Nut Beds Flora, Clarno Formation, Oregon." *Palaeontographica Americana* 58: 1–205.

Manchester, S. R., W. S. Judd, and B. Handley. 2006. "Foliage and Fruits of Early Poplars (Salicaceae: *Populus*) from the Eocene of Utah, Colorado, and Wyoming." *International Journal of Plant Sciences* 167 (4): 897–908.

Manchester, S. R., and E. Zastawniak. 2007. "Fruit with Perianth Remains of *Chaneya* Wang & Manchester (Extinct Rutaceae) in the Upper Miocene of Sośnica, Poland." *Acta Palaeobotanica* 47 (1): 253–59.

———. 1877. "Notice of Some New Vertebrate Fossils." *American Journal of Science*, ser. 3, 14: 249–56.

Martin, A. J., G. M. Vasquez-Prokopec, and M. Page. 2010. "First Known Feeding Trace of the Eocene Bottom-Dwelling Fish *Notogoneus osculus* and Its Paleontological Significance." *PLoS ONE open access on-line journal* 5:e10420: 1–8.

Mayr, G. 2000. "A New Raptor-like Bird from the Lower Eocene of North America and Europe." *Senckenbergiana lethaea* 80 (1): 59–65.

———. 2004. "Phylogenetic Relationships of the Early Tertiary Messel Rails (Aves, Messelornithidae)." *Senckenbergiana lethaea* 84: 319–24.

———. 2005. "The Postcranial Osteology and Phylogenetic Position of the Middle Eocene *Messelastur gratulator* Peters, 1994—a Morphological Link between Owls (Strigiformes) and Falconiform Birds?" *Journal of Vertebrate Paleontology* 25 (3): 635–45.

———. 2008. "Phylogenetic Affinities of the Enigmatic Avian Taxon Zygodactylus Based on New Material from the Early Oligocene of France." *Journal of Systematic Paleontology* 6: 333–44.

———. 2009. *Paleogene Fossil Birds*. Berlin: Springer-Verlag.

Mayr, G., and M. Daniels. 2001. "A New Short-Legged Landbird from the Early Eocene of Wyoming and Contemporaneous European Sites." *Acta Palaeontologica Polonica* 46 (3): 393–402.

Mayr, G., and A. Manegold. 2004. "The Oldest European Fossil Songbird from the Early Oligocene of Germany." *Naturwissenschaften* 91: 173–77.

Mayr, G., C. Mourer-Chauviré, and I. Weidig. 2004. "Osteology and Systematic Position of the Eocene Primobucconidae (Aves, Coraciiformes *sensu stricto*), with First Records from Europe." *Journal of Systematic Palaeontology* 2 (1): 1–12.

Mayr, G., and I. Weidig. 2004. "The Early Eocene Bird *Gallinuloides wyomingensis* a Stem Group Representative of Galliformes." *Acta Palaeontologica Polonica* 49 (2): 211–17.

McGrew, P. O. 1975. "Taphonomy of Eocene Fish from Fossil Basin, Wyoming." *Fieldiana* 33 (14): 257–70.

McMurran, D. M., and S. R. Manchester. 2010. "*Lagokarpos lacustris*, a New Winged Fruit from the Paleogene of Western North America." *International Journal Of Plant Sciences* 171 (2): 227–34.

McNeill, J., F. R. Barrie, H. M. Burdet, V. Demoulin, D. L. Hawksworth, K. Marhold, D. H. Nicolson, J. Prado, P. C. Silva, J. E. Skog, J. H. Wiersema, and N. J. Turland. 2006. *International Code of Botanical Nomenclature*. *Regnum Vegetabile* 146, A. R. G. Gantner Verlag KG.

Meyer, H. W. 2003. *The Fossils of Florissant*. Washington, DC: Smithsonian Institution Press.

Meylan, P. A. 1987. "The Phylogenetic Relationships of the Soft-Shelled Turtles (Family Trionychidae)." *American Museum of Natural History Bulletin* 186: 1–101.

Micklich, N., and G. Klappert. 2001. "*Masillosteus kelleri*, a New Gar (Actinopterygii, Lepisosteidae) from the Middle Eocene of Grube Messel (Hessen, Germany)." *Kaupia, Darmstädter Beiträge zur Naturgeschichte* 11: 73–81.

Miller, R. R. 1959. "Origin and Affinities of the Freshwater Fish Fauna of Western North America." In "Zoogeography," edited by C. L. Hubbs, 187–222. *American Association for the Advancement of Science, Publication* 51.

Mook, C. C. 1959. "A New Species of Crocodile of the Genus *Leidyosuchus* from the Green River Beds." *American Museum Novitates* 1933: 1–6.

Nelson, J. S. 2006. *Fishes of the World*. 4th ed. Hoboken, NJ: John Wiley & Sons.

Nesbitt, S. J., D. T. Ksepka, and J. A. Clarke. 2011. "Podargiform Affinities of the Enigmatic Fluvioviridavis platyrhamphus and the Early Diversification of Strisores ("Caprimulgiformes" + Apodiformes)." *PLoS One, 6* (11): 10.

Newberry, J. S. 1883. "Brief Descriptions of Fossil Plants, Chiefly Tertiary, from Western North America." *U.S. National Museum, Proceedings* 5: 502–14.

Olson, S. 1977. "A Lower Eocene Frigatebird from the Green River Formation (Pelecaniformes: Frigatidae)." *Smithsonian Contributions to Paleobiology* 35.

————. 1987. "An Early Eocene Oilbird from the Green River Formation of Wyoming (Caprimulgiformes: Steatornithidae)." *Documents des Laboratoires de Géologie de Lyon* 99: 57–69.

————. 1989. "Aspects of Global Avifaunal Dynamics during the Cenozoic." In *Acta XIX congressus internationalis ornithologici*, edited by H. Ouellet. Ottawa: University of Ottawa Press.

————. 1992. "A New Family of Primitive Landbirds from the Lower Eocene Green River Formation of Wyoming." In "Papers in Avian Paleontology Honoring Pierce Broadcorb," edited by K. E. Campbell. *Science Series of the Natural History Museum of Los Angeles County* 36: 127–36.

Olson, S., and A. Feduccia. 1980. "*Presbyornis*, and the Origin of the Anseriformes (Aves: Charadriomorphae)." *Smithsonian Contributions to Zoology* 323: 1–24.

Olson, S., and H. Matsuoka. 2005. "New Specimens of the Early Eocene Frigatebird *Limnofregata* (Pelecaniformes: Fregatidae), with the Description of a New Species." *Zootaxa* 1046: 1–15.

Osborn, H. F. 1902. "American Eocene Primates, and the Supposed Rodent Family Mixodectidae." *American Museum of Natural History, Bulletin* 16: 169–214.

Packard, A. S. 1880. "Fossil Crawfish from the Tertiaries of Wyoming." *American Naturalist* 14: 222–23.

Pitman, J. K., F. W. Pierce, and W. D. Grundy. 1989. "Thickness, Oil-Yield, and Kriged Resource Estimates for the Eocene Green River Formation, Piceance Creek Basin, Colorado." *U.S. Geological Survey Oil and Gas Investigations Chart* OC-132.

Plato. 380 BC. *Politeia* [later translated as *The Republic* in English].

Powell, I. 1934. "'Compleat Angler' Fishes for Fossils: Fish Fossil Hunting in the Rocky Mountains of Wyoming." *National Geographic Magazine* 66 (2): 251–58.

Rafinesque, C. S. 1820. "Ichthyologia Ohiensis [Part 8]." *Western Review and Miscellaneous Magazine* 3 (3): 165–73. Also published in same year in *Ichthyologia Ohiensis, or natural history of the fishes inhabiting the river Ohio and its tributary streams, preceded by a physical description of the Ohio and its branches.* Lexington, KY: printed for the author by W. G. Hunt.

Regier, J. C., J. W. Shultz, A. Zwick, A. Hussey, B. Ball, R. Wetzer, J. W. Martin, and C. W. Cunningham. 2010. "Arthropod Relationships Revealed by Phylogenomic Analysis of Nuclear Protein-Coding Sequences." *Nature* 463 (7284): 1079–83.

Ride, W. D. L., H. G. Cogger, C. Dupuis, O. Kraus, A. Minelli, F. C. Thompson, and P. K. Tubbs. 1999. *International Code of Zoological Nomenclature.* 4th ed. London: International Trust for Zoological Nomenclature, Natural History Museum, London.

Rieppel, O., and L. Grande. 1998. "A Well-Preserved Fossil Amphiumid from the Eocene Green River Formation of Wyoming." *Journal of Vertebrate Paleontology* 18: 700–708.

————. 2007. "The Anatomy and Relationships of the Fossil Varanid Lizard *Saniwa ensidens* Leidy, 1870 Based on a Newly Discovered Complete Skeleton." *Journal of Paleontology* 81: 643–65.

Romer, A. S., and F. M. Fryxell. 1928. "*Paramiatus gurleyi*, a Deep Bodied Amiid Fish from the Eocene of Wyoming." *American Journal of Science*, ser. 5, 16: 519–27.

Rose, K. D. 2006. *The Beginning of the Age of Mammals.* Baltimore: Johns Hopkins University Press.

Rose, K. D., and W. v. Koenigswald. 2005. "An Exceptionally Complete Skeleton of *Palaeosinopa* (Mammalia, Cimolesta, Pantolestidae) from the Green River Formation, and Other Postcranial Elements of the Pantolestidae from the Eocene of Wyoming (USA)." *Palaeontographica* Abt. A, Band 273: 55–96.

Rosen, D. E., and C. Patterson. 1969. "The Structure and Relationships of Paracanthopterygian Fishes." *Bulletin of the American Museum of Natural History* 141, art. 3: 359–474.

Scotese, C. R., 2004. "Cenozoic and Mesozoic Paleobiogeography: Changing Terrestrial Biogeographic Pathways." In *Frontiers in Biogeography*, edited by L. Heaney and M. V.

Lomolino, 9–26. Sinauer Associates.

Scott, W. B., and E. J. Crossman, 1973. "Freshwater Fishes of Canada." *Fisheries Research Board of Canada, Ottawa, Bulletin* 184.

Scudder, S. H. 1890. "The Tertiary Insects of North America." *U.S. Geologic and Geographic Survey of the Territories, Report* 13: 1–734.

Shelubsky, M., 1951. "Observations on the Properties of Toxin Produced by Microcystis." *Verhandlungen der Internationalen Vereinigung fur Theoretische und Angewandte Limnologie* 11: 362–66.

Shufeldt, R. W. 1913. "Further Studies of Fossil Birds with Descriptions of New Species." *Bulletin of the American Museum of Natural History* 33: 285–306.

Silcox, M. T., J. I. Block, D. M. Boyer, and P. Houde. 2010. "Cranial Anatomy of *Paleocene Labidolemur kayi* (Mammalia: Apatotheria), and Relationships of the Apatemyidae to Other Mammals." *Zoological Journal of the Linnean Society*, no. doi:10.1111 /j.1096-3642.2009.00614.

Simmons, N. B., and J. H. Geisler. 1998. "Phylogenetic Relationships of *Icaronycteris, Archaeonycteris, Hassianycteris,* and *Palaeochiropteryx* to Extant Bat Lineages, with Comments on the Evolution of Echolocation and Foraging Strategies in Microchiroptera." *Bulletin of the American Museum of Natural History* 235: 1–182.

Simmons, N. B., K. L. Seymour, J. Habersetzer, and G. F. Gunnell. 2008. "Primitive Early Eocene Bat from Wyoming and the Evolution of Flight and Echolocation." *Nature* 451: 818–21.

Smith, J. B. L. 1949. "The Sea Fishes of Southern Africa." *South Africa, Central News Agency,* 1949 (*non vidi*).

Smith, M. E., A. R. Carroll, and B. S. Singer. 2008. "Synoptic Reconstruction of a Major Ancient Lake System: Eocene Green River Formation, Western United States." *GSA Bulletin* 120: 54–84.

Smith, M. E., K. R. Chamberlain, B. S. Singer, and A. R. Carroll. 2010. "Eocene Clocks Agree: Coeval $^{40}Ar/^{39}Ar$, U-Pb, and Astronomical Ages from the Green River Formation." *Geology* 38: 527–30.

Sterli, J. 2010. "Phylogenetic Relationships among Extinct and Extant Turtles: The Position of Pleurodira and the Effects of the Fossils on Rooting Crown-Group Turtles." *Contributions to Zoology* 79 (3): 93–106.

Stevens, P. F. 2008. Angiosperm Phylogeny Website. Version 9. http://www.mbot.org/MBOT /APweb.

Sullivan, S. P., L. Grande, A. Gau, and C. McAllister. 2012. "Taphonomy in North America's Most Productive Freshwater Fossil Locality: Fossil Basin, Wyoming." *Fieldianna Life and Earth Sciences* 5: 1–4.

Tanner, V. M. 1925. "Notes on the Collection of Fossil Fishes Contained in the University of Utah Collection, with the Description of One New Species." *Bulletin of the University of Utah* 15 (6): 1–13.

Tarduno, J. A., D. B. Brinkman, P. R. Renne, R. D. Cottrell, H. Scher, and P. Castillo. 2011. "Evidence for Extreme Climatic Warmth from Late Cretaceous Arctic Vertebrates." *Science* 282: 2241–44.

Taverne, L. 2009. "New Insights on the Osteology and Taxonomy of the Osteoglossid Fishes *Phareodus, Brychaetus,* and *Musperia* (Teleostei, Osteoglossomorpha)." *Bulletin Van Het Koninkluk Belgisch Instituut Voor Natuurwetenschappen Aardwetenschappen* 79: 175–90.

Uyeno, T., and R. R. Miller. 1963. "Summary of the Late Cenozoic Freshwater Fish Records for North America." *Occasional Papers of the Museum of Zoology, University of Michigan, Ann Arbor* (631): 1–34.

Wang, Y., and S. R. Manchester. 2000. "*Chaneya*, a New Genus of Winged Fruit from the Tertiary of North America and Eastern Asia." *International Journal of Plant Sciences* 161 (1): 167–78.

Warren, L. 1998. *Joseph Leidy: The Last Man That Knew Everything*. New Haven, CT: Yale University Press.

Weidig, I. 2003. "Fossil Birds from the Lower Eocene Green River Formation (USA)." PhD diss., Johann-Wolfgang-Goethe-Universitat, Frankfurt.

———. 2006. "The First New World Occurrence of the Eocene Bird *Plesiocathartes* (Aves: ?Leptosomidae)." *Paläontologische Zeitschrift* 80: 230–37.

———. 2010. "New Birds from the Lower Eocene Green River Formation, North America." *Records of the Australian Museum* 62: 29–44.

Wetmore, A. 1926. "Fossil Birds from the Green River Deposits of Eastern Utah." *Annals of the Carnegie Museum* 16: 391–402.

Whitfield, R. P. 1890. "Observations on a Fossil Fish from the Eocene Beds of Wyoming." *Bulletin, American Museum of Natural History* 3: 117–20.

Whitlock, J. A. 2010. "Phylogenetic Relationships of the Eocene Percomorph Fishes *Priscacara* and *Mioplosus*." *Journal of Vertebrate Paleontology* 30: 1037–48.

Wilf, P. 2000. "Late Paleocene–Early Eocene Climate Changes in Southwestern Wyoming: Paleobotanical Analysis." *GSA Bulletin* 112 (2): 292–307.

Wilf, P., C. C. Labandeira, K. R. Johnson, and B. Ellis. 2006. "Decoupled Plant and Insect Diversity after the End-Cretaceous Extinction." *Science* 313 (5790): 1112–15.

Wing, S. L., and L. J. *Hickey*. 1984. "The Platycarya Perplex and the Evolution of the Juglandaceae." *American Journal of Botany* 71: 388–411.

Yahya, H. 2006. "Atlas of Creation." Norwich, UK: Bookwork.

Zachos, J. C., G. R. Dickens, and R. E. Zeebe. 2008. "An Early Perspective on Greenhouse Warming and Carbon-Cycle Dynamics." *Nature* 451: 279–83.

Zhang, M., J. Zhou, and D. Qin. 1985. "Tertiary Fish Fauna from Coastal Region of Bahia Sea." *Institute of Vertebrate Paleontology and Paleoanthropology*, Memoir 17: 1–136. [In Chinese with English summary.]

Taxonomic Index

Arthropoda, 53, 57–58
arthropods. *See* Arthropoda
†Asineopidae, 108, 151–54, 182
†*Asineops*, 151–54, 176, 179
†*Asterotrygon*, 13, 98, 99, 101–6, 176, 180
†*Asterotrygon maloneyi*, 13, 98, 99, 101–6, 176, 180
†Asterotrygonidae, 97, 98, 99, 101–6, 182
astronium, 308, 333, 335
Astronium sp., 333, 335
Atractosteus, 112, 114–17, 176–80
†*Atractosteus atrox*, 115, 117, 176
†*Atractosteus simplex*, 114–16, 177, 179
Aves, 183, 184, 215–57, 383, 384
†*Axestemys byssinus*, 195–98
Azolla, 288, 289
Azollaceae, 288

†Baenidae, 190–94
†*Bahndwivici ammoskius*, 208, 209
Banksia, 310, 312
†*Baptemys*, 200–202
bats, 260, 274–78, 354
beaked sandfishes, 108, 146–49
†*Bechleja rostrata*, 86, 87
bees, 73, 75
beetles, 69–71
Bibionidae, 78, 79
Bignoniaceae, 308, 316
Biomphalaria, 90–92
birds. *See* Aves
birthworts, 296–98
Bivalvia, 90, 93
blue-green algae, 53, 55, 56
†*Boavus idelmani*, 204, 205
bony tongues, 108, 127–34
†*Borealosuchus wilsoni*, 210, 211, 213
bowfins, 108, 119–24, 176, 178, 179
†Brontotheriidae, 260, 270, 271
Buprestidae, 69–71
butterflies, 76–77

caiman, 210, 212, 213
Calopterygidae, 62, 63
Caprimulgiformes, 218
Carabidae, 69–71
†*Cardiospermum coloradensis*, 336–38
cartilaginous fishes. *See* Chondrichthyes
catalpa, 316
catfishes, 7, 180
Catostomidae, 7, 180
cattails, 296, 303
Caudata, 184, 186, 187

†*Cedrelospermum*, 327, 328
†*Cedrelospermum nervosum*, 327, 328
†*Celericolius acriala*, 245, 247
Ceratophyllaceae, 296, 303–5
Ceratophyllales, 296
Ceratophyllum, 296, 303–5
Ceratophyllum muricatum, 303–4
†*Chaneya*, 340
Chelicerata, 57–59
Chelydridae, 191
Chironomidae, 81, 82
chironomids, 81, 82
Chiroptera, 260, 274–78
†*Chisternon undatum*, 190–93
Chlorophytes, 282
Chondrichthyes, 97
Chordata, 53, 96
cicadas, 67
Cicadidae, 67
†*Cimolestidae*, 260–63
Cinnamomum, 297
Cixiidae, 68, 69
clams, 90, 93
climbing fern, 286, 289, 290
Clupeidae, 18, 108, 141–45, 176–79, 352, 355–57
Clupeiformes, 108
†*Cockerellites*, 52, 140, 157, 159, 163, 177, 179
†*Cockerellites liops*, 52, 169–71
Coenagrionidae, 62, 63
Coleoptera, 62, 69–71
Coliidae, 218, 245, 247
Coliiformes, 218, 245
Colocasia, 298
Comptonia, 332
†Condylarthra, 260, 269, 270
Coniferophyta, 282, 291–93
conifers. *See* Coniferophyta
Convulvaceae, 340
Coraciiformes, 218
courols, 218, 244, 245
crab spiders, 58, 59
crane flies, 79, 80
crayfishes, 84, 85, 87
crickets, 62, 65, 66
crocodile lizards, 208, 209
crocodilians, 184, 210–13
Crocodylomorpha, 184, 210–13
†*Crossopholis*, 15, 108–12, 176, 178, 180
†*Crossopholis magnicaudatus*, 15, 108–12, 176, 178, 180
Crustacea, 84
Cryptodira, 184
"cuckoo-rollers," 218, 244, 245

Raptor, 254, 255
ray-finned fishes. *See* Actinopterygii
reptiles. *See* Reptilia
Reptilia, 183, 184, 189, 215
reticulated beetles, 70, 71
†*Rhus longipetiolata*, 333, 334
†*Rhus nigricans*, 333, 334
river mussels, 93
rollers, 243, 244
Rosales, 308, 327–30
Rosids, 308
royal moth, 76, 77

†*Sabelites powelli*, 300, 301
salamanders, 184–87
Salicaceae, 321, 323
†*Salmilidae*, 218, 237–40
Salticidae, 58, 59
†*Salvinia preauriculata*, 289
Salviniaceae, 286, 288, 289
†*Sandcoleidae*, 218, 244–45
†*Saniwa ensidens*, 205–8
Sapindaceae, 308, 333, 336, 337
Sapindales, 308, 333
Sapindus sp., 336, 337
Sarcopterygii, 183, 184
Saturniidae, 76, 77
Saxifragales, 308, 314, 315
Schizaeales, 286, 289, 290
Sciaridae, 79, 81
Scoliidae, 73, 74
scorpion wasps, 72, 73
seed-shrimp, 57, 87, 88
Serpentes, 205
Shinisauridae, 208, 209
shrimp, 86, 87
Simaroubaceae, 308, 338–40
snails, 90–92
snakes, 205
soapberry, 336–38
soldier flies, 79, 80
Sphecoidea, 73
spiders, 57–59
spiny-rayed fishes, 151
spoonbills, 218, 234, 235
Squamata, 184
staghorn ferns, 286–88
Steatornithidae, 218, 226, 227
Sterculia, 340, 341
†*"Sterculia" coloradensis*, 341
Sterculiaceae, 308
stingrays, 32, 98–106, 176–80
stink bugs, 66, 67

stone flies, 64, 66
Stratiomyidae, 79, 80
suckers, 7, 180
sumac, 308, 333, 334, 336
sunbittern, 218, 230, 231
†*Swartzia wardelli*, 323, 324
"sweet fern," 330, 332
sweetgum, 314, 315
swifts, 218, 226–30

Testudines, 184, 190–203
tetrapods, 183, 184
Thomisidae, 58, 59
Threskiornithidae, 218, 234, 235
tinamou-like birds, 218–22
Tipulidae, 79, 80
tree of heaven, 338–39
Tricoptera, 62, 77–78
Trionychidae, 190, 191, 195–200
trout-perches, 154–57
true bugs, 62, 66–69
†*Tsoabichi greenriverensis*, 210, 212, 213
tube flowers, 298–300
tube-sheep, 269–70
Turaco?, 218, 235, 236
turtles, 190–203
†*Tynskya eocaena*, 248, 252
Typha, 303
Typhaceae, 296, 303

Ulmaceae, 308, 327
Unionidae, 90, 92, 93
Uticaceae, 327, 328

Vertebrata, 95, 96
vertebrates. *See* Vertebrata
Vespoidea, 73
Viviparidae, 90–92
Viviparus sp., 90–92

walnut, 330
wasps, 62, 72–74
water striders, 68, 69
waterfowl, 218, 225, 323–25
"wave beast," 270
weevils, 70, 71
wolf spider, 58, 59

zelkovas, 327
†*Zygodactylidae*, 218, 252–54
†zygodactylids, 218, 252–54
Zygoptera, 62, 63

Subject Index